HIGH RESOLUTION MORPHODYNAMICS AND SEDIMENTARY EVOLUTION OF ESTUARIES

Coastal Systems and Continental Margins

VOLUME 8

The titles published in this series are listed at the end of this volume.

High Resolution Morphodynamics and Sedimentary Evolution of Estuaries

Edited by

Duncan M. FitzGerald
Boston University, MA, U.S.A.

and

Jasper Knight
University of Exeter, UK

Springer

A C.I.P. Catalogue record for this book is available from the Library of Congress.

ISBN-10 1-4020-3295-1 (HB)
ISBN-13 978-1-4020-3295-0 (HB)
ISBN-10 1-4020-3296-X (e-book)
ISBN-13 978-1-4020-3296-7 (e-book)

Published by Springer,
P.O. Box 17, 3300 AA Dordrecht, The Netherlands.

www.springeronline.com

Cover illustration: View of Nauset Inlet, a small estuarine system located along the outer coast of Cape Cod, Massachusetts.

Printed on acid-free paper

Printed in the Netherlands.

Table of Contents

Chapter 1

TOWARDS AN UNDERSTANDING OF THE MORPHODYNAMICS AND SEDIMENTARY EVOLUTION OF ESTUARIES

Jasper Knight[1] and Duncan M. FitzGerald[2]
[1]*Department of Geography, University of Exeter, Rennes Drive, Exeter, Devon, EX4 4RJ, UK, email j.knight@exeter.ac.uk*

[2]*Department of Earth Sciences, Boston University, Boston, MA 02015, USA*

1. INTRODUCTION

Estuaries are found along many of the world's coastlines irrespective of geological setting, energy regime, and depositional environment (Perillo, 1995a). They also represent one of Earth's most dynamic sedimentary environments because they lie at the interface of the terrestrial and marine spheres, and evolve in response to the interaction of fluvial, coastal (tidal) and marine (wave) processes. The genetic classification of estuaries has focused on the interaction of processes in these fluvial, coastal, and marine environments (e.g. Perillo, 1995b; Elliott and McLusky, 2002), although in practice the processes influencing estuary morphodynamics vary along the length of the estuary, with tidal state, and over different time-spans. Estuaries are therefore not homogeneous sedimentary systems: their fluvial, coastal and marine environmental regimes are all subject to change in their intrinsic characteristics and their interactions over different scales of time and space, particularly in response to changes in climate and relative sea-

D.M. FitzGerald and J. Knight (eds.),
High Resolution Morphodynamics and Sedimentary Evolution of Estuaries, 1-9.

level (RSL) (Uncles, 2002). It can be argued, therefore, that the estuaries found along present-day coasts worldwide are both environmentally-sensitive and geologically-transient phenomena.

It is this sensitivity and transient nature that, in part, make the study of estuaries so important and interesting. Estuary and associated coastal valley-fill sediment successions contain a record of change in the erosional and depositional processes of their fluvial, coastal and marine-associated components. Marginal estuarine and coastal valley-fill sediment successions also record the signatures of transgressive and regressive RSL phases and related changes in coastal sediment depositional patterns.

There is a large literature on the morphological characteristics of estuaries and their sedimentary evolution. Notable monographs include those by Dyer (1973, 1986) and Perillo (1995b) and that on tidal inlets by Aubrey and Weishar (1988). These works focus in particular on descriptive studies of individual estuaries, and include conceptual physical models developed to explain estuarine hydraulics, morphology, sedimentary processes and facies distributions in response to a range of external forcing factors. Most of these established physical models stress the traditional view that estuaries are long-term sediment repositories, trapping fluvial sediment as well as bedload sediment from marine sources. More recent studies based on high-resolution field data, however, show that estuaries are more sedimentologically dynamic, often exporting sand to the nearshore or inner shelf (FitzGerald *et al.* 2000; and summarized by Uncles, 2002). Estuaries that discharge sand are dominated by flood events that overpower normal estuarine circulation and tide-induced sediment transport patterns. This more modern work presents a paradigm shift in the way in which estuaries, and the sedimentary systems of land-sea margins more generally, should be observed, monitored and modeled.

Studying the morphodynamics and sedimentary evolution of estuaries is fraught with difficulty. Despite offering an esthetically pleasing and physically diverse environment, data collection in estuaries is often difficult because of poor accessibility; safety problems of traversing exposed tidal flats; treacherous tidal currents and shifting patterns of intertidal creeks; instrumentation problems across the land-sea interface; issues of scale; and the high cost of water-based research. At best, studies can offer only a limited spatial and temporal shapshot of estuary morphodynamic behavior, and make quantitative assessments of sediment fluxes between certain portions of the estuary (Uncles, 2002). Much of estuarine behavior, and response to external forcing factors, therefore remain unknown.

2. RECENT ADVANCES IN COASTAL AND MARINE SCIENCE

Recent methodological and technical advances in field data collection and analysis have transformed estuarine studies from descriptive and area-based to quantitative and based on integration of datasets from different sources and on different spatial and temporal scales (Pye and Allen, 2000; Williams *et al.*, 2003). These more quantitative investigations have also helped in the definition and classification of estuarine systems (Elliott and McLusky, 2002). These methodological and technical advances include:

1. Remote sensing of the morphology of estuarine and coastal environments is very useful for regional-scale mapping and, when repeated and the images rectified, can indicate temporal changes in these environments. Remote sensing methods include vertical and oblique aerial photography from aircraft (error of ± 10 m), elevation mapping by radar and lidar (error of ± 0.3 m), and high-resolution satellite imagery (error of ± 10 m). These techniques are useful because they can aid accurate geomorphic mapping in both terrestrial and shallow-water environments (Jones, 1999; Rainey *et al.*, 2003). Diverse datasets on different scales can be integrated most successfully using a geographical information system (GIS) package such as ArcView.
2. Marine geophysical techniques are also useful in rapid field mapping of surface and subsurface sediment types and differentiation of sediment bodies. Field data can be collected digitally and post-processed to remove noise or error produced by, for example, vessel heave. Side-scan sonar used in water depths as shallow as only a few metres can differentiate between sediments with different acoustic backscatter characteristics, which are a function of sediment grain size density of the reflective medium (Briggs *et al.*, 2002; Davis *et al.*, 2002). Resolution can be varied to suit individual surveys using single or multibeam equipment, and with different input frequency and swath width (Jones, 1999). Sub-bottom acoustic units can be imaged in shallow water-depths using Chirp, boomer or sparker seismic profiling equipment. Penetration into the sediment profile, and vertical resolution of seismostratigraphic boundaries, can be optimized by varying input seismic frequency (Jones, 1999). The boundaries of acoustic units derived from closely-spaced seismic profile lines can be used to reconstruct the three-dimensional geometry and estimate the volume of sedimentary units. This is important in identifying unit boundary relations, morphostratigraphic development of nearshore sediment wedges, and may be important in

estimates of marine aggregate reserves or near-surface gas traps. Side-scan sonar and sub-bottom profiling data can be integrated effectively within programs such as Surfer or within a GIS. These field data types can be ground-truthed when coupled with surface sediment sampling (by Van Veen or bucket grabs) or matched against the stratigraphy of marine cores, respectively. In some intertidal environments, especially in well-drained sand and gravel sediments, internal sedimentary structures and bounding surfaces can be imaged using ground-penetrating radar (GPR). Offshore bathymetry can be measured quickly and accurately ($\pm$ 0.1 m resolution) using echo-sounding when these point data are kriged.

3. Onshore and offshore data collection in the field involves the use of a range of equipment designed to give speedier access to all parts of the study area, in all conditions, and with greater reliability. Equipment includes all-terrain vehicles (ATVs), hovercrafts and shallow-draught boats. Accurate field mapping in the x, y and z planes using a differential global positioning system (dGPS) enables rapid data collection, having a low degree of error (usually $\pm$ 0.03 m), and can be imported directly into digital terrain model (DTM) packages. In addition, a range of other field equipment can be used directly in the supratidal, intertidal and subtidal zones to monitor changes in bed morphology, surface sediments, and water physical characteristics such as temperature, salinity, dissolved oxygen, etc. Instrumentation includes acoustic doppler current profilers (ADCPs), current meters and tide gauges. These instruments can be deployed and the data collected digitally and downloaded straight to PC. This aids numerical data analysis and as input into quantitative models.
4. Sediments recovered through coring (usually box, gravity or piston cores in shallow water) can be examined in several ways. Physical properties measured includes sedimentary structures, grain size, lithology and heavy mineral analysis, core magnetometry and x-ray analysis. Dating core components may be through accelerator mass spectrometry (AMS) ^{14}C dating of organic fractions, or measurement of excess radioisotopes (^{210}Pb, ^{134}Cs, ^{137}Cs) in the < 63 µm fraction (e.g. Wheeler *et al.*, 1999). Estuary sediments may also contain microfaunal or floral components which can be examined using transfer functions to derive estimates of changes in salinity and other estuarine parameters. Linked to changes in core physical characteristics, different sedimentary facies and environments can be reconstructed. In addition, these stratigraphic elements can also be linked using techniques such as Markov chain analysis and principal component analysis (PCA). On a larger scale, this analysis of

ground-truth data can be used in a sequence stratigraphic context to reconstruct systems tracts and in facies modeling.

5. Finally, data on coastal forcing factors such as RSL change and onshore and offshore wind and wave climates are of better quality and more readily available. Monitoring and analysis of present-day tide gauge data, field investigation and dating of RSL index points in the geological record, and geophysical modeling have produced a better understanding of RSL change on different scales, and thus their likely effects on coastal sediment systems. High-resolution climate data from fixed ocean buoys, satellites, and field-based automatic weather stations (AWSs) is easily linked to concurrent monitoring of estuary morphodynamics, thus helping to identify, for example, coastal forcing by large storms (e.g. Orford *et al.*, 1999). Advances in understanding these forcing factors may be somewhat offset by human activity within estuaries in changing sediment supply (through armoring, river channelisation and reclamation) and sediment movement (through dredging).

2.1 A Holistic Approach to Estuarine Studies

Modern methods of field data collection, analysis, integration and interpretation, outlined above, emphasize the significance of estuaries as a dynamic interface between terrestrial and marine environments (Uncles, 2002). Data integration using historical maps and modern field surveys provides long- and short-term perspectives on estuary evolution. Estuaries are also important because of the close relationship between their morphodynamic behavior and human activity (Pye and Allen, 2000). A holistic approach to estuarine studies should therefore consider estuaries as multi-use systems (Nordstrom, 2000):

1. As part of a sediment system. Estuaries form part of coastal and nearshore sediment systems in which sediment is circulated between temporary onshore and offshore storage areas as a result of wind and water transport processes. Changes in any one component of this system results in sediment oversupply and deficit, leading to morphodynamic changes and environmental stress over different spatial and temporal scales.
2. As a coastal resource. Associated with the presence and development of estuaries are other landscape components such as sensitive coastal features (beaches, sand dunes, saltmarsh, intertidal flats), unique flora and fauna, and aspects of landscape heritage including archeological features.

3. As human-use systems. Estuaries often form natural harbors, the entrance to ports, or waterways downstream from major cities. Navigation may be maintained by dredging or parts of the estuary stabilized by reclamation. Estuaries may also be used for a range of human activities including commercial fishing, oyster farming, aggregate extraction, waste disposal and dumping, tourism and recreation. Estuarine sediments, including contaminants, can record the history of regional-scale human activities.

Clearly, such multi-use systems are sensitive to a range of human and environmental variables on different scales. The focus of this book is to examine in more detail some of these components.

3. AIMS AND STRUCTURE OF THIS BOOK

This book does not intend to be all-encompassing; rather, it seeks to raise some issues of the morphodynamics and sedimentary evolution of estuaries, including the ways in which they are (or should be) observed, monitored, modeled and managed. Significantly, this book highlights the role of high-resolution data collection in the field and through remotely-sensed (geophysical) methods. These data should be integrated with baseline monitoring and integration with historical datasets (e.g. aerial photographs and maps) on different scales, as through the use of a GIS. Throughout, the use of multi-proxy indicators of changes in estuary environments reinforces the fact that estuaries are multi-use, multi-dimensional systems. Papers in this book offer a new approach to nearshore and estuary studies with an emphasis on multidisciplinary techniques and data integration.

The book is organized into three main themes, which are not mutually exclusive.

Remote-sensing and geophysical techniques are examined in three papers. McDowell *et al.* (chapter 2) use integrated CHIRP sub-bottom profiler and side-scan sonar techniques to investigate late Pleistocene and Holocene sediment dynamics of the Northern Ireland coast. Nitsche *et al.* (chapter 3) describe results from a project aimed at mapping benthic habitats of the Hudson River estuary (New York State, USA). Geophysical data were integrated with multiple acoustic sensor data to produce an automated classification scheme. Gonzalez *et al.* (chapter 4) use a temporal record of aerial photos to identify land-cover changes within the Guadiana River (Iberia). Land-cover changes are quantified using a modified Markov chain analysis within a GIS.

Sediment dynamics and fluxes are examined in six papers. Pendleton and FitzGerald (chapter 5) describe the changes in hydrodynamics and sediment fluxes, including flood-ebb dominance, following spit breaching at New

Inlet (Massachussets, USA). Perillo *et al.* (chapter 6) investigate suspended sediment fluxes in the Bahía Blanca River estuary (Argentina) during flood and ebb cycles, including identifying points of flow separation. Ward and Bub (chapter 7) investigate temporal variations in hydrological parameters and suspended sediment dynamics in Great Bay estuary (New Hampshire, USA). French *et al.* (chapter 8) consider the sediment dynamics and morphological evolution of the Blyth estuary (England) within the context of long-term channel modification and reclamation. Fenster *et al.* (chapter 9) describe the sediment dynamics of Merrymeeting Bay (Maine, USA) in response to varying flood-ebb conditions. FitzGerald *et al.* (chapter 10) summarize studies of New England estuaries (northeast USA) and argue that spring freshets are hydrodynamically important in the seaward transport of coarse sediment.

The multiscale morphodynamic evolution of estuaries is investigated in six papers. Burningham (chapter 11) examines the mesoscale evolution of a tidal inlet in County Donegal (Ireland) and identifies possible coastal forcing by episodic storms and variations in the North Atlantic Oscillation. Two papers explore the sensitivity of the Mississippi River estuary in coastal Louisiana (USA). Stone *et al.* (chapter 12) present storm wind and wave data to demonstrate the importance of cold fronts as agents of shoreline change. The paper by Roberts *et al.* (chapter 13) discusses the effects of cold fronts on bayhead delta development. Duck (chapter 14) describes the sediment and bedform dynamics of the Tay River estuary (Scotland) in response to estuary front formation. Buynevich and FitzGerald (chapter 15) describe the relationship between river sediment discharge and barrier evolution along the coast of Maine (USA). Finally, the paper by Belknap *et al.* (chapter 16) discusses how RSL position and riverine sediment fluxes contributed to the formation of the now-submerged Penobscot paleodelta, Maine (USA).

4. ESTUARIES AND THE FUTURE

Future changes in the external environment (including RSL, storm surge frequency, hurricane frequency, wave height) are likely to exert a strong influence on the morphodynamics and functioning of coastal and estuarine sediment systems (Pethick, 2001). Estuaries and associated geomorphic features will take the first impact of these changes, such as storm and hurricane landfall. Estuarine systems, at the interface of the physical and human environments of developed coastlines (Nordstrom, 2000), are also well placed to respond dynamically to changes in morphology and sediment budgets associated with dredging, reclamation and channelisation. Understanding the morphodynamics and sedimentary evolution of estuaries is therefore fundamental to predictions of estuary response to future changes in environmental systems and human development in the coastal zone.

REFERENCES

Aubrey, D.G. and Weishar, L. (eds) 1988. *Hydrodynamics and Sediment Dynamics of Tidal Inlets*. Lecture Note on Coastal and Estuarine Studies Vol. 29. Springer-Verlag, New York.

Briggs, K.B., Williams, K.L., Jackson, D.R., Jones, C.D., Ivakin, A.N. and Orsi, T.H. 2002. Fine-scale sedimentary structure: implications for acoustic remote sensing. *Marine Geology*, 182, 141-159.

Davis, A., Haynes, R., Bennell, J. and Huws, D. 2002. Surficial seabed sediment properties derived from seismic profiler responses. *Marine Geology*, 182, 209-223.

Dyer, K.R. 1973. *Estuaries: a physical introduction*. Wiley, London. 140pp.

Dyer, K.R. 1986. *Coastal and Estuarine Sediment Dynamics*. Wiley, Chichester. 342pp.

Elliott, M. and McLusky, D.S. 2002. The need for definitions in understanding estuaries. *Estuarine Coastal and Shelf Science*, 55, 815-827.

FitzGerald, D.M., Buynevich, I.V., Fenster, M.S. and McKinlay, P.A. 2000. Sand dynamics at the mouth of a rock-bound, tide-dominated estuary. *Sedimentary Geology*, 131, 25-29.

Jones, E.J.W. 1999. *Marine Geophysics*. Wiley, Chichester. 466pp.

Nordstrom, K.F. 2000. *Beaches and Dunes of Developed Coasts*. Cambridge University Press, Cambridge. 352pp.

Orford, J.D., Cooper, J.A.G. and McKenna, J. 1999. Mesoscale temporal changes to foredunes at Inch Spit, south-west Ireland. *Zeitschrift für Geomorphologie, N.F.*, 43, 439-461.

Perillo, G.M.E. (ed) 1995a. *Geomorphology and Sedimentology of Estuaries*. Developments in Sedimentology 53, Elsevier, Amsterdam. 471pp.

Perillo, G.M.E. 1995b. Definitions and geomorphologic classifications of estuaries. In: Perillo, G.M.E. (ed) *Geomorphology and Sedimentology of Estuaries*. Developments in Sedimentology 53, Elsevier, Amsterdam. 17-47.

Pethick, J. 2001. Coastal management and sea-level rise. *Catena*, 42, 307-322.

Pye, K. and Allen, J.R.L. (eds) 2000. *Coastal and Estuarine Environments: sedimentology, geomorphology and geoarchaeology*. Geological Society, Special Publication 175. Geological Society, London. 435pp.

Rainey, M.P., Tyler, A.N., Gilvear, D.J., Bryant, R.G. and McDonald, P. 2003. Mapping intertidal estuarine sediment grain size distributions through airborne remote sensing. *Remote Sensing of Environment*, 86, 480-490.

Uncles, R.J. 2002. Estuarine physical processes research: Some recent studies and progress. *Estuarine Coastal and Shelf Science*, 55, 829-856.

Wheeler, A.J., Orford, J.D. and Dardis, O. 1999. Saltmarsh deposition and its relationship to coastal forcing over the last century on the north-west coast of Ireland. *Geologie en Mijnbouw*, 77, 295-310.

Williams, J.J., O'Connor, B.A., Arens, S.M., Abadie, S., Bell, P., Balouin, Y., van Boxel, J.H., Do Carmo, A.J., Davidson, M., Ferreira, O., Heron, M., Howa, H., Hughes, Z., Kaczmarek, L.M., Kim, H., Morris, B., Nicolson, J., Pan, S., Salles, P., Silva, A., Smith, J., Soares, C. and Vila-Concejo, A. 2003. Tidal inlet function: Field evidence and numerical simulation in the INDIA project. *Journal of Coastal Research*, 19, 189-211.

Chapter 2

HIGH-RESOLUTION GEOPHYSICAL INVESTIGATIONS SEAWARD OF THE BANN ESTUARY, NORTHERN IRELAND COAST

J. Lyn McDowell[1], Jasper Knight[2]* and Rory Quinn[1]

[1]*Coastal Studies Research Group, School of Environmental Sciences, University of Ulster, Coleraine, BT52 1SA, Northern Ireland, UK*

**Author for correspondence (j.knight@exeter.ac.uk)*

1. INTRODUCTION AND AIMS

The coast of Ireland, located on the paraglacial shelf of the north-east Atlantic (Carter, 1990), is well placed to respond dynamically to external forcing factors in the marine and onshore environments. These factors include eustatic changes in relative sea-level (RSL) driven by glacial cycles on 3rd and 4th order (Milankovitch) time-scales; changes in shelf, nearshore, coastal, estuarine, dune and fluvial sediment storage and supply; changes in North Atlantic wind and wave climates; and the effects of high-magnitude events such as storms, storm surges and sea floods (Delaney and Devoy, 1995). In addition, formerly paraglacial coasts and shelves in particular are subject to a range of environmental factors impacting on present-day shelf stratigraphy and sediment dynamics. These factors include sediment supply to continental shelves, and 4th order (glacioisostatic-driven) changes in RSL

D.M. FitzGerald and J. Knight (eds.),
High Resolution Morphodynamics and Sedimentary Evolution of Estuaries, 11-31.

(Barnhardt *et al.*, 1997; Plag *et al.*, 1996; Syvitski, 1991; Kelley *et al.*, 1989). In western Britain, late Devensian (Wisconsinan; ~ 25-13 kyr BP) ice spread outwards from dispersal centres in upland areas of Scotland, Wales, northern England, and northern and western Ireland (Bowen *et al.*, 1986). This ice spread generally radially onto adjacent lowlands and offshore shelves in the North Atlantic, North Sea and Irish Sea (including the Northern Ireland coast) which were dry due to 4th order eustatically-driven RSL fall (Bridgland, 2002). These ice sheets carried glacially-eroded sediment onto the present-day continental shelf. During glacioisostatic RSL rise (from ~ 13-8 kyr BP) shelf sediment was mobilised as the land-sea interface migrated landwards. Evidence for this generally transgressive RSL phase comes from the presence of submerged lowstand deltas, shoreline notches and beaches on these paraglacial coasts which are onlapped and overstepped by transgressive intertidal to subtidal sands. This is a common feature of most paraglacial shelves (e.g. Cooper *et al.*, 2002; Barnhardt *et al.*, 1997; Shipp *et al.*, 1991; Kelley *et al.*, 1989). Slowed eustatic RSL rise towards the mid-Holocene highstand (~ 7-5 kyr BP) acted to reduce nearshore accommodation space, and allow for the formation of coastal and nearshore sediment wedges. Stabilised RSL since the mid-Holocene has resulted largely in reduced onshore sediment transport, and reworking of sediment between onshore (dune), nearshore (subtidal) and intertidal (estuarine) sinks, and the infilling of back-barriers, lagoons and estuaries.

The above description of dynamic sediment response to RSL forcing over time-scales of 10^3-10^5 years can be described with reference to the formation of regressive, lowstand, transgressive and highstand systems tracts (Vail *et al.*, 1991). The systems tracts concept refers to the formation and preservation of discrete coastal and nearshore 3-dimensional sediment bodies under particular RSL stages. These sediment bodies and associated systems tracts can be identified through the use of sub-bottom geophysical techniques in the marine environment which are able to image stratal (systems tract) boundaries.

In order to investigate mesoscale sediment dynamics and the evolution of sediment bodies on the paraglacial coast of Northern Ireland, an integrated geophysical investigation was carried out using side-scan sonar and sub-bottom profiling techniques, linked to ground-truthed surface sediment sampling. Together, these techniques provide insight into the character, disposition and history of 3-dimensional sediment bodies in the nearshore zone. This paper presents preliminary results of this investigation. In detail this paper has two main aims. These are to (1) investigate the present-day sediment dynamics of the Bann estuary, Northern Ireland coast, through repeat side-scan sonar surveys; and (2) investigate the post-glacial evolution of the area through integration of high-resolution marine geophysical data (Chirp sub-bottom profiler, bathymetry and side-scan sonar).

1.1 Regional physical setting

Geologically the north coast of Ireland generally comprises flat-bedded Tertiary basalt beds which overlie karstified Cretaceous-age chalk of the Ulster White Limestone series (Wilson, 1972). The basalt generally extends a few kilometres offshore, often forming a flat, scoured platform (Cooper *et al.*, 2002), and Mesozoic sediments are found up to 15 km offshore (Fyfe *et al.*, 1993). The basalts are cut by minor north-south faults which intersect the coast at right angles, controlling its indented nature (Roberts, 1976). The major northeast-southwest Tow Valley fault, a Caledonian-age lineament, cuts the basalt series further inland and intersects the coast at Ballycastle.

The River Bann, Northern Ireland's longest river (Wilcock, 1982), discharges near Portstewart into the Malin Sea (North Atlantic ocean) through a funnel-shaped estuary which is bounded by basalt headlands (4 km apart) (Fig. 1). The River Bann has a seasonally-varying fluvial discharge of 60-250 m^3s^{-1} (Carter and Rihan, 1976). During ebb tides, discharge peaks at 2000 m^3s^{-1} (Carter and Rihan, 1976), which means there is a large tidal prism compared to river outflow. Presently, the tidal limit is located at Mountsandel, 8 km upstream from the river mouth (termed the Barmouth), but this limit has migrated considerably during the Holocene as a result of changes in RSL, river discharge, tidal range and estuary configuration (Battarbee *et al.*, 1985; Carter, 1993a). The lower Bann valley is underlain entirely by basalt as far as Lough Neagh, and is thus not structurally controlled. However, the presence of overdeepened linear sections beneath the present river channel may suggest glacial and subsequent fluvial erosion took place along lines of weakness, possibly intra- or sub-basalt faults (Carter, 1993a). Late Pleistocene and Holocene changes in RSL also acted to overdeepen, infill and rejuvenate sections of the lower Bann (Carter, 1993a).

The Malin Sea lies open to the high wave-energy, swell-dominated North Atlantic (significant wave height of 2.5 m). The region comprises a shallow shelf (generally up to 80 m depth) with localised trenches up to 200 m depth around Rathlin Island and in the North Channel. Tidal current ellipsoids for the southern Malin Sea display a strong rectilinear nature for surface currents, with a degree of spreading towards the seafloor. In the Portstewart region, the current ellipses display a rotary tidal flow, with a peak spring flow on the flood cycle of 0.55 ms^{-1} orientated at 90-100° (Lawlor, 2000). The Bann estuary itself experiences decreased flows as it is influenced by both the ebb tidal delta of Lough Foyle (Tuns Bank) and discharge from the River Bann. The corresponding spring tide ebb flow is orientated at 270-

280°. Maximum neap tidal current velocities for the Bann estuary approach 0.35-0.45 ms^{-1} at the surface and 0.1-0.2 ms^{-1} at the river bottom (Lawlor, 2000). Spring-neap M_2 tidal range is 3.10 m.

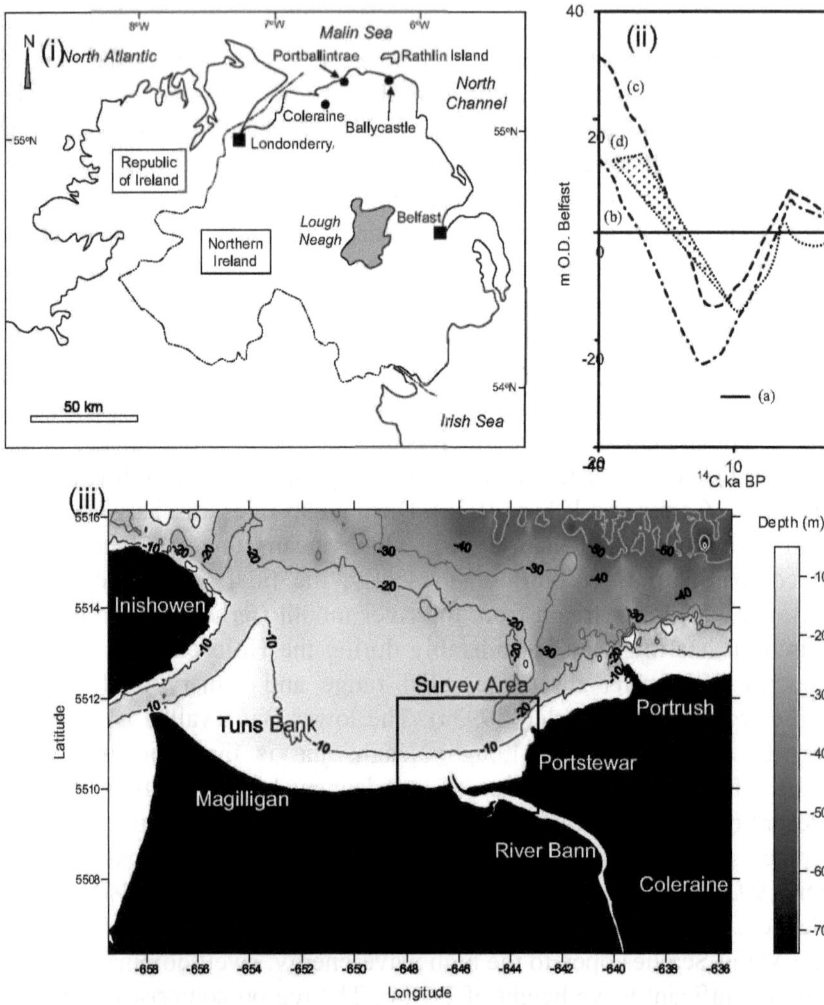

Figure 1 (i) Location of the study area; (ii) Sea-level curves for the north coast of Ireland: ^{14}C ky BP-present. Curves b and c are modelled curves for Ballycastle and Inishowen Head adjacent to the study area (Lambeck, 1996). Curve d (Carter, 1982), based upon field evidence, is complemented by an interpreted sea-level lowstand of –30m denoted by (a) in the diagram (Cooper et al., 2002); (iii) Regional setting of the Bann Estuary and the area under investigation (boxed).

The generalised distribution of offshore surface sediments has been described by Pendlebury and Dobson (1976) and Lawlor (2000) from a combination of side-scan sonar and surface sediment sampling. Offshore sand waves were described by Carter and Kitcher (1979). Off Portstewart, the planar seafloor is dominated by sand with locally developed gravel patches and exposed bedrock. A series of megaripples and concentrations of asymmetrical sand waves to the north and east of the study area, located in water depths ranging from 50-150 m (Lawlor, 2000), indicate sediment transport directions from the north, north-west and north east towards the Bann estuary.

1.2 Previous work

A range of field morphological, sedimentary and dating evidence provides information on RSL history and, indirectly, past sediment dynamics in the Bann estuary area (Carter, 1982; Wilson and McKenna, 1996; Wilson and Orford, 2002). This evidence includes raised beaches (Stephens, 1963; Carter, 1982), raised clifflines, notches and platforms (McKenna, 2002), buried intertidal peats or paleosols (Hamilton and Carter, 1983; Wilson, 1991, 1994; Wilson and McKenna, 1996), sand dunes (Carter and Wilson, 1990; Wilson, 1991) and estuarine sediments (Battarbee *et al.*, 1985). Sea-level changes during the late Pleistocene and Holocene are broken into a number of different phases (Fig. 1ii). The reconstructed RSL changes in this region are based on dated index points from marine, estuarine and terrestrial sediments (Wilson and Orford, 2002) which are subject to dating and elevation error, and error due to changes in tidal range, storminess and exposure over time, which may have an uncertain relationship to RSL. Between ~ 18-11 kyr BP RSL fell rapidly as isostatic rebound of the land outpaced eustatic RSL rise, culminating in a RSL lowstand of possibly as much as –30 m OD between about 11-10 kyr BP (Cooper *et al.*, 2002). From this period, RSL rose rapidly to a mid-Holocene highstand of +2 to +3 m OD at 6 kyr BP from which time it has declined steadily to the present day (Carter, 1982). Modern tide-gauge data indicate slight RSL regression (Wilson and Orford, 2002).

Using onshore geomorphic, sedimentary and dating evidence, Wilson and McKenna (1996) proposed a three-stage model for the Holocene evolution of the Bann estuary using more precise RSL data. (1) Around 9 kyr BP, when RSL was –6 to –8 m OD and the coastline up to 1 km seaward of its present position, the River Bann meandered through an estuarine landscape of proto-dunes and lagoons. (2) RSL rise to the mid-Holocene highstand saw the formation of a funnel-shaped coastal re-entrant with subtidal sand and gravel shoals, and active cliff erosion. (3) Around 4 kyr BP, with RSL at –2 to –3 m OD, beach ridges and dunes formed on an emergent Portstewart Strand, which constrained the location of the Barmouth. The dating and RSL control on this proposed model remains an issue, since neither are as yet fully evaluated, and the model has not been compared to offshore field evidence. In addition, the model poses a number of questions concerning long-term changes in tidal range, sediment supply, interaction of the fluvial and marine environments as the Bann estuary changes shape, and controls on, for example, the formation and dynamics of subtidal shoals and the evolution of Portstewart Strand. Some of these issues were raised in other works (e.g. Battarbee *et al.*, 1985; Cooper *et al.*, 2002) and are not discussed here.

2. METHODOLOGY

A series of repeat high-resolution marine acoustic surveys was conducted in the Bann estuary between 2000 and 2002. The surveys included side-scan sonar, Chirp sub-bottom profiler and single-beam echo-sounder surveys of a 5.3 x 4.6 km area in water depths between 5 m and 30 m (Fig. 2).

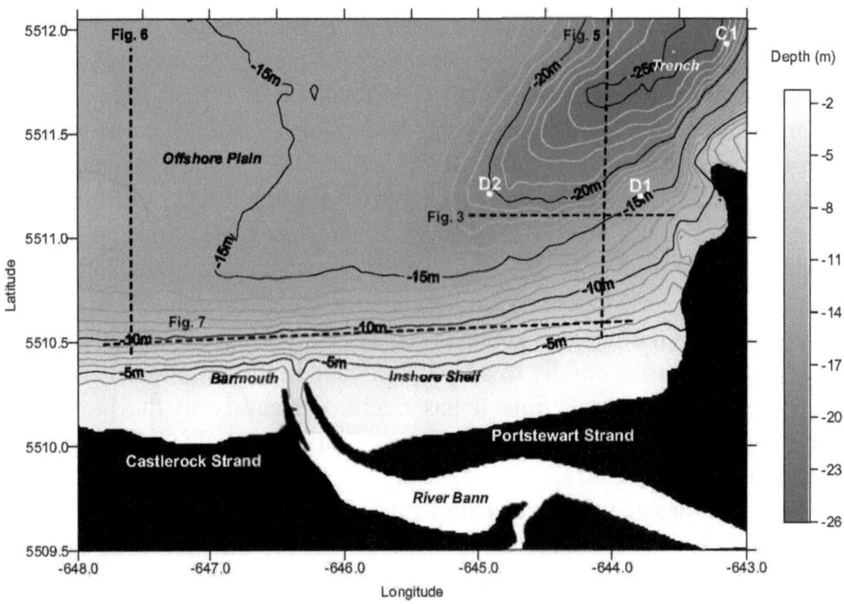

Figure 2. Contoured results of the bathymetric survey of the study area. The location of the side-scan sonographs, sub-bottom profiles and sonograph extracts (C1, D1, D2) presented in the main text are indicated.

In excess of 200 km of trackline data were acquired during surveys over the three-year period. Positional data (WGS-84 Ellipsoid) for all surveys were provided by a *Litton Marine* LMX-400 series GPS, with real-time differential corrections broadcast by the General Lighthouse Authorities. GPS data were corrected for towfish layback; total positional error is estimated to be about ± 15 m.

The composition and morphology of the seafloor was mapped using an *EdgeTech* Model 272-TD dual-frequency (100/500 kHz) side-scan system. Bathymetric data were collected using an *AutoHelm* SeaTalk 50kHz single beam echo-sounder with a vertical resolution of a few decimetres. The bathymetric data were gridded and contoured to produce 2- and 3-dimensional contour plots and digital terrain models of the study area. Sub-surface architecture was investigated using an *EdgeTech* SB-216 Chirp sub-bottom profiler operating at 2-10 kHz. Post-processing of the sub-bottom data involved heave compensation, the application of a time-varied gain (TVG) algorithm and band-pass filtering to increase the signal-to-noise ratio of the Chirp data. Depth conversion of the time-sections was based on a single compressional-wave velocity of 1500 ms^{-1}.

The side-scan sonar data were ground-truthed by a programme of grab-sampling. The interpretation of the geophysical data was enhanced by previously published bathymetric data (Hydrographic Office, 1986), offshore geophysical surveys (Lawlor, 2000; Cooper *et al.*, 2002) and onshore terrestrial mapping (McCabe *et al.*, 1994).

3. RESULTS AND INTERPRETATION

3.1 Morphology

The morphology of the Bann estuary (based upon the results of the bathymetric survey) is characterised by an inshore shelf between 0 and 10 m water depth, giving way to an offshore plain of average 15 m depth in the western sector of the survey area. A distinct bathymetric 'trench' is located in the north-eastern region of the study area, reaching a maximum of 27 m depth north of Portstewart Head. The inshore shelf is dissected by the River Bann channel immediately adjacent to the Barmouth (Fig. 2).

3.2 Substrate Type and Dynamics

The substrate in the area is sub-divided into three acoustic facies (SS-I to SS-III), identified on the basis of their backscatter characteristics (Figs. 3, 4).

SS-I, the dominant facies throughout the study area, is located in water depths between 3-25 m and is characterised by a moderate backscatter surface, and smooth uniform tone returns (Fig. 3) with bedforms developed locally (Fig. 3ii). A series of sediment samples from this facies indicates the substrate comprises fine sand (0.125-0.250 mm).

SS-II is located predominantly on the southern margins of the trench area in 12-22 m water depth, although one area of SS-II is located on the northern margin of the trench at a depth of 25 m (Fig. 4). This acoustic facies is characterised by moderate to high backscatter returns, with moderate tonal variation, locally developed shadows and bedforms (Fig. 3ii). Sediment samples indicate that this facies comprises gravel.

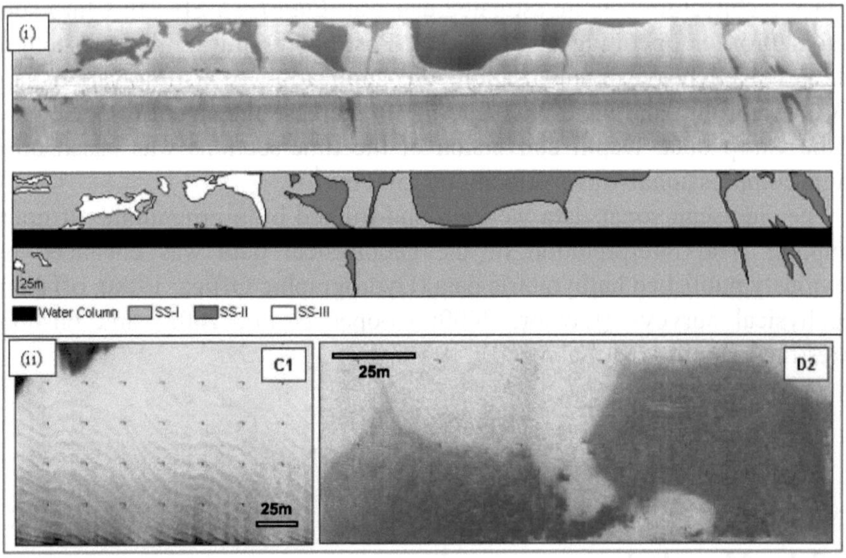

Figure 3. Side-scan sonar from west-east track. (i) 500 kHz type-sonograph and interpretation of the three dominant substrate facies identified in the study area; (ii) Bedforms developed in the sand (C1) and gravel (D2) facies. Refer to Figure 2 for the locations of the sonographs.

SS-III, located within the trench and on the western margin of Portstewart Head (Fig. 4), is characterised by high backscatter returns and a rough surface texture and is predominantly confined to water depths between 15-25 m (Fig. 3). Individual high-backscatter targets are strewn on the exposed surface, averaging 0.8 m diameter. This facies is interpreted as either an exposed bedrock or glacial diamict (till) surface (which have very similar acoustic signatures), and possibly may contain both components. The high-backscatter surface targets are interpreted as strewn clasts of cobble to boulder size which were derived either from erosion of a bedrock surface or winnowing of exposed diamict.

(i)

(ii)

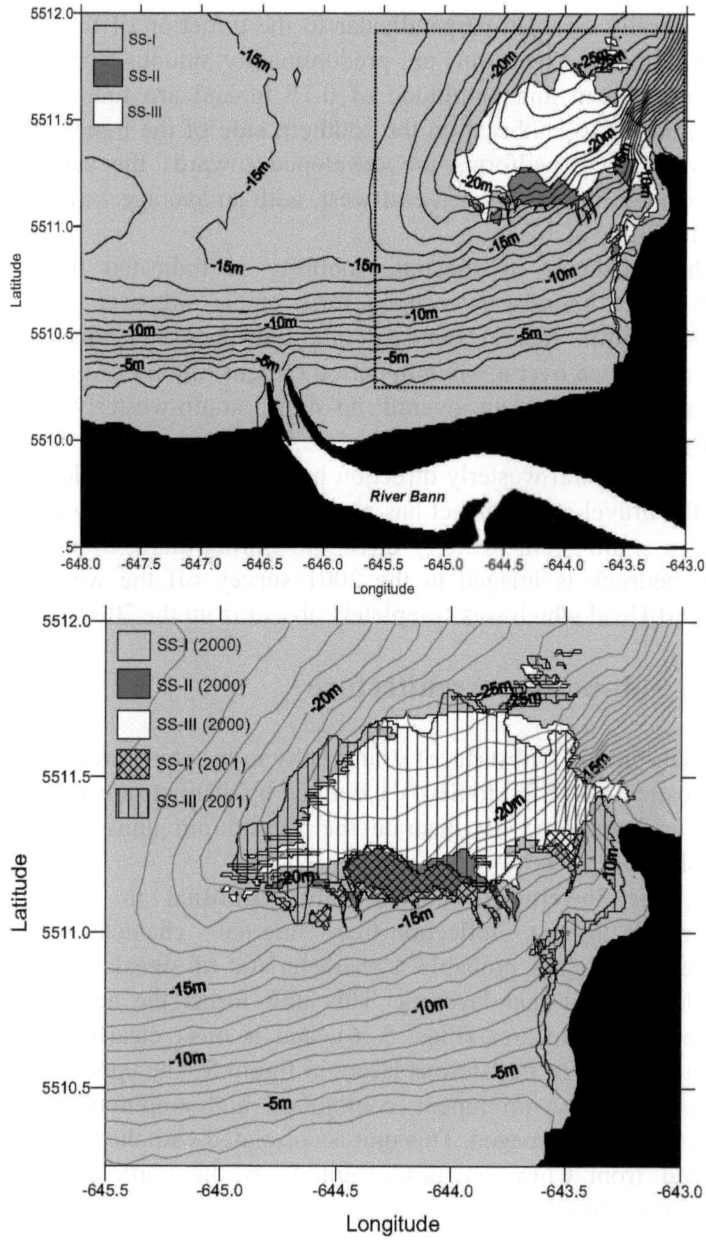

Figure 4. (i) Substrate map of the study area compiled from the side-scan sonar data (2001). The blank area in the River Bann was not surveyed; (ii) Enlarged view of the substrate map depicting the changes in the boundary positions identified from the side-scan sonar surveys of 2000 and 2001.

Although the dominant bed type within the study area is planar sand, bedforms are developed locally in both the sand and gravel facies, with crests generally oriented perpendicular to the direction of tidal currents. In the sand facies, the bedforms are predominantly sinuous ripples of average wavelength 6.0 m and amplitude of 0.75 m and are found in particular aligned northwest-southeast on the southern side of the trench (Fig. 3ii). In the gravel facies, bedforms are developed towards the sand contact as straight-crested ripples aligned east-west, with an average wavelength of 1.0 m and amplitude of 1.5 m.

Further evidence of substrate mobility is indicated by comparison between facies boundaries mapped from repeat side-scan sonar surveys. Figure 4ii shows facies boundary migration between successive repeat surveys conducted over a 9-month period (August 2000 to April 2001) in the trench area. There is an overall up-slope, south-westerly trend in the boundary migrations. The sand-bedrock/diamict contact has migrated upslope in a general westerly direction by 114 m. Towards the south of the trench, the gravel-sand contact has also migrated up-slope by an average of 37 m in a south-south-westerly direction. Furthermore, a large section of exposed bedrock is imaged in the 2001 survey off the western shore of Portstewart Head which was completely absent from the 2000 survey.

3.3 Sub-bottom Architecture

The sub-bottom (subsurface) stratigraphy in the study area is divided into four acoustic units (SB-I to SB-IV), defined by distinct seismic facies. Three profiles are presented to illustrate the sub-bottom units and their inter-relationships (Figs. 5-7).

SB-I, the lowermost acoustic unit identified in the profiles, is characterised by a reflection-free internal character (acoustically transparent), which is probably a characteristic of signal absorption rather than a lack of internal layering. This unit forms the acoustic basement throughout the field area (Figs. 5, 6), and is most clearly imaged in the trench, north-west of Portstewart Head, in figure 5. The upper surface of this unit is marked by a prominent, continuous, high-amplitude reflector. The lower surface is not imaged. This unit is correlated with the bedrock reflector interpreted from Chirp profiles acquired off the north coast of Ireland (Cooper *et al.,* 2002).

SB-II, characterised by a high degree of internal backscatter, directly overlies SB-I (Figs. 5, 6). Internally, the unit is structureless to chaotic in nature, with pronounced topographic expression (maximum relief of 3 m) of the uppermost surface in the inshore region. Offshore, the unit has a planar upper surface at an average depth of 25 m. Where distinguishable internal reflectors are present, they are planar and dip steeply in an offshore direction. In places, the surface of the unit is characterised by distinct hyperbolic reflectors. The full thickness of the unit is difficult to ascertain due to signal attenuation, however it exceeds 4 m in places. Scattering of the acoustic energy in the form of hyperbolic reflectors is indicative of gravel-rich beds or diamicts (Stoker *et al.,* 1997). Furthermore, where SB-II is exposed on the seafloor, the unit is characterised by the boulder-rich substrate (SS-III) described above.

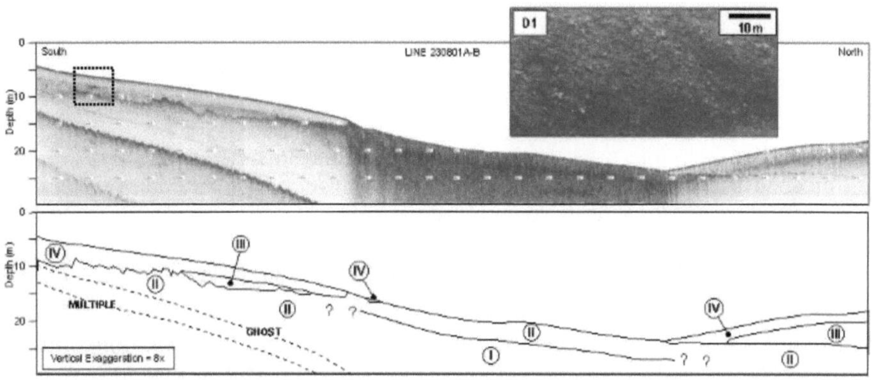

Figure 5. Interpreted Chirp sub-bottom profile 230801A-B showing the positions of acoustic units SB-I to SB-IV, and inset sonograph D1 showing high backscatter returns, interpreted as surface boulders. See Figure 2 for locations of the geophysical data. The boxed area in the upper section shows an area of high acoustic impedance, interpreted as a lowstand peat deposit

An extract of sonar data (D1) is shown on the Chirp profile in figure 5, illustrating the boulder-strewn surface. This interpretation is further enhanced by the onshore sequence at Portballintrae, 15 km east of the study area, where bedrock is directly overlain by diamict in an emergent shallow marine sequence (McCabe *et al.,* 1994). SB-III is characterised by moderate internal backscatter. The upper surface is planar and laterally continuous, whilst both the upper and lower surfaces are characterised by high amplitude reflections, revealing a high density and/or velocity contrast between SB-III and the underlying SB-II and overlying SB-IV.

SB-III reaches a maximum thickness of 5 m, although it is typically 3 m thick, forming a wedge which generally thickens in an offshore direction. Unit SB-III is characterised by a reflection-free internal configuration, implying a massive, homogeneous deposit with a uniform lithology, such as marine muds. In places, the upper surface of SB-III is diffuse (Fig. 5), indicating a gradational boundary between it and the overlying SB-IV. This may be an expression of an increase in the sand component (coarsening-up sequence) towards the top of SB-III.

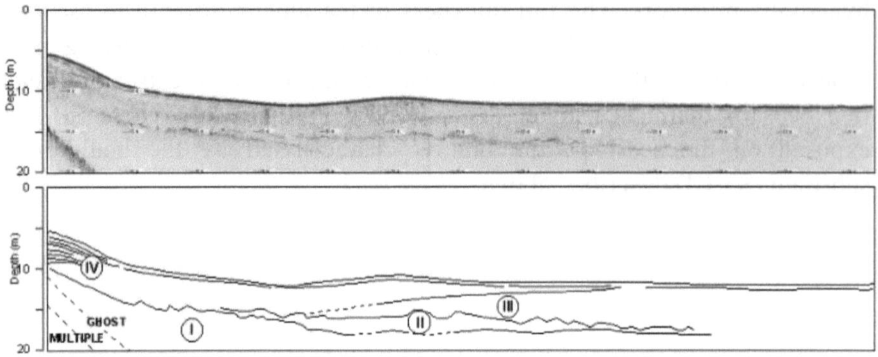

Figure 6. Interpreted south to north Chirp profile 230801E-F showing the positions of acoustic units SB-I to SB-IV (vertical exaggeration x 8). See Figure 2 for location.

SB-IV is subdivided into two sub-units (SB-IVi and SB-IVii; Fig. 7). SB-IV is typically 3 m in thickness (forming sheet or drape deposits offshore) but reaches 6 m in the inshore region of the study area. SB-IVi, concentrated in the inshore region, is characterised by a series of clinoformal, offshore-dipping parallel and sub-parallel continuous reflectors.

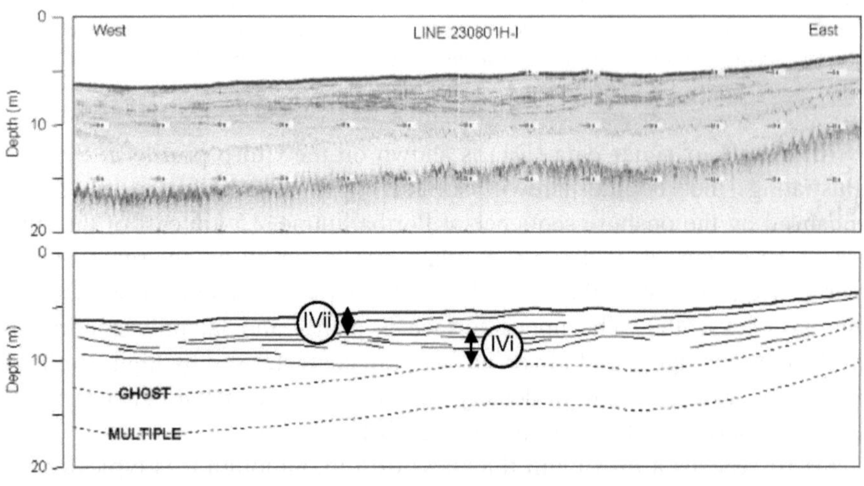

Figure 7. Interpreted Chirp profile 230801H-I showing the positions of acoustic units SB-IVi and SB-IVii. See Figure 2 for location of the profile.

Locally, the internal reflectors are of variable continuity, amplitude and frequency (Fig. 7) and are developed as channel fills and lenses (< 3 m deep), draped on erosional surfaces. Some of the channel fills are capped by a high amplitude reflector (see the boxed section in the Chirp profile in Fig.

5). Such 'brightspot' reflections are often indicative of buried peat horizons (Bacharach *et al.,* 1998).

SB-IVii has the geometry of a massive sheet deposit. Unit SB-IVii is interpreted as sandy sediment on the basis of its backscatter properties, its high reflectivity, attenuation of the acoustic signal in the Chirp profiles, and correlation with the sand unit SS-I interpreted from the side-scan data.

4. DISCUSSION

Geophysical surveying off the north coast of Northern Ireland reveals mobile surface marine bedforms and a buried succession of tabular sedimentary units which have distinctive acoustic characteristics (Table I).

Table I *Summary of the side-scan and sub-bottom units identified in the study area, with their interpretations.*

Side-scan facies	*Sub-bottom unit*	*Thickness (m)*	*Lithology*	*Interpretation*
SS-I	SB-IV	3-6	Sand	Late Holocene marine sand (mobile)
SS-II	Not imaged	?	Gravel	Late Holocene gravel (mobile)
Not imaged	SB-III	3-5	Massive muds	Glaciomarine/marine mud
SS-III	SB-I/SB-II	< 3	Bedrock/ diamict	Diamict lag deposit (SB-II) and glacially scoured bedrock platform (SB-I)

These sedimentary units are also laterally extensive both across the study area (Cooper *et al*., 2002) and onshore the adjacent coast (McCabe *et al.*, 1994). Surface bedforms, imaged in the side-scan sonar data, have their crestlines orientated at right angles to the direction of strongest currents, characteristic of in-phase nearshore ripples developed on a uniform substrate. Changes in the position of the rippled sand (SS-I) and gravel substrates (SS-II) over the time-period of field observations (18 months) likely indicate changes in the strength of bottom currents. For example, the observed westward shift in the location of facies boundaries (Fig. 4ii) may reflect a decrease in the strength of east-going bottom currents, thus a decrease in eastward sediment transport. Other components, such as the strength of the west-going returning gyre west of Portstewart Head, may also partly determine the position of these facies boundaries and the degree to which bedrock/diamict surfaces are uncovered. Surface sediment movement may also have a strong onshore-offshore component driven by seasonal changes in wave regime and storminess.

Determining the elevation of sediment bodies or sedimentary boundaries is central to reconstructing RSL changes. At Portballintrae, McCabe *et al.* (1994) described an emergent late-glacial sediment succession comprising a glacially-eroded bedrock platform overlain by glaciomarine diamict, wave-disturbed and rhythmically-bedded sand/mud couplets, and interbedded gravel and sand. This sediment succession, found onshore above a present-day beach, closely matches that interpreted offshore from acoustic data presented in this paper. The mismatch in elevation between onshore and offshore evidence was noted by McCabe *et al.* (1994) and explained by differential isostatic effects in the Irish Sea region (McCabe, 1997; Knight, 2001). Likewise, spatial variability in RSL index points in Northern Ireland has been discussed by Carter (1982), Shaw and Carter (1994) and Wilson and Orford (2002). Additionally, the flat-bedded Tertiary basalts of the north coast of Ireland may have exerted a strong geological control on the elevation of bedrock shore platforms, thus of supposed RSL indicators (McKenna, 2002). Relating the elevation of specific sediment bodies or stratigraphic boundaries to RSL, therefore, may be rather difficult. It is therefore more useful to consider the general relationship of these sediment units to transgressive or regressive RSL stages using the systems tract approach, and associated changes in coastal and nearshore marine environments. These aspects are discussed below.

4.1 Event Sequence and Environmental Interpretation

The sub-bottom acoustic facies identified in this study, and their stratigraphic positions, can be used to reconstruct a history of environmental change for the north coast of Ireland.

1. During the last northward late Devensian ice advance off the north coast of Ireland and into the Malin Sea (associated with the Heinrich event 1 ice-rafting event at 14.5 ^{14}C kyr BP; McCabe *et al.*, 1998) ice scoured and smoothed the bedrock platform (SB-I), observed as the acoustic basement in this and other studies (e.g. Cooper *et al.*, 2002). The depth to which this glacial scouring took place is uncertain. Glacigenic landforms developed in bedrock are developed up to 100 m OD only 2 km onshore, but along the River Bann infilled bedrock palaeovalleys are present at –30 m OD (Battarbee, 1973). This elevation is identical to overdeepened and infilled bedrock palaeovalleys located 40 km to the west around the River Foyle (Bazley *et al.*, 1997), and corresponds to a prominent erosional surface in an adjacent offshore borehole (Bazley *et al.*, 1997) and the hypothesised RSL lowland (Cooper *et al.*, 2002). A –30 m OD surface may only be one of many such surfaces off the north coast; from the sub-bottom data presented in this study (Figs. 5-7) seaward-dipping bedrock platforms are also evident

at –24 m and –16 m OD. It is also possible that the lowermost bedrock platform is of earlier age, formed at the time of the last glacial maximum (~ 19-21 kyr BP) when ice from Scotland and Ireland extended farther offshore (Fyfe *et al.*, 1993). Such platforms may have been reoccupied several times by later ice advances.

2. Ice advance during Heinrich event 1 also resulted in the deposition of a subglacial diamict (till) (SB-II), evidenced by its structureless and chaotic acoustic signature. The surface relief of this unit may indicate subglacial streamlining. These characteristics correspond directly with streamlined till sections seen overlying bedrock onshore (McCabe *et al.*, 1994, 1998). The high-backscatter surface reflector of SB-II may suggest that the till is overconsolidated and/or has an erosional surface contact. Such subglacial sediment is not commonly recorded farther offshore (Binns *et al.*, 1974; Fyfe *et al.*, 1993).

3. During onshore ice retreat following Heinrich event 1 (~ 14-13 kyr BP) massive glaciomarine and marine mud was deposited (SB-III). The association of the retreating late Devensian Irish ice margin with high RSL is well documented (e.g. McCabe, 1997; Knight, 2001), and may suggest that the ice margin off the north coast was destabilised by marine flooding of the flat Malin Sea region during glacioisostatic depression of the land (Knight, 2003). The absence of internal reflectors within SB-III suggests that ice-rafted dropstones are not present within this unit. The well-defined upper termination of this unit at –12 m OD (Fig. 5) may indicate the position of contemporaneous RSL. Although this unit is laterally widespread, its variable thickness (3-5 m thick) and character of its uppermost surface suggest that this massive mud is transitional to the overlying sandy unit SB-IVi. The gradational upward transition to the overlying acoustic unit may reflect a change from muddy to sandy sediment (from marine to coastal in character) in the period after 13 kyr BP. Additionally, the uppermost surface of SB-III in nearshore areas is often sharp, indicating erosion during RSL fall (see Section 4.2), contrasting with the gradational upper transition of SB-III seen in deeper water, indicating more continuous sedimentation. The total preserved age of this unit is therefore likely greater in deeper than in shallower water and probably extends into the early Holocene.

4. Unit SB-IVi is present only in inshore areas to –8 m OD and lies draped over unit SB-III, forming broad channel fills (Fig. 7). The distinctive layered stratigraphy of SB-IVi, found especially between –6 m and –8 m OD, suggests alternating muddy and sandy layers (few tens of cm thick). Based on observed clinoforms, the calculated maximum geometry of individual channels is 3 m deep and 8 m in width, but overall channel sequences are up to 200 m across. This strongly suggests that channels were formed and infilled by a meandering proto-River Bann in an open estuarine

or intertidal environment. This interpretation is supported by sedimentary evidence at the Barmouth (Carter and Wilson, 1990) which shows freshwater marls and peats at –6 m OD. A peat sample from this level is radiocarbon-dated to 8960 ± 110 (Beta-34315) (Carter and Wilson, 1990). Progradation of these fluvio-estuarine sediments is evidenced by the seaward-dipping clinoforms seen in figure 6.

5. Late Holocene marine sands (SB-IVii) are distributed uniformly across the study area, draping earlier units and separated by both sharp (erosional) and gradational contacts from unit SB-IVi. These sediments, part of the present-day rippled sand (SS-I) and gravel substrate (SS-II), likely date from after the mid-Holocene RSL maximum which peaked at 6.5 kyr BP (Wilson and Orford, 2002). Radiocarbon dating indicates that these sediments are highly reworked and homogenised (Kershaw, 1986).

4.2 Systems Tract Analysis of the Sedimentary Succession

Examining the acoustic units described above within the context of the systems tract approach highlights the response of offshore sediment systems to changes in RSL, and identify periods of time in which these sediment systems were most active or inactive. Figure 8 is a schematic time-space diagram illustrating the major environmental changes offshore Northern Ireland from 15 kyr BP to present, and the periods of formation of SB-I to SB-IV.

Final planation of unit SB-I and deposition of unit SB-II relate to the last glacial advance at ~ 14.5 kyr BP. It is possible that SB-II was deposited by basal ice during both ice advance and retreat phases, draped by glaciomarine/marine distal muds (unit SB-III). Regionally, unit SB-III corresponds with the late Devensian-age Jura Formation, a glaciomarine diamict, identified offshore western Scotland in the Main Sea and Hebrides Sea (Fyfe *et al.*, 1993) where it overlies glaciated bedrock (SB-I), supporting the model presented here. Rapid late-glacial RSL transgression (transgressive systems tract) accompanying ice retreat is due to marine flooding of an isostatically-depressed basin (Eyles and McCabe, 1989). High RSL at this time is marked by the raised marine succession at Portballintrae (McCabe *et al.*, 1994), discussed previously. The highstand systems tract (HST) is evidenced by late-glacial raised beaches found across northeastern Ireland especially around 15-20 m OD (Stephens, 1963), and by a raised gravel beach ridge at 9.7 m OD on Rathlin Island, radiocarbon-dated to 12,320 ± 80 yr BP (Beta-48580) (Carter, 1993b). There is no clear offshore signature of this HST.

The – 30 m OD lowstand identified by Cooper *et al.* (2002) is not observed on our data which do not penetrate to that depth. Sea-level regression from the late-glacial highstand to the –30 m lowstand, however, is marked by an erosional surface over the nearshore components of SB-III (Fig. 5). Because offshore components of this acoustic unit do not show this erosional upper surface, we suggest that SB-III continued to accumulate during both regressive and transgressive phases (Fig. 8). The transition from SB-III to SB-IVi (marine to estuarine sedimentation) during continued RSL transgression marks an important change in depositional environment, likely related to a slowdown in the rate of transgression (at about 9 kyr BP) which allowed for fluvial progradation and estuary development, which reducted nearshore accommodation space. At this time, sedimentation likely matched the rate of RSL rise towards the mid-Holocene highstand of +2 to +3 m OD, and is marked around the north coast of Ireland by dated evidence for estuary and back-barrier infilling, and sand dune and intertidal peat development (Shaw and Carter, 1994; Wilson and McKenna, 1996). In the Bann estuary, this is also evidenced by repeated intertidal channel infilling and migration at around –8 to –6 m OD (SB-IVi). The highstand at around 6 kyr BP is marked by a distinctive surface across SB-IVi, which may be either an erosional or flooding surface. Estuary and ebb-tide delta progradation (Fig. 6) continued as RSL regressed from this highstand position to be at about –2 to –3 m OD at 4 kyr BP (Wilson and McKenna, 1996), although modelling evidence does not support this later lowstand (Lambeck, 1996).

The draped, uniform nature of the present offshore substrate (SB-IVii) suggests it has been reworked from pre-existing sediments under a period of oscillating RSL, and with a contribution from both onshore and offshore sources. Onshore, this unit overlies the fluvial unit SB-IVi, whereas offshore it overlies the marine unit SB-III (Fig. 8).

Throughout the period 15 kyr BP to present, there is not a close match between trangressive and regressive RSL phases and periods of time in which the distinctive sub-bottom acoustic facies accumulated (Fig. 8). For example, units SB-III and SB-IV accumulated under both transgressive and regressive RSL phases, and their deeper water components appear more insensitive to RSL changes than their shallow water components. The clear upper and lower stratigraphic boundaries of SB-IVi suggest a marked change in nearshore sediment dynamics which appears to be almost independent of RSL stage. The main geomorphic elements of the Bann estuary region were emplaced at this period between 9-6 kyr BP. The reasons why this time-period should be so dynamic in contrast to previous and subsequent periods are unknown. Likely reasons may include fluvial and nearshore sediment supply, climate change, wind, wave and tidal regimes, and human activity (e.g. Wilson and Braley, 1997). It may also be the case that more subtle changes in RSL, when linked to sediment supply, were more important in nearshore sediment transport in the mid-Holocene than were the larger RSL changes of the late-glacial period.

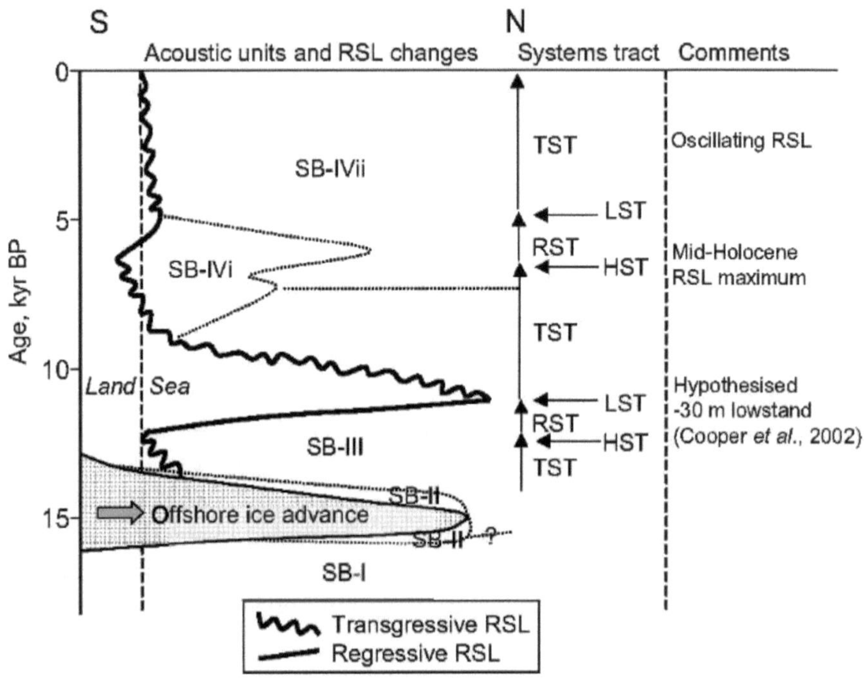

Figure 8. Schematic time-space diagram illustrating changes in RSL and the formation of different sediment packages off the River Bann estuary region, and formation of different systems tracts: transgressive (TST), lowstand (LST), regressive (RST) and highstand (HST). Postulated boundaries between different sediment packages are shown by the dotted lines

Off the north coast of Ireland, patterns of nearshore sediment responses to external forcing by RSL changes are similar to responses recorded on other paraglacial coasts. For example, an erosional unconformity, formed off the Gulf of Maine during RSL fall to a postglacial lowstand, was followed by RSL transgression and the deposition of fluvial-estuarine sediments (Kelley *et al.*, 1989; Barnhardt *et al.*, 1997). Similar transgressive-regressive RSL histories, and nearshore facies architecture, may therefore characterise a range of paraglacial shelves (Plag *et al.*, 1996).

REFERENCES

Bacharach, R., Dvorkin, J. and Nur, A. 1998. High-resolution shallow-seismic experiments in sand, Part II: Velocities in shallow unconsolidated sand. *Geophysics,* 63, 1233-1240.

Barnhardt, W.A., Belknap, D.F. and Kelley, J.T. 1997. Stratigraphic evolution of the inner continental shelf in response to late Quaternary relative sea-level change, northwestern Gulf of Maine. *Geological Society of America Bulletin*, 109, 612-630.

Battarbee, R.W. 1973. *A pollen and diatom study of the late-Flandrian sediments of Lough Neagh.* Unpublished PhD thesis, New University of Ulster.

Battarbee, R.W., Scaife, R.G. and Phethean, S.J. 1985. Palaeoecological evidence for sea-level change in the Bann estuary in the Early Mesolithic period. In: Woodman, P.C. (ed) *Excavations at Mount Sandel 1973-77.* Northern Ireland Archaeological Monographs No 2. HMSO, Belfast. 111-120.

Bazley, R.A.B., Brandon, A. and Arthurs, J.W. 1997. *Geology of the country around Limavady and Londonderry.* Geological Survey of Northern Ireland Technical Report GSNI/97/1.

Binns, P.E., Harland, R. and Hughes, M.J. 1974. Glacial and postglacial sedimentation in the Sea of the Hebrides. *Nature*, 248, 751-754.

Bowen, D.Q., Rose, J., McCabe, A.M. and Sutherland, D.G. 1986. Correlation of Quaternary glaciations in England, Ireland, Scotland and Wales. *Quaternary Science Reviews,* 5, 299-340.

Bridgland, D.R. 2002. Fluvial deposition on periodically emergent shelves in the Quaternary: example records from the shelf around Britain. *Quaternary International*, 92, 25-34.

Carter, R.W.G. 1982. Sea level changes in Northern Ireland. *Proceedings of the Geologists' Association*, 93, 7-23.

Carter, R.W.G. 1990. Coastal processes in relation to geographic setting, with special reference to Europe. *Senckenbergiana Maritima*, 21, 1-23.

Carter, R.W.G. 1993a. Geology, hydrology and land-use of Lough Neagh and its catchment. In: Wood, R.B. and Smith, R.V. (eds) *Lough Neagh.* Kluwer, Amsterdam. 11-33.

Carter, R.W.G. 1993b. Age, origin and significance of the raised gravel barrier at Church Bay, Rathlin Island, County Antrim. *Irish Geography*, 26, 141-146.

Carter, R.W.G. and Kitcher, K.J. 1979. The geomorphology of offshore sand bars on the north coast of Ireland. *Proceedings of the Royal Irish Academy*, 79B, 43-61.

Carter, R.W.G. and Rihan, C.L. 1976. The River Bann mouth bar. *Irish Geography*, 9, 121-123.

Carter, R.W.G. and Wilson, P. 1990. Portstewart Strand and the Bann estuary. In: Wilson, P. (ed) *North Antrim and Londonderry.* Field Guide No. 13, Irish Association for Quaternary Research, Dublin. 18-23.

Cooper, J.A.G., Kelley, J.T., Belknap, D.F., Quinn, R. and McKenna, J. 2002. Inner shelf seismic stratigraphy off the north coast of Northern Ireland: new data on the depth of the Holocene lowstand. *Marine Geology*, 186, 369-387.

Delaney, C. and Devoy, R. 1995. Evidence from sites in western Ireland of Late Holocene changes in coastal environments. *Marine Geology*, 124, 272-287.

Eyles, N. and McCabe, A.M. 1989. The Late Devensian (<22,000 BP) Irish Sea Basin: The sedimentary record of a collapsed ice sheet margin. *Quaternary Science Reviews*, 8, 307-351.

Fyfe, J.A., Long, D. and Evans, D. 1993. *United Kingdom offshore regional report: the geology of the Hebrides-Malin sea area.* HMSO, London, for the British Geological Survey.

Hamilton, A.C. and Carter, R.W.G. 1983. A mid-Holocene moss bed from eolian dune sands near Articlave, Co Londonderry. *Irish Naturalists' Journal*, 21, 73-75.

Hydrographic Office. 1986. *Admiralty Chart 2499: River Foyle to Derry.* Hydrographic Office, Taunton.

Kelley, J.T., Belknap, D.F. and Shipp, R.C. 1989. Sedimentary framework of the southern Maine inner continental-shelf – influence of glaciation and sea-level change. *Marine Geology*, 90, 139-147.

Kershaw, P.J. 1986. Radiocarbon dating of Irish Sea sediments. *Estuarine, Coastal and Shelf Science*, 23, 295-303.

Knight, J. 2001. Glaciomarine deposition around the Irish Sea Basin: some problems and solutions. *Journal of Quaternary Science*, 16, 405-418.

Knight, J. 2003. Evaluating controls on ice dynamics in the north-east Atlantic using an event stratigraphy approach. *Quaternary International*, 99-100, 45-57.

Lambeck, K. 1996. Glaciation and sea-level change for Ireland and the Irish Sea since Late Devensian/Midlandian time. *Journal of the Geological Society of London*, 153, 853-872.

Lawlor, D.P. 2000. *Inner shelf sedimentology off the North Coast of Northern Ireland.* Unpublished DPhil thesis, University of Ulster.

McCabe, A.M. 1997. Geological constraints on geophysical models of relative sea-level change during deglaciation of the western Irish Sea Basin. *Journal of the Geological Society of London*, 154, 601-604.

McCabe, A.M., Carter, R.W.G. and Haynes, J.R. 1994. A shallow marine emergent sequence from the northwestern sector of the last British ice sheet, Portballintrae, Northern Ireland, *Marine Geology*, 117, 19-34.

McCabe, A.M., Knight, J. and McCarron, S.G. 1998. Evidence for Heinrich event 1 in the British Isles. *Journal of Quaternary Science*, 13, 549-568.

McKenna, J. 2002. Basalt cliffs and shore platforms between Portstewart (Co Derry) and Portballintrae (Co Antrim). In: Knight, J. (ed) *Field Guide to the Coastal Environments of Northern Ireland.* University of Ulster, Coleraine. 157-164.

Pendlebury, D.C. and Dobson, M.R. 1976. Sediment and macrofaunal distribution in the eastern Malin Sea, as determined by side-scan sonar and sampling. *Scottish Journal of Geology*, 11, 315-332.

Plag, H.P., Austin, W.E.N., Belknap, D.F., Devoy, R.J.N., England, J., Josenhans, H., Peacock, J.D., Petersen, K.S., Rokoengen, K., Scourse, J.D., Smith, D.E. and Wingfield, R.T.R. 1996. Late Quaternary relative sea-level changes and the role of glaciation upon continental shelves. *Terra Nova*, 8, 213-222.

Roberts. J.C. 1976. The joint and fault patterns of the north coast of counties Antrim and Londonderry between Murlough Bay and Castlerock. *Proceedings of the Royal Irish Academy*, 76B, 619-628.

Shaw, J. and Carter, R.W.G. 1994. Coastal peats from northwest Ireland: implications for late-Holocene relative sea-level change and shoreline evolution. *Boreas*, 23, 74-91.

Shipp, R.C., Belknap, D.F. and Kelley, J.T. 1991. Seismic-stratigraphic and geomorphic evidence for a postglacial sea-level lowstand in the northern Gulf of Maine. *Journal of Coastal Research*, 7, 341-364.

Stephens, N. 1963. Late-glacial sea levels in northeast Ireland. *Irish Geography*, 4, 345-359.

Stoker, M.S., Pheasent, J.B. and Josenhans, H. 1997. Seismic methods and interpretation. In: Davies, T.A., Bell, T., Cooper, A.K., Josenhans, J., Polyak, L., Solheim, A., Stoker, M.S. and Stravers, J.A. (eds) *Glaciated Continental Margins: An Atlas of Acoustic Images*. Chapman and Hall, London. 9-27.

Syvitski, J.P.M. 1991. Towards an understanding of sediment deposition on glaciated continental shelves. *Continental Shelf Research*, 11, 897-937.

Vail, P.R., Audemard, F., Bowman, S.A., Eisner, P.N. and Perez-Cruz, C. 1991. The stratigraphic signatures of tectonics, eustasy and sedimentology – an overview. In: Einsele, G., Ricken, W. and Seilacher, A. (eds) *Cycles and Events in Stratigraphy*. Springer-Verlag, Berlin. 617-659.

Wilcock, D.N. 1982. Rivers. In: Cruickshank, J.G. and Wilcock, D.N. (eds) *Northern Ireland Environment and Natural Resources*. Queen's University, Belfast, and New University of Ulster. 43-71.

Wilson, H.E. 1972. *Regional Geology of Northern Ireland.* HMSO, Belfast.

Wilson, P. 1991. Buried soils and coastal aeolian sands at Portstewart, Co. Londonderry, Northern Ireland. *Scottish Geographical Magazine*, 107, 198-202.

Wilson, P. 1994. Characteristics, age and significance of buried podzols in the Grangemore sand dunes, Co Londonderry. *Irish Naturalists' Journal*, 24, 475-480.

Wilson, P. and Braley, S.M. 1997. Development and age structure of Holocene coastal sand dunes at Horn Head, near Dunfanaghy, Co. Donegal. *The Holocene,* 7, 187-197.

Wilson, P. and McKenna, J. 1996. Holocene evolution of the River Bann estuary and adjacent coast, Northern Ireland. *Proceedings of the Geologists' Association*, 107, 241-252.

Wilson, P. and Orford, J. 2002. Relative sea-level changes. In: Knight, J. (ed) *Field Guide to the Coastal Environments of Northern Ireland.* University of Ulster, Coleraine. 11-15.

Chapter 3

A SEABED CLASSIFICATION APPROACH BASED ON MULTIPLE ACOUSTIC SENSORS IN THE HUDSON RIVER ESTUARY

Frank O. Nitsche, Suzanne Carbotte, William Ryan and Robin Bell
Lamont-Doherty Earth Observatory of Columbia University

1. INTRODUCTION

Due to the increasing demand for clean water, recreation areas, and healthy ecosystems the management of watersheds has become an issue of increasing importance. This demand has increased the need for detailed inventories of the present state of watersheds and the understanding of related processes. In 1996, the New York State Department of Environmental Conservation (DEC) initiated an effort to map the benthic habitat of the Hudson River Estuary as part of a larger Hudson River Action Plan. This project includes extensive mapping using sidescan sonar, sub-bottom profiling, single and multi-beam bathymetry, as well as collecting ground truth data with sediment cores, grab samples, and sediment profiling imagery (SPI). The goal of the project is the creation of a comprehensive data set that includes detailed interpretive maps of sediment distribution, grain size, bed forms, and benthic habitats (Ladd *et al.*, 2002).

During an initial pilot study (1997-2000) four sections of the Hudson River totaling ~60 km were selected, surveyed, and interpreted. Based on the results of this study the entire ~240 km of the lower Hudson River from Manhattan to Troy will be mapped (Fig. 1).

The pilot study produced over 30 Gbyte of data that was interpreted by visual analysis to derive a qualitative sediment classification based on sidescan sonar backscatter amplitudes, reflection seismic character, and

D.M. FitzGerald and J. Knight (eds.),
High Resolution Morphodynamics and Sedimentary Evolution of Estuaries, 33-55.

bottom samples. Significantly more data will be collected when the remaining parts of the estuary are mapped. Automatic or semi-automatic classification procedures are needed to make the interpretation process more efficient and quantitative.

Difficulty of managing the large data volumes that derive from modern acoustic surveying techniques has led to the development of automatic classification methods (e.g. Pace, 1983; Reut *et al.*, 1985; Blondel and Murton, 1996; Blondel, 2001). Such methods include classification based on the texture of sidescan images (e.g. Reed and Hussong, 1989), the spectral analysis of sidescan data (e.g. Reut *et al.*, 1985; Pace and Gao, 1988; Tamsett, 1993), the grazing angle of multi-beam swath bathymetry backscatter (e.g. Hughes Clarke, 1994), the sea-floor morphology (e.g. Mitchell and Hughes Clarke, 1994; Fox, 1996; Herzfeld and Higginson, 1996), and the waveform of echo sounder or sub-bottom profiler data (e.g. Pace, 1983; Hamilton *et al.*, 1999; Hines *et al.*, 2001). Although there are studies using, for example, different frequencies of the same system (e.g. Ryan and Flood, 1996), most of these approaches are based on parameters derived from a single system.

The Hudson River Benthic Mapping Project produced multiple co-located data sets for different acoustic systems that can be exploited for automatic classification. The different acoustic systems employed include sidescan sonar, seismic profiling, single-, and multi-beam bathymetry. Sidescan sonar and seismic data are collected simultaneously with both systems installed on the same boat. Therefore, data from both systems yield the same navigation and survey conditions. Simultaneous data collections from multiple devices has recently become more common due to modern developments in the size and handling of such systems (e.g. Roberts, 2001).

This paper describes our approach of an open classification procedure which combines data from multiple sensors. Based on data collected during the Hudson River Benthic Mapping Project we derived different parameters that appeared useful for later classification. A comparison and correlation with ground truth information, including samples from cores, grabs, and images yields insight on the meaning of different parameters and their use for classification. Selected parameters have been combined as layers comparable to optical bands of multi-spectral satellite images and supervised as well as unsupervised classification techniques common in remote sensing have been tested.

2. SETTING OF THE LOWER HUDSON RIVER

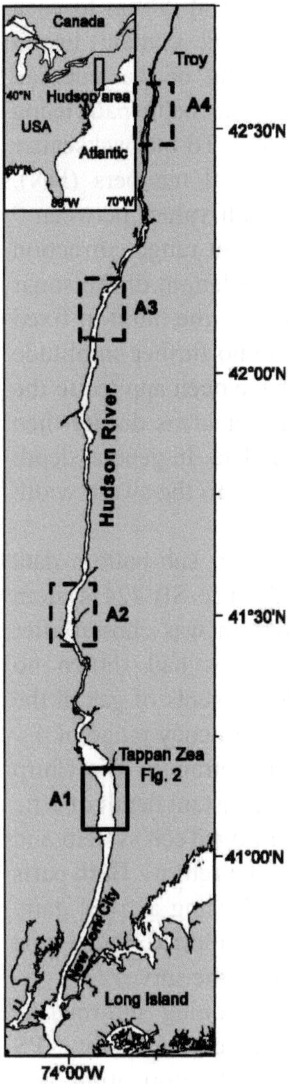

The study area is the Tappan Zee part of the ~240 km long lower Hudson River Estuary between Manhattan and Troy (Fig. 1), which is dominated by tides and shows decreasing influence of the salt water intrusion to the north. Water depths of the study area range between 10-15 m in the main channel and 2-4 m for the subtidal flats which occupy most of the Tappan Zee area.

The Hudson valley was mainly shaped by glacial erosion. Its Holocene sedimentary history consists of several main stages (Newman *et al.*, 1969): (i) After retreat of the ice at the end of the last glaciation, glacial lakes were formed when the melt-water runoff was blocked by the terminal moraines; (ii) after draining of these lakes the Hudson valley changed into a fluvial environment; (iii) with ongoing sea-level rise marine water flooded the Hudson valley and the present estuarine environment was established. Seismic and sampling studies suggest variations in flow regime and salinity throughout the estuarine stage (Newman *et al.*, 1969; Carbotte *et al.*, in prep.).

The present sediment cover within the saline mid-portion of the estuary consists mainly of silt with high amount of organic material and clay whereas the amount of sand increases towards both the north and south (Newman *et al.*, 1969; Coch and Bokuniewicz, 1986).

Figure. 1. Overview map of the lower Hudson River between Manhattan and Troy. Encircled areas have been surveyed during the first phase of the project. The southernmost encircled area (solid line) represents the Tappan Zee, which is the focus of this study.

3. DATASET

This study is based on simultaneously collected sidescan sonar and sub-bottom profiler data. Sidescan data were collected using an EdgeTec DF-

1000 dual-frequency system (100 kHz and 384 kHz) which was towed from the bow of the boat at a constant depth of 1.5 m to avoid disturbances from the boat's wake as well as collisions in the shallow parts of the Hudson River. The maximum swath-range of this system was set at 200 m to each side. Previous tests had shown that the 100 kHz data contain useful information up this range and would enable overlapping of the sidescan tracks that allowed substitution of the high amplitude near-nadir path using the neighboring track during mosaic generation. The raw 16-bit backscatter amplitudes are reduced by linear scaling to 8-bit digital numbers (DN), which can easily be represented in grayscale images with values between 0 and 255. Further processing consisted of layback and slant range correction before final mosaics are created. Source level and pulse length of the sonar system were kept constant throughout the survey. Besides the build-in fixed time-variant gain (TVG) of 60 dB to 300 ms (100 kHz) no further amplitude corrections for bottom slope, depth, or beam pattern have been applied to the data, because detailed bathymetry was only available for areas deeper then 4 m and not recorded simultaneously with the sidescan data. In general depth variation (2-20 m) are small in the survey area compared to the swath width (200 m).

Simultaneously with the sidescan data, single-channel sub-bottom data were collected using a Chirp sub-bottom profiler EdgeTec SB-424 system with a sweep of 4 - 16 kHz. This high-frequency system was chosen after previous tests of systems with lower frequency ranges had shown no improved sub-surface penetration of 2-5 m due to the presents of gas in the sediment in many parts of the river bottom. The high frequency range of 4 – 16 kHz provided the best resolution for this limited penetration. The Chirp sonar was towed behind the boat or along the side at a constant depth of 1 m. Correlation of the chirp sweep was handled inside the EdgeTech system and raw output data consists of real and imaginary parts of the returns. Both parts were combined by calculating the envelope and scaled using a fixed gain. Neither TVG nor ping-to-ping amplitude balancing was applied. Pulse type and power output of the source were kept constant during the survey.

All data were stored digitally together with positioning information derived from a differential GPS system and have been corrected for GPS antenna position, tow cable (layback) and tides. Tidal information was obtained from fixed NOAA and USGS tide gauges as well as gauges specifically deployed for the time of the survey along the Hudson River. Water level was reduced to NADV88 datum and interpolated between these stations. The areas were covered in north-south as well as in east-west direction to provide insonification from orthogonal angles. The grid spacing was 95 m for the N-S lines and 185 m for the E-W lines to ensure significant overlap of the sidescan swath.

4. DERIVING PARAMETER FOR CLASSIFICATION

To test the classification approach we concentrate on the Tappan Zee area, which consists of extended subtidal flats (1-4 m deep) and a <15 m deep main channel (Fig. 2). For this area all acoustic and ground truth data had been analyzed and manually interpreted (Fig. 2b). For example, oyster beds are found on the marginal flats and confirmed by samples. Recent sediments are located within the channel and in an L-shaped area on the flats. These recent sediments are identified through the presents of a ^{137}Cesium (Cs) peak in gravity cores that is related to nuclear testing during 1950-1960 (Simpson *et al.*, 1976). Hence we have a very good understanding of the distribution of sediment types of the river bottom in this area (Carbotte *et al.*, in prep.).

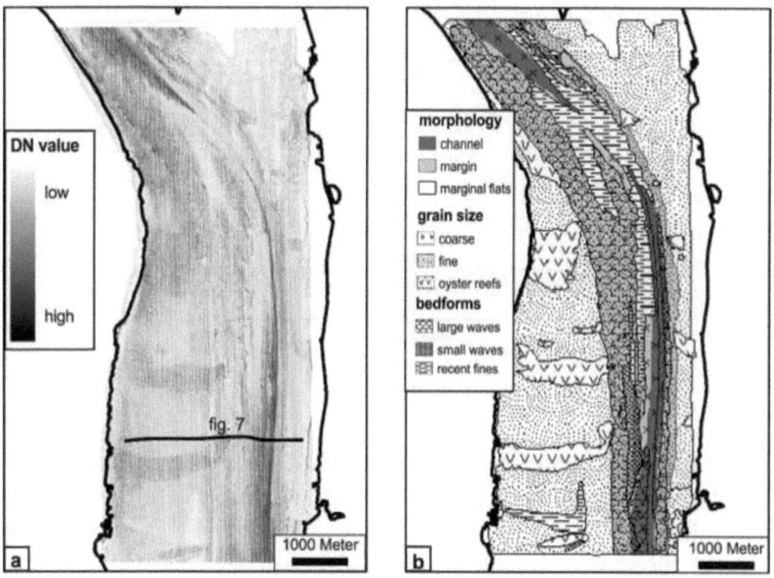

Figure. 2. (a) Sidescan sonar mosaic of the Tappan Zee area (Fig. 1). Dark gray represents high backscatter, while light gray represents low backscatter values. (b) Manually derived interpretation based on all available information including sidescan sonar, seismic and bathymetry data.

4.1 Sidescan sonar data

Most parameters used in this approach are derived from sidescan sonar data. The signal strength in sidescan imagery results from many factors including changes in sea-floor roughness, grain size, and compaction (Pace, 1983; Blondel and Murton, 1996).

Amplitudes of sidescan data can vary with the angle of transmission and receiving of the energy. This variation in amplitudes is caused by the beam pattern of the system and the variation of backscatter and direct back-reflection with distance to the nadir (Blondel and Murton, 1996). Many of these factors are depending on the transmission angle or the angle between the sea-floor and the direction of insonification called the grazing angle (Fig. 3). Therefore, several authors proposed to use the grazing angle instead of the distance from the nadir to compare different data sets to reduce such system related variations of the signal (Hughes Clarke, 1994; Keeton and Searle, 1996).

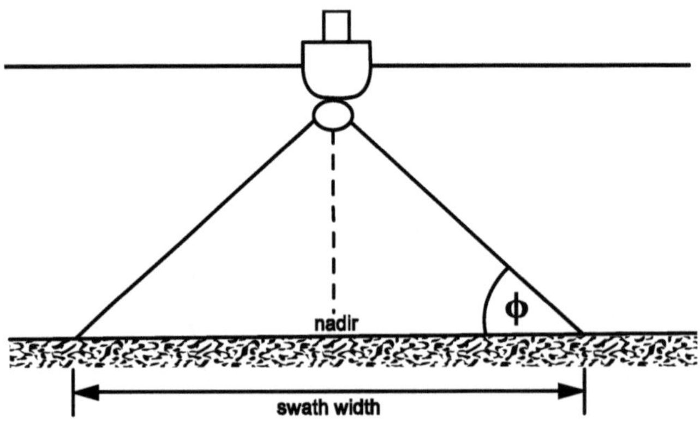

Figure. 3. Sketch to illustrate the grazing angle Φ.

For this study grazing angle is calculated assuming a horizontal bottom, because no bathymetric correction data were acquired with this system. Due to the shallow water depth and relatively gentle slopes this assumption should be a fair approximation for most of the survey area.

Hughes Clarke (1994) demonstrated that parameters derived from backscatter amplitudes vs. grazing angle vary between different bottom types. These parameters include average backscatter amplitudes for different ranges of grazing angles, standard deviation, a coefficient of variation, and the slope of the amplitude fall off.

Following this approach, we calculated the average amplitude response over different ranges of grazing angles. However, the ranges chosen in this study are different from those used by Hughes Clarke (1994), because our data have been collected in very shallow water (mostly <20 m) and yield mainly small angle values (Fig. 4).

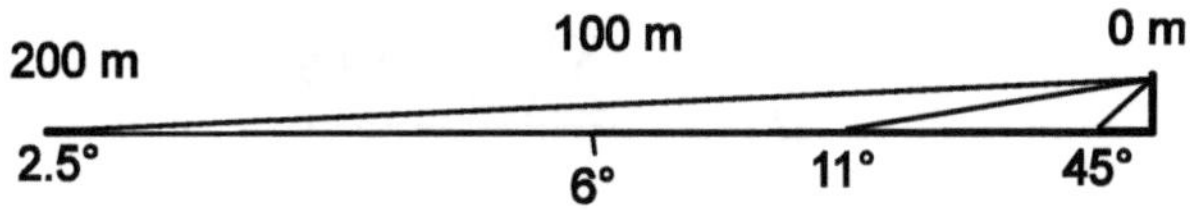

Figure. 4. Sketch to illustrate the relation of grazing angle and range for the shallow water settings of this survey.

The amplitude vs. grazing angle function is calculated over an average of 10 successive pings combining the port and starboard soundings. With a survey velocity of 5 kn this represents a distance of ~8.5 m in direction of the ship motion. Single values are calculated for each combination of 10 pings representing ~400 m of river bottom perpendicular to the survey track. Therefore it is likely that some pings cross the boundary of different substrate and thus the calculated parameter value represents an average of these different bottom types.

4.2 Calculation of means and related parameters

Figure 5 shows an example of backscatter amplitude vs. grazing angle for a single ping. The amplitudes are relative values represented by digital numbers (DN) ranging between 0 and 255. We calculated the total mean μ_{all} between angles θ of 3° and 70° as

$$\mu_{all} = \frac{\sum_{\theta>3}^{\theta<70} DN(\theta)}{n_{3..70}} \quad (1)$$

where n is the number of values in this range. Angles smaller then 3° were excluded because these far offset data yield a low signal-to-noise ratio. Angles larger then 70° were excluded because they are influenced by the back reflection of the vertical incident rays, though very few samples actually have angles larger than 70°. This range was used for all subsequent calculations of other parameters. To quantify amplitude differences between low and high grazing angles we calculated two means with restricted range of grazing angles in addition to the total mean amplitude

$$\mu_{10} = \frac{\sum_{\theta>3}^{\theta<10} DN(\theta)}{n_{3..10}} \quad (2)$$

$$\mu_{20..70} = \frac{\sum_{\theta>20}^{\theta<70} DN(\theta)}{n_{20..70}} \quad (3).$$

The first mean contains values for outer range samples with grazing angles between 3° and 10°. This range contains most data points. For angles <3° the noise level increases strongly. The second average includes inner range samples between 20° and 70° and usually contains fewer values due to the shallow water depths surveyed. Other ranges of angles were tested but the selected two ranges appeared most promising results for the classification.

Hughes Clarke (1994) showed that the variation of the amplitude values with grazing angle could be indicative of different sediment types. To quantify this variation we calculated the standard deviation σ_{all} as

$$\sigma_{all} = \sqrt{\frac{\sum_{\theta>3}^{\theta<70} \left(DN(\theta) - \mu_{all}\right)^2}{n_{all} - 1}} \quad (4)$$

and the coefficient of variation CV as

$$CV = \frac{\sigma_{all}}{\mu_{all}} \quad (5).$$

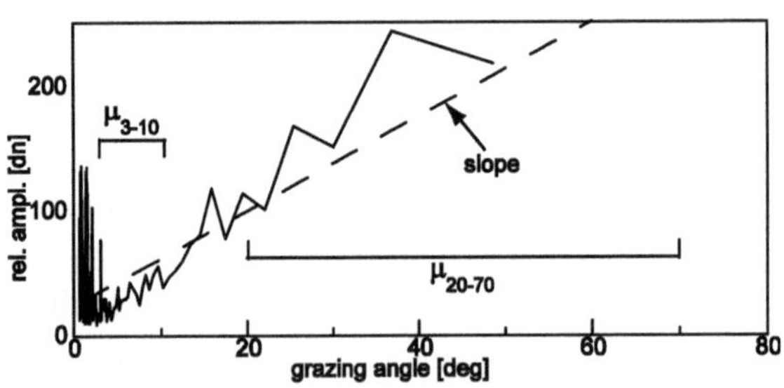

Figure. 5. Relative amplitudes of one sidescan sonar ping against grazing angle. Different grazing angle ranges are marked as used for calculation. Dashed line represents the calculated slope for this ping.

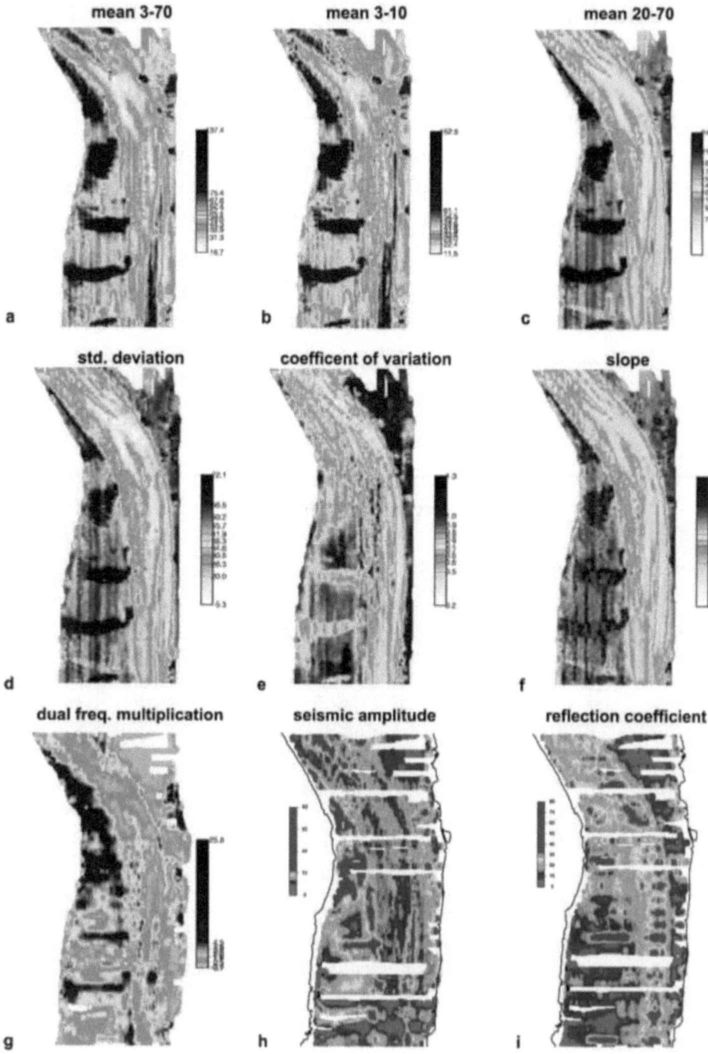

Figure. 6. Examples of calculated parameters for north-south profiles of 100 kHz sidescan data. Data have been gridded with 60 m grid size. Different means as outlined in figure 5 are shown in (a), (b) and (c). (d) and (e) represent standard deviation and coefficient of variation of (a). The slope is shown in (f). (g) shows the mean of the multiplication of 100 kHz and 384 kHz data. Seismic reflection amplitude and coefficient are displayed in (h) and (i). Note that the color scales are different and adjusted to the value range of each data set. The same parameters have been calculated for 384 kHz data and for both frequencies of the EW-profiles.

Another possible classifier of different sediment types is the amplitude decay of the grazing angle curve (Hughes Clarke, 1994). Therefore, we calculated the slope of the grazing angle curve by performing a least-square fit to the data (Fig. 5). Figure 6 (a-f) yields examples of the different parameters calculated for North-South oriented profiles of 100-kHz sidescan data. The same parameter were calculated for 384 kHz data and for East-West oriented profiles.

Ryan and Flood (1996) demonstrated the potential of distinguishing different sediment types by multiplying the sidescan data collected at two different frequencies. Because we simultaneously collected two sets of sidescan data at 100 kHz and 384 kHz we were able to calculate this parameter by multiplying each sample of both datasets and calculating the mean values after Ryan and Flood (1996) (Fig. 6g).

4.3 Seismic data

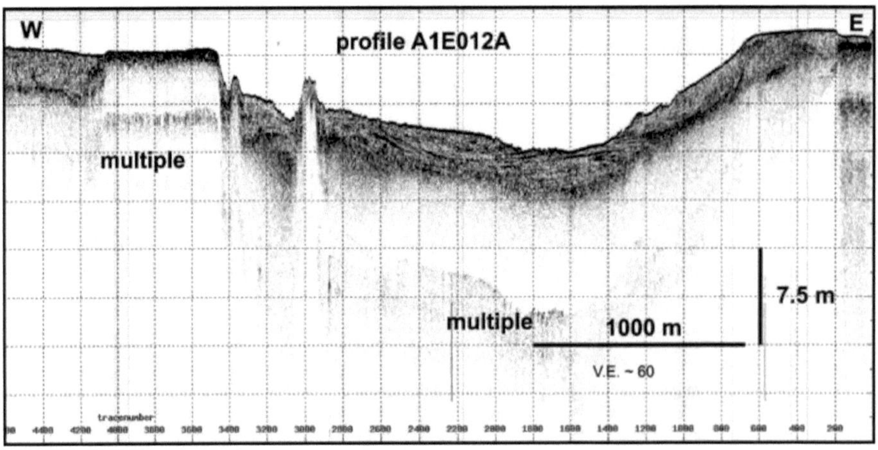

Figure. 7. Example of seismic data collected during the survey (Fig. 2). Relative reflection amplitudes were derived from the river bottom reflection. Vertical exaggeration is ~1:60.

In addition to the parameters derived from the sidescan data we calculated parameters based on the seismic data (Fig. 7) including river bottom amplitude and reflection coefficient. A bottom detect algorithm based on an amplitude threshold criteria was used to determine the travel time of the river bottom reflection. The reflection amplitude was determined by averaging amplitude values of 5 samples (~0.2 ms) on each side of the detected bottom to allow for variations and uncertainty of the bottom detect algorithm. No trace-to-trace amplitude balancing had been performed on the data and, therefore, relative amplitude differences between traces were still preserved. Furthermore, we estimated a reflection coefficient. Instead of using the quotient between the amplitudes of the source signal which is unknown and the river bottom return we used the quotient between the amplitude of the river bottom return and the first multiple, assuming the water-air surface is a perfect reflector. The generally good sea-stage limited scattering of the energy at the water-air interface. The amplitude of the multiple is estimated by averaging the values at a travel-time two times of the bottom reflection. For some parts of the data where the energy of the multiples is interfering with energy of real reflections this approach might not produce the correct amplitude of the multiple.

5. CORRELATION WITH GROUND TRUTH DATA

One of the major advantages of this study is the high number of ground truth data available for the test area. As part of the project 69 shallow cores (< 2 m length), 55 grab samples and 75 sediment profile imaging (SPI) data have been collected for the Tappan Zee area. These samples have been analyzed for grain size and habitat description and structured into bottom types. These visual and sample data permit acoustic parameters to be related to sediment properties.

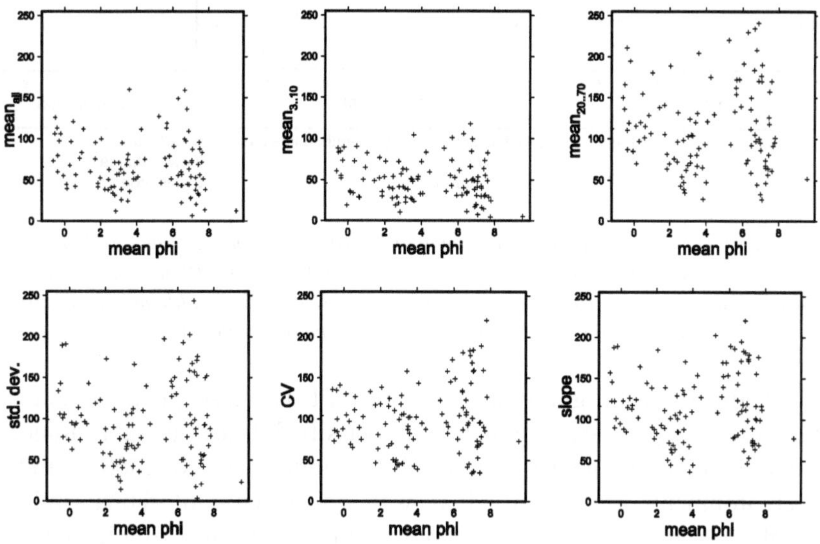

Figure. 8. DN values of different parameters derived from sidescan sonar 100 kHz line compared to mean grain size as phi value. Note that the lower values of the mean$_{3..10}$ are related to the lower amplitudes of the smaller grazing angles (compare with fig. 5).

Figure 8 shows the different parameters derived from 100 kHz sidescan data plotted against mean grain size of the core tops and grab samples. Although all plots exhibit considerable scatter, some parameters like total mean and mean$_{3..10}$, yield a trend of lower DN values with increasing phi, i.e.

smaller grain size. In contrast, $mean_{20..70}$, CV and the standard deviation vary strongly. In general, the correlation of DN-values of single parameters and grain sizes is poor.

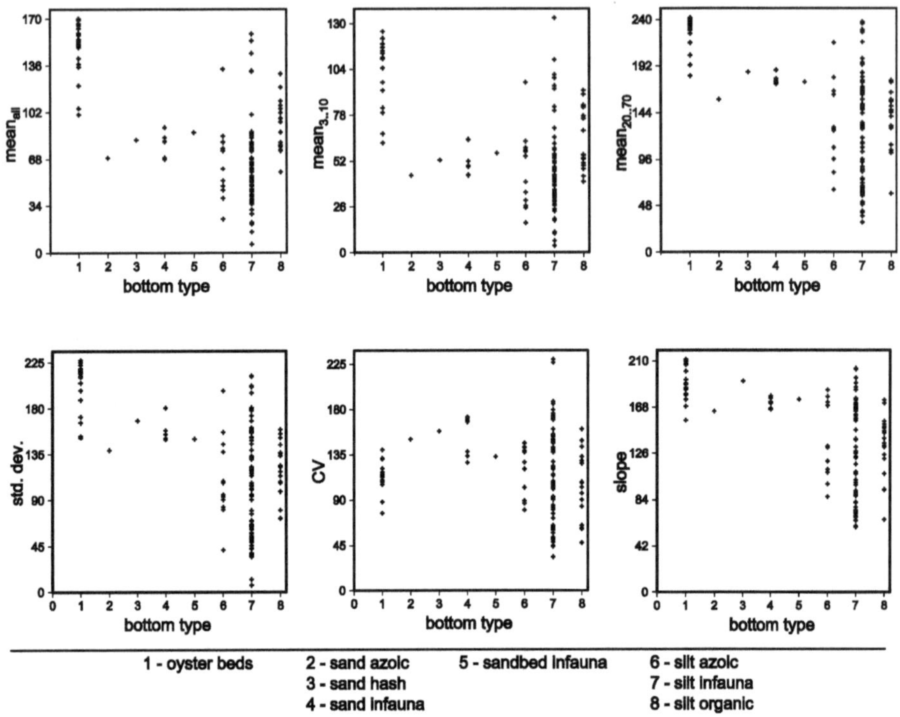

Figure. 9. DN values of different parameters derived from sidescan sonar 100 kHz line compared to bottom type classification made by the SPI data. Most SPI bottom types are more constrained to certain DN ranges than the mean grain size of Fig. 8, although the assignment of certain DN-values to a certain bottom type is not unique.

In addition to the grain size data the SPI data also yield ground truth information. These imaging data are classified based on appearance of benthic fauna in the images and visual descriptions (Iocco, 2000). Sand and silt environments are identified as well as presence of fauna (infauna), general organic material (organic), absence of life forms (azoic), shell hash, and oyster beds. These bottom types are compared to the different acoustic parameters in figure 9. The majority of SPI sites have been identified as silty habitats (types 7, 8) that cover a wide range of DN values and overlap with most other habitat classes. The oyster beds (type 1) cover a more distinct range of usually high DN values, but they still overlap with sandy and silty values. The sandy habitats (types 2-4) are apparently restricted to a limited range of medium DN values. But the number of sandy habitat samples is small compared to the other classes consisting of only one to five samples.

However, although most DN values cannot be uniquely assigned to certain bottom types of the SPI classification, the different values appear more separated from each other than the grain size range in Figure 8.

6. CLASSIFICATION APPROACH

Before the actual classification, we normalized all parameters to a range of 0-255 to ensure that all parameters have the same impact on the process. Furthermore, we combined all parameters derived from the N-S sidescan profiles of both frequencies. Applying a principle component analysis these 12 parameters were reduced to four by selecting the four most relevant components. The same procedure was performed for the E-W sidescan profiles. This additional analysis reduced the total numbers of layers and allowed the selection of principal components that were less affected by the striping pattern visible in Fig. 6. The selected principal components (PCx, where x represents the order of the component, i.e. x=1 being the most significant component) were combined together with parameters derived from the seismic and from the multiplication of the two frequencies. Table I lists the selection of parameters used here.

Table I. List of selected parameters for classification

Seismic amplitudes (N-S)	PC1 N-S sidescan	PC1 E-W sidescan
Seismic amplitudes (E-W)	PC2 N-S sidescan	PC2 E-W sidescan
Multiplied sidescan freq. (N-S)	PC4 N-S sidescan	PC3 E-W sidescan
Multiplied sidescan freq. (E-W)	PC5 N-S sidescan	PC4 E-W sidescan

To perform the actual classification based on the multiple parameters, we established algorithms commonly used in analysis and classification of satellite images. The different parameter layers we have calculated are used instead of the optical bands of satellite images.

Each of these n parameter layers represents a dimension in n-dimensional space (Fig. 10). The aim of the classification process is to divide this space into different regions that represent individual classes. Similar classification approaches have used with parameters derived from single systems (e.g. Fox, 1996; Hamilton *et al.*, 1999).

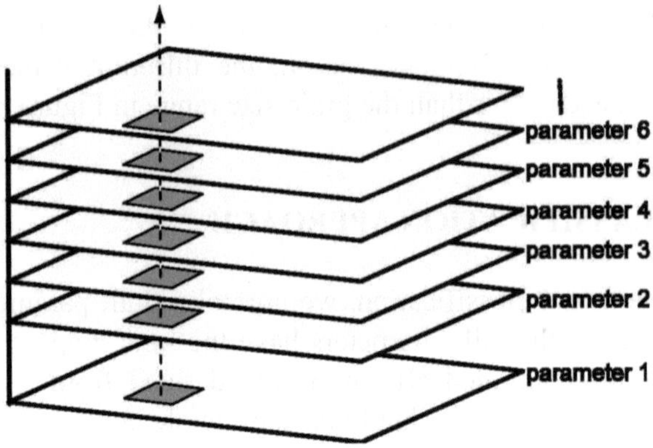

Figure. 10. Each of the different parameters represents a layer of an n-dimensional space. The location of one point in this space is given by the value of a cell in each layer.

We applied supervised and unsupervised classification algorithm (Schowengerdt, 1997). Algorithms for both are commonly implemented in image analysis and Geographic Information Systems (GIS) software. For the supervised classification, training areas for which the nature are known are identified and the reference signature, i.e. the combination of the different parameters, of the related classes is determined. Subsequently, all other points are classified by assigning them to the class that has the most similar signature. This classification is done by applying a maximum likelihood algorithm, which calculates a probability function for each layer that describes the probability of a point belonging to a certain class (Schowengerdt, 1997). Each point is then assigned to the class that has the highest probability for this point. This method requires that good a priori knowledge from ground truth like cores or other sources is available, and, therefore, the nature of different classes is known.

In contrast, unsupervised classification algorithms are used if no a priori information is available. In this case, a clustering algorithm is typically used to divide the data space in clusters of points with similar parameter values, i.e. they are close together in the n-dimensional parameter space (Schowengerdt, 1997). One common algorithm for deriving the clusters is the k-mean approach (Duda and Hart, 1973). The ArcInfo implementation used here is a variation of k-means based on the ISODATA algorithm (Ball and Hall, 1967). The number of classes used for the clustering can be chosen freely. Tests with different numbers of classes should be run to check the significance of the different classes. This method does not apply any knowledge of the real physical meaning of different classes. Thus, the output classes need to be manually assigned to physical bottom types.

7. RESULTS

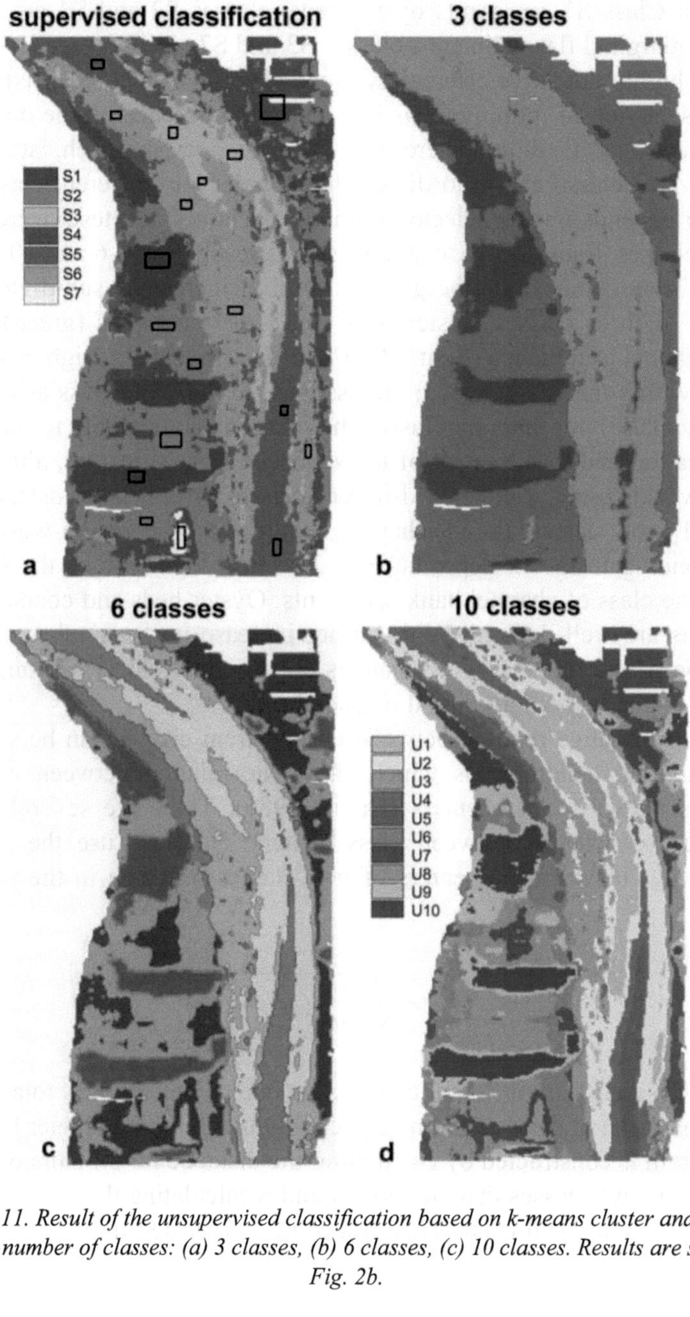

Figure. 11. Result of the unsupervised classification based on k-means cluster analysis for different number of classes: (a) 3 classes, (b) 6 classes, (c) 10 classes. Results are similar to Fig. 2b.

Supervised classification was performed on the Tappan Zee data by selecting training areas for seven different classes based on the initial manual interpretation (Fig. 11a). The result is similar to the original interpretation (Fig. 2b): Class S1 represents oyster beds, classes S2 and S4 are different types of marginal flat sediments, classes S3 and S7 represent recent deposits, class S5 highly reflective channel axis, and class 6 channel bank sediments.

The selection of training areas had a strong influence on the outcome of supervised classification. Therefore, in a second approach, we applied unsupervised classification to divide the data set into different classes. The outcome depends on the selected number of classes and tests were run for 2 to 20 classes. Figure 11 (b, c, and d) shows results for 3, 6 and 10 different classes. They demonstrate the general trend of increasing sub-dividing. For the case of three classes, oyster beds (red), marginal flats (green) and the main channel sediments (light blue) are discriminated. High backscatter regions within the channel (red) are assigned to the same class as the oyster beds. By classifying into six classes the image is comparable to the manual interpretation and to the result of the supervised classification, although the unsupervised results yield two different classes on the flats instead of one and did not detect the L-shaped feature at the south-west corner corresponding to recent deposits (Fig. 2b). Instead it assigns the L-shaped area to the class of channel bank sediments. Oyster beds and coarse channel sediments are well defined. With further increase in the number of classes the shape of the major classes changes only slightly (Fig. 11d) and classes are subdivided without altering the general image.

The quantitative relation between the different classes can be visualized by calculating dendrograms which show the relation between classes as distances between class centers (Davis, 1986). There are several ways to calculate the distance between class centers. Here we use the Euclidian distance d_{mn} between the centers of two classes n and m in the parameter space

$$d = \sqrt{\sum_{i=1}^{N} (\mu_{ni} - \mu_{mi})^2} \qquad (6)$$

where n and m are different classes; i is one parameter of N total; μ_{mi} and μ_{ni} are the mean of class m and n respectively in the i-th parameter layer. The dendrogram is constructed by calculating the distance for all pairs of classes, merging the two classes that are closest and recalculating the distance values using the new merged class. This procedure is repeated until all classes are merged.

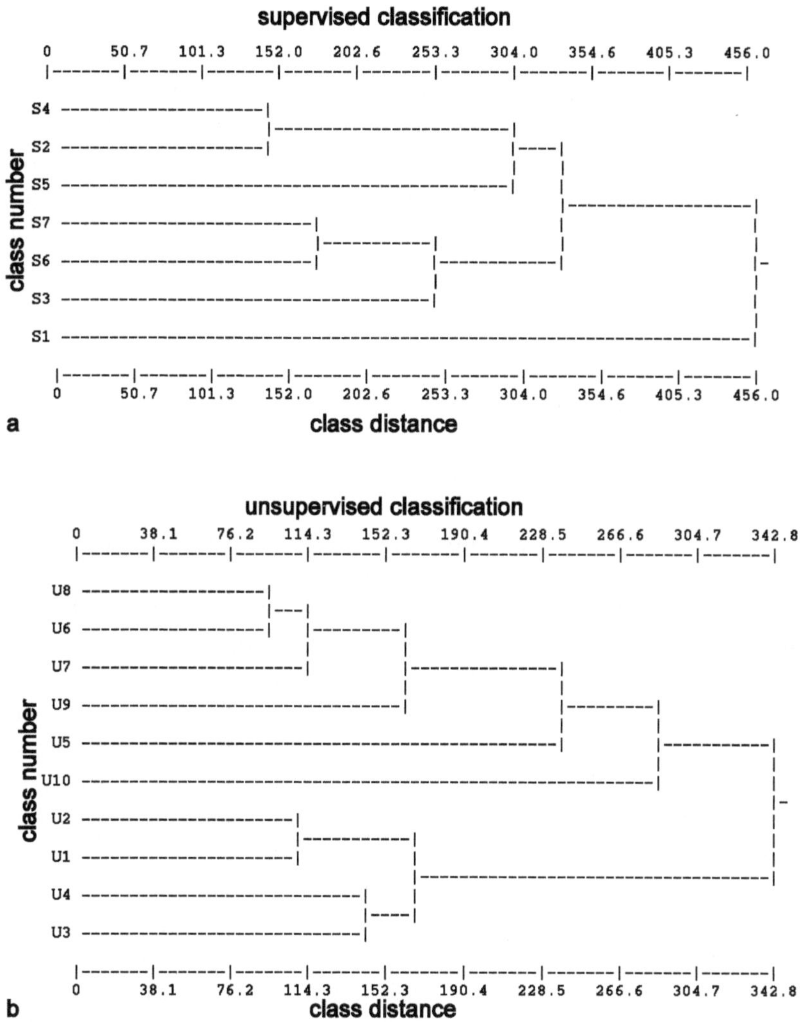

Figure. 12. Dendrograms showing the differences between classes obtained from (a) the supervised classification and (b) the unsupervised classification. The value of the x-axis represents the Euclidian distance between two classes.

The resulting dendrogram of the supervised classification indicates the close relation of the two margin flat classes S2 and S4 (Fig. 12a) as well as the similarity between the channel bank sediment classes S6 and S7, which is a low backscatter region interpreted as thin layer of recent deposits. Both, S6 and S7, are closer in character to the marginal flat recent sediment class (S3) then to the strongly reflective sediments in the channel (S5) or the compacted marginal flats classes (S2, S4). The most distinct class is S1 which represents the oyster beds.

The dendrograms of Figure 12b correspond with the unsupervised classification with 10 classes (Fig. 11d). Classes U6, U7, and U8 are all closely related and all correspond with marginal flat sediment. Class U9 is related to these classes and appears to be an intermediate class between U10 (oyster beds) and the flats. Classes U1, U2, U3, and U4 are closely related and are mostly located in the channel/channel bank area where fine-grained sediments are found. Class U5, which corresponds to high reflective channel sediments, is more closely related to the marginal flat sediments and the oyster beds then to the finer channel sediment.

8. DISCUSSION

Application of both supervised and unsupervised classification led to similar discrimination of major bottom types in the Tappan Zee test area for the acoustic parameters used in this study. Once the data have been collected and processed these parameters can be calculated more or less automatically and converted into the layers used for classification. Manual identification of training areas is necessary for supervised classification. Training areas may be difficult to define with confidence if ground truth information is sparse. In contrast, the clustering algorithm of the unsupervised classification requires the number of classes to be identified and the interpreter also needs to decide, which the relevant classes are and what their physical meaning is.

If the signature-class relation for certain river bottom types is available beforehand, this procedure could be fully automated. Such automation would require that the survey system setup as well as the environmental and bottom conditions are the same as within the reference area. In most cases this is not a valid assumption and new training areas need to be defined for different surveys. The need for adjustment to different survey conditions or general changes of the environment is repeatedly emphasized in reports on different classification approaches (Reed and Hussong, 1989; Hughes Clarke, 1994). The same reference values could be used for monitoring changes within a certain area if the same system setup was used for the repeat surveys.

Even if the classification process is not fully automated, the quantitative classification approach presented here greatly aids bottom terrain identification. The resulting classes can be used as templates for a more refined manual interpretation, and the calculated parameters reveal the extent of areas with different acoustic properties. The combination of different parameters from multiple sensors reveals differences in river bottom response to the different sensors that might not be detected otherwise, because it is difficult to compare the individual data sets. The combination of sidescan data from profiles of different look direction (east-west and north-south) reveals bottom features whose backscatter amplitudes are dependent on the direction of the insonification. In case of the Tappan Zee area the classification differentiates sediment classes on the marginal flats (classes U5, U6, U7) which appear similar visually on the original sidescan mosaic (Fig. 2a).

The relation of the different classes to the physical environment and habitats remains subjective. The classes can be related by comparing them with the previous interpretation.

Although several previous studies relate their acoustic classification results with grain size (e.g. Reut *et al.*, 1985; Tamsett, 1993; Cholwek *et al.*, 2000) it seems to be only of limited value for this study. The different parameters calculated here do not allow a clear separation into grain size categories as e.g. silt, sand, and gravel. This lack of separation might be the result of special conditions of the Hudson Estuary where sediments are dominated by silt and coarser material is sparse. Therefore, the variation in grain size is small and the differences in acoustic backscatter seen in this area are likely to reflect other factors like compaction, mineral composition, as well as variation in roughness due to bioturbation, burrowing, benthic vegetation or current activity. For example, the consolidated silts and clays on the marginal flat have different acoustic character than recently deposited silts and clays that are less consolidated and still contain a high amount of water. Furthermore, mean grain size is only of limited value to characterize sediments. The same mean grain size could reflect different sample types from well sorted sediment to sediment with a large variance. Outliers could also shift the mean significantly.

The bottom type classification based on SPI data is more likely to include some of these factors like bioturbation and bottom vegetation directly or indirectly. This is probably the reason why the different SPI bottom types are better separated by the individual parameters (Fig. 9) than the main grain size. On the other hand, the bottom types are still not likely to include all factors. For example the wide range of DN values representing the silt azoic and silt infauna bottom types indicates that these types include different sediment types like consolidated as well as non-consolidated sediments. Thus, it might be possible and useful to apply simple classification schemes where distinct bottom types are present, but in this case of the estuarine environment of the Hudson River the relation between acoustic classes and sediment parameters appears to be more complex.

The resolution of this classification method is limited and heterogeneous. We are averaging the sidescan data over an area of ~400 m perpendicular to the track line and 8-10 m along track (average over 10 pings). Furthermore, the classification approach combines information of different lateral resolution. The sidescan data provide complete lateral coverage and the calculated parameters represent an average of the sediments between the tracks. In contrast, the seismic amplitude data reflect only bottom types along the tracks with no information between two lines. All acoustic parameters are gridded at 60 m interval which is a compromise to account for the different resolution of the parameters. Therefore, smaller variations might not been represented correctly in the classes. Nevertheless, the larger regions >300 m could be classified using this.

The calculated parameters are not independent from each other and contain some amount of redundancy, because several parameters are derived from one set of sidescan sonar data. Most likely some parameters are more relevant for the final classification than others are. We did not investigate these relationships in detail and expect that redundancy will be separated by the principle component analysis.

In some of the parameters in figure 6 an artificial looking striping pattern can be identified. These stripes are coincident with profile tracks. They probably represent a survey footprint caused by small amplitude variations due to changes of sea state, altitude above riverbed, and speed of the boat. Some parameters like $mean_{3..10}$ and CV are more sensitive to these variations than others. The principal component analysis partially separates this striping pattern from major changes in the data and enables those components where the striping is dominant to be excluded.

The data used for this study have undergone only minor processing. No corrections for beam pattern, variations in depth or slope have been applied due to the lack of necessary information. At the time this survey was carried out the application of such a classification method was not intended. However, due to the limited depth variations of the survey area and the uniform system setup these corrections are likely to have only minor influence for this particular survey area. The results correspond well with the manual interpretation of different bottom types, and major classes are distinguished in areas of the same depth, e.g. on the flats or in the channel. For areas with greater depth or stronger relief better corrections for geometry and system parameter will be more important. Such corrections will also be necessary for comparing classes of different areas and for establishing system-independent classification. With the system setup used here the described method will provide classes that are different from each other, but their signatures cannot be transferred directly to other areas or used with data collected by other systems. Such a signature transfer requires a well-calibrated system that would allow to correct for system parameters as well as correction for bathymetry effects.

However, the described method of combining different data sets and classifying them is independent from the system used and will work with other sets of input parameters. Additional parameter like e.g. seabed slope and azimuth derived from multi-beam bathymetry, and lateral texture variation of sidescan images can be used as well. Systems with calibrated and controlled source output and a known angular response function allow other parameters like e.g. spectral analysis of sidescan data (Pace and Gao, 1988) or wave form analysis of reflected single- or multi-beam data (Hamilton *et al.*, 1999) to be used also. This is likely to enhance the classification abilities of the method. Improved geometrical and radial correction of sidescan and seismic data might also improve the classification potential of the parameters used in this study given that the necessary information is available. In fact, it is likely that better correction are necessary for areas with stronger relief for this method to provide good results. What parameters are available or useful depends on the deployed systems, the local geology, and the target of the survey.

Present and future advances in sonar systems in terms of size, costs, and handling will make it more common to simultaneously collect data with different systems to study an area or address scientific questions. This trend is likely to increase in the future. Therefore, co-located and simultaneously collected input data as used here will be frequently available for the classification.

9. CONCLUSIONS

This study describes an approach to classify shallow water sediments by calculating parameters from different acoustic survey systems instead of using data from a single system only. Based on sidescan and seismic sub-bottom profiler data different classifiers were calculated.

Supervised and unsupervised classification methods led to results comparable to the manual interpretation. Limited manual interaction is still necessary to select training areas for the supervised classification and to relate obtained classes to different environments. But speed and quality of the interpretation process is highly improved. Additionally, the calculated parameters provide new insights into the data as they enhance features and distinguish classes which are not clearly distinguishable by eye in the raw data.

For this study limited processing was sufficient to distinguish major classes. The used parameters and procedure are likely to yield similar good results in other estuaries that contain similar extended shallow flats with low relief. However, for more general application in other areas with higher relief improved corrections should be applied.

The discussed method of different layers is flexible and can be easily adjusted or extended to other parameters than shown here. This is especially important as simultaneous measurements with different systems on integrated platforms become more common. Future work should investigate the use of additional parameters, especially those based on multi-beam bathymetry. And survey design should account for the possible later application of classification procedures.

This method can be used in many regions, although new training areas need to be created for each survey area for the supervised classification. For repeat surveys in the same area assuming the same systems and setup are used the discussed approach could provide a practical tool for quantitative and efficient analysis of change in bottom type.

Further understanding of the relation of different physical differences of the sediment and the sonar response is needed to improve the physical understanding of the classes.

10. ACKNOWLEDGMENTS

We thank Chris Small for providing insights into supervised classification and Chris Bertinato for help with implementing the algorithms. The grain size data were made available by Cecilia McHugh. John Hughes Clarke and an anonymous reviewer provided comments that helped to improve the manuscript. This work was supported by the Cooperative Institute for Coastal and Estuarine Environmental Technology (CICEET) grant. Further funding was provided by New York State department of Environmental Conservation for this project from the Environmental Protection Fund through the Hudson.

REFERENCES

Ball, G. and Hall, D. 1967. A clustering technique for summarizing multivariate data. *Behavioral Science*, 12, 153-155.

Blondel, P. 2001. New techniques of acoustic seabed classification at ocean margins. *EOS Trans. AGU, Fall Meet. Suppl.*, OS31B-0416.

Blondel, P. and Murton, B.J. 1996. *Handbook of seafloor sonar imagery*. John Wiley and Sons, Chichster, UK, 314 pp.

Carbotte, S.M., Bell, R.E., Ryan, W.B.F., Chillrud, S. and McHugh, C.M.G. in prep. Climate Driven Estuary Process Linked to Oysters Productivity and Harvesting. *Science*.

Cholwek, G., Bonde, J., Li, X., Richards, C. and Yin, K. 2000. Processing RoxAnn sonar data to improve its categorization of lake bed surficial substrates. *Marine Geophysical Researches*, 21, 409-421.

Coch, N.K. and Bokuniewicz, H.J. 1986. Oceanographic and geologic framework of the Hudson system. In: K. Coch Nicholas and J. Bokuniewicz Henry (ed) *Sedimentation in the Hudson system; the Hudson River and contiguous waterways*. Northeastern Geology. Rensselaer Polytechnic Institute, Department of Geology, Troy, NY, 96-108.

Davis, J.C. 1986. *Statistics and data analysis in geology*. Wiley, New York, 646 pp.

Duda, R.D. and Hart, P.E. 1973. *Pattern classification and scene analysis*. John Wiley & Sons, New York, NY, 482 pp.

Fox, C.G. 1996. Objective classification of oceanic ridge-crest terrains using two-dimensional spectral models of bathymetry. *Marine Geophysical Researches*, 18, 707-728.

Hamilton, L.J., Mulhearn, P.J. and Poeckert, R. 1999. Comparison of RoxAnn and QTC-view acoustic bottom classification system performance for the Cairns area, Great Barrier Reef, Australia. *Continental Shelf Research*, 19, 1577-1597.

Herzfeld, U.C. and Higginson, C.A. 1996. Automated geostatistical seafloor classification - principles, parameters, feature vectors, and discrimination criteria. *Computers and Geosciences*, 22, 35-52.

Hines, P., Osler, J. and MacDougald, D. 2001. Normal incidence classification of seabed and sub-bottom sediments using acoustic backscatter measurements from 1-10kHz. *EOS, Trans. AGU, Fall Meet. Suppl.*, 82(47), OS31B-0425.

Hughes Clarke, J. 1994. Toward remote seafloor classification using the angular response of acoustic backscattering: a case study from multiple overlapping GLORIA data. *IEEE Journal of Oceanic Engineering*, 19, 112-127.

Iocco, L.E. 2000. *Benthic habitats of selected areas of the Hudson River, NY, based on Sediment Profile Imagery*. Final report to NYS DEC, Technology planning and managment corporation, NOAA Coastal Services Center, Charleston, SC.

Keeton, J.A. and Searle, R.C. 1996. Analysis of Simrad EM12 multibeam bathymetry and acoustic backscatter data for seafloor mapping, exemplified at the Mid-Atlantic Ridge at 45°N. *Marine Geophysical Researches*, 18, 663-688.

Ladd, J.W. et al. 2002. Mapping the Hudson Estuary's submerged lands. *Clearwaters*, 32, 5-7.

Mitchell, N.C. and Hughes Clarke, J.E. 1994. Classification of seafloor geology using multibeam sonar data from the Scotian Shelf. *Marine Geology*, 121, 143-160.

Newman, W.S., Thurber, D.H., Zeiss, H.S., Rokach, A. and Musich, L. 1969. Late Quaternary geology of the Hudson River Estuary: a preliminary report. *Transactions of the New York Academy of Sciences*, 31, 548-569.

Pace, N.G. 1983. Acoustic classification of the seabed. In: R.A. Geyer (ed) *CRC Handbook of geophysical exploration at sea*. CRC Press, Inc., Boca Raton, Florida, 211-218.

Pace, N.G. and Gao, H. 1988. Swath seabed classification. *IEEE Journal of Oceanic Engineering*, 13, 83-90.

Reed, T.B., IV and Hussong, D. 1989. Digital image processing techniques for enhancement and classification of SeaMARC II side scan sonar imagery. *Journal of Geophysical Research*, 94B, 7469-7490.

Reut, Z., Pace, N.G. and Heaton, M.J.P. 1985. Computer classification of sea beds by sonar. *Nature*, 314, 426-428.

Roberts, H.H. 2001. New geologic data from the complex estuarine environments of Louisiana; ground truth "calibrated" acoustic surveys in shallow water. *Geological Society of America, 2001 annual meeting, Abstracts with Programs*, A274.

Ryan, W.B.F. and Flood, R.D. 1996. Side-looking sonar backscatter response at dual frequencies. *Marine Geophysical Researches*, 18, 689-705.

Schowengerdt, R.A. 1997. *Remote Sensing - Models and methods for image processing*. Academic Press, San Diego, 522 pp.

Simpson, H.J., Olsen, C.R., Trier, R.M. and Williams, S.C. 1976. Man-made radionuclides and sedimentation in the Hudson River estuary. *Science*, 194, 179-183.

Tamsett, D. 1993. Sea-bed characterization and classification from the power spectra of side-scan sonar data. *Marine Geophysical Researches*, 15, 43-64.

Chapter 4

ANALYSIS OF LAND-COVER SHIFTS IN TIME AND THEIR SIGNIFICANCE

An example from the mouth of the Guadiana Estuary (SW Iberia)

Ramon Gonzalez, João M.Alveirinho Dias and Óscar Ferreira
CIACOMAR-Universidade do Algarve, Avenida 16 de Junho s/n, 8700-311 Olhão, Portugal, rgonzale@ualg.pt

1. INTRODUCTION

Over the past 10 to 15 years the use of Geographical Information Systems (GIS) in the analysis of coastal and marine systems has expanded dramatically. One of the reasons for this is the fact that GIS is an ideal means to analyse and visualize the complex spatial and temporal evolution of morphologically complex areas (Bartlett, 1999). Most environmental data in dynamic coastal areas are complex and show variations in location, depth, and time (Li and Saxena, 1993; Kemp, 1997), raising the problem of finding suitable ways to represent environmental phenomena (Lucas, 1999).

What is true for the coast in general applies particularly to estuaries and their mouths. Estuaries are constantly changing and evolving on a range of time scales (e.g. Dyer, 1973). The deposition and erosion of sedimentary bodies in estuaries is dependent on many factors such as sediment supply (both from rivers and littoral drift), local morphology, wave and tidal climate, storm frequency, floods, and, increasingly, human interventions.

In order to predict the future evolution of estuaries and to understand their present underlying dynamics it is necessary to analyse the past evolution of these areas. However, a lack of precise data on environmental parameters often makes this task difficult, or even impossible. In many cases the only existing data are maps and, in some cases, aerial photographs.

This paper presents a GIS-aided method to extract environmental information about the past behaviour of highly dynamic sedimentary areas

D.M. FitzGerald and J. Knight (eds.),
High Resolution Morphodynamics and Sedimentary Evolution of Estuaries, 57-82.

(such as estuaries) from vertical aerial photographs, by determining the significance and magnitude of shifts in land-cover areas.

To determine the significance of land-cover shifts, a variation of Markov chain analysis (De Raaf *et al.*, 1965; Miall, 1973; Powers and Easterling, 1982; Harper, 1984) is used. The Markov chain analysis approach as applied to sedimentary depositional systems assumes that vertical facies successions are not random but are subject to 'natural laws'. As a consequence of Walther's law, which states that vertical facies shifts are also found laterally (Walther, 1894), lateral changes in facies must be equally not random.

This concept of facies relationships will be tested on a temporal sequence of historic vertical aerial photographs from the Guadiana Estuary in SW Iberia, using land-cover areas as proxies for facies belts. The environmental parameters influencing land-cover areas determine the speed and magnitude at which shifts and changes occur. Consequently, by carefully analysing and quantifying these land-cover shifts some deductions about the magnitude and the timing of the impact of environmental changes can be made.

2. REGIONAL SETTING

The Guadiana Estuary is situated in the SW of the Iberian Peninsula, forming part of the Guadiana River basin (Fig. 1). The river is 810 km long, and its basin is 66,960 km^2, the fourth largest in size of the Iberian Peninsula. Of this area, 55,260 km^2 (83%) are located in Spain, and 11,700 km^2 (17%) in Portugal.

2.1 Hydrological setting

The Guadiana River flow volume is marked by unusually large seasonal changes, as well as changes associated with dry and wet years (Loureiro *et al.,* 1986). The regional climate is classified as semi-arid, with the exception of July and August (arid) and November to January (temperate-humid) (Morales, 1995).

The minimum measured flow volume is in the region of just 6 m^3s^{-1}, while the maximum reached during floods can be as high as 3000 m^3s^{-1}, and more. Since the year 680 a total of 128 floods were considered to be catastrophic (Ortega and Garzón, 1997).

The hinterland of the Guadiana River Basin is extremely dry, especially during the summer. A series of large barrages and dams have been built to alleviate shortfalls in irrigation, and to provide water and electricity for Spanish and Portuguese cities. In 1990, about 70% of the Guadiana drainage basin was regulated by some kind of barrage (Morales, 1995), most recently by the inauguration of the *Alqueva* dam in February 2002, which almost doubled the amount of water stored in dam lakes throughout the river basin (Gonzalez *et al.*, 2001). The result of this regulation was a dramatic reduction of flow volume in the lower Guadiana River during droughts and winters (when most water is retained in dam lakes). As a consequence the annual flash floods, which usually occurred in winter, have strongly diminished in magnitude in recent years.

The coast adjacent to the estuary is meso-tidal, with average tidal amplitudes of *c.* 2 m, reaching a maximum of 3.4 m at spring tides. Tidal currents at the river mouth reach 0.6 ms^{-1} (direction 340°) during peak flood tide, and 1.2 ms^{-1} (direction 140°) during peak ebb tide (Instituto Hidrográfico, 1998). At high tide, saline coastal waters penetrate 40-50 km into the hinterland of the estuary (Rocha *et al.*, 2002).

The offshore coastal wave regime is primarily dominated by waves from the W and SW (approximately 50% of occurrences). SE waves also have a significant influence with *c.* 25% of occurrences. Average offshore significant wave height is about 0.92 m, with an average period of 8 s (Costa, 1994). The net resulting yearly littoral drift is from W to E.

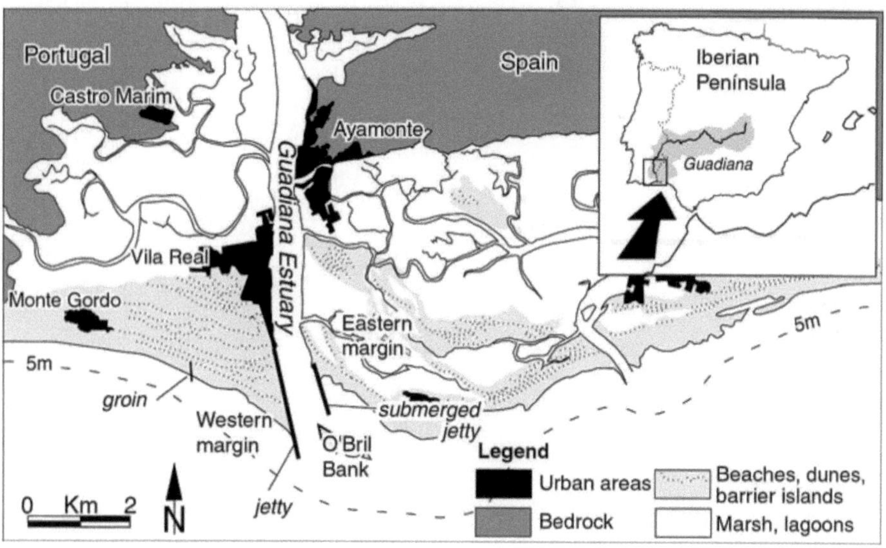

Figure 1. Main geomorphological elements of the lower Guadiana estuary and delta (after Morales, 1995)

2.2 Geomorphological setting

The river only crosses an open coastal plain in its last 7 km (Fig. 1). This coastal plain is part of an old deltaic plain, dominated by marsh systems formed by the river (Morales, 1997).

The mouth of the estuary remains to this day a highly dynamic area, with considerable movement of sediments and associated morphological changes (Gonzalez *et al.*, 1999, 2001). The coastline only reached a position close to the present one about 200 years ago (Morales, 1997).

The mouth of the Guadiana Estuary can be split into three main morphological elements (Fig. 1): the western and eastern estuary margins, and the O'Bril sand bank.

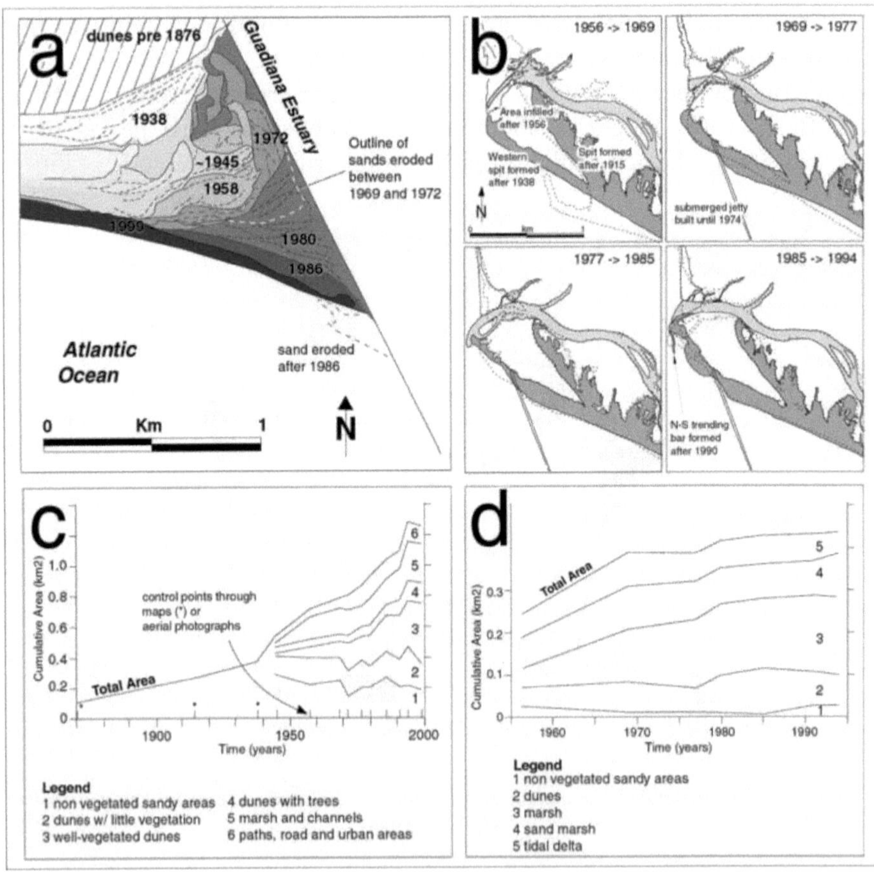

Figure 2. Summary of the morphosedimentary evolution and cumulative curves of the evolution of some land-cover areas for both sides of the estuarine mouth based on the analysis of maps and/or aerial photographs (from Gonzalez et al.*, 2001): a) & c) western margin (since 1876); b) & d) eastern margin (since 1956).*

Large dune belts, with a maximum width of 3 km characterize the western margin. Since 1870 this margin has prograded by about 950 m, accumulating an estimated 4.6 x 10^6 m^3 of sediment (Gonzalez *et al.* 1999, 2001) (Figs. 2a,c). About 20-25% of this accumulation was caused by the construction of jetties confining the estuary mouth margins in 1974, leading to the trapping of the entire littoral drift in the following years (Gonzalez *et al.* 1999, 2001). A remarkable dune system developed on this western margin, enclosing a small marsh area.

On the eastern margin, a long spit growing into the main estuary channel defines the eastern limit of a barrier island that forms the front of the Guadiana Delta. Large lagoons and marshes dominate the back of these islands (Figs. 2b,d). Although the coastline showed average erosion rates of about 3 myr^{-1} (Gonzalez *et al.,* 2000), the total area of the eastern estuary margin increased as the spit prograded into the main river channel (Fig. 2b). The marsh systems in the back of the westward growing spit have shown considerable infilling with fine-grained sediments since the 1950s (Gonzalez *et al.*, 2001).

The O'Bril Sand Bank located in front of the main estuary channel (Fig. 1) is the remnant of a much larger sand body that existed in the area prior to the construction of the jetties, and was a problem for navigation in the past (Weinholtz, 1978). Most of it is now found to the east of the main estuary channel, where it might merge with the eastern margin barrier islands in the future (Gonzalez *et al.*, 1999).

Important anthropogenic elements influencing the estuarine mouth are the jetty on the western margin, with a length of 2,040 m, a 900 m long submerged jetty on the eastern margin, and a groin with a length of 150 m about 1,700 m west of the western jetty, all built between 1972 and 1974 (Weinholtz, 1978; Dias, 1988).

2.3 Sediment supply

Sediment is brought into the area from littoral drift, and from the river basin. Estimated values of the littoral drift range from 180,000 m^3yr^{-1} (Cuena, 1991; Gonzalez *et al.*, 2001) to 300,000 m^3yr^{-1} (CEEPYC, 1979). The amount of sediment brought in from the river basin has not yet been accurately determined, and has been affected by several factors during the past century: 1) deforestation due to mining; 2) agricultural practices favouring erosion in the river basin hinterland, mainly during the 1930s and 1940s; 3) the building of dams in the river basin from approximately 1955 to 1965, reducing both the supply of water, and of most sediments except for the finest fractions; and 4) the building of the jetties between 1972 and 1974, causing large-scale accumulation of sand against the western estuary margin

- particularly between 1974 to 1986 - and resulting in the development of a large dune system.

3. METHODS

3.1 Image treatment and production of land-cover maps

The evolution of the Guadiana Estuary mouth as discussed here was studied mainly with the help of vertical aerial photographs. A number of photographs only show either the western, Portuguese side, or the eastern, Spanish side of the estuary. Table I shows a complete list of used aerial photographs.

Unfortunately the time of the year some photographs were taken is unknown. However, it is very likely that most if not all pictures were taken between March and October. The coastline usually recedes during winter storms between December and February, and progrades during the summer. An earlier analysis (Gonzalez *et al.*, 2001) indicates that the standard deviation from the average trend of movement of the shoreline in the past 60 years is of not more than 3 to 5 m on the Portuguese side, and 5 to 9 m on the Spanish side. This deviation can be attributed amongst other factors also to seasonal variations of the coastline. The distribution of vegetation in the area of the Guadiana estuary on the scale of aerial photographs is not significantly affected by seasonal changes.

Table I. *Aerial photographs used for this study*

Date	Scale	Visible area of margin	Format
194? (Date uncertain, most likely 1945)	1:20,000	Portugal	Black and white
July 1956	1: 33,000	Spain	Black and white
1958	1: 20,000	Portugal	Black and white
1969	1: 25,000	Both margins	Black and white
1972	1: 6,000	Portugal	Black and white
1977	1: 18,000	Both margins	Black and white
October 1980	1: 20,000	Both margins	Black and white
March 1985	1: 20,000	Both margins	Black and white
March 1986	1: 20,000	Both margins	Black and white
October 1991	1: 20,000	Both margins	Black and white
May 1994	1: 20,000	Both margins	Black and white
October 1999	1: 20,000	Both margins	Colour

All aerial photographs were scanned at 300 pixels/inch, translating to a resolution of about 3 m per pixel at a scale of 1:33,000, and about 0.5 m per pixel at a scale of 1:6,000. The photographs were geo-referenced with ERMapper 6.0 using a series of geographic tie points. The projection used is UTM (Datum Lisbon, Castelo de S. Jorge). Comparisons of known points (e.g. houses) on all photos showed that the maximum overlap error between such points is generally below 5 m, and for some photographs up to 9 m on the eastern margin (where fewer accurate tie points exist). This is below the error given by Dolan *et al.* (1991) for the overlap of aerial photographs.

The criteria for distinction of the areas were shape, shade of grey (or colour where available), texture, and context. The land-cover areas identified on the scanned photographs were digitised with the help of MapInfo5.5, a GIS based program. Table II shows mapped geomorphologic elements and their definition.

Table II. *Mapped geomorphologic elements*

Name	Description
Sub- and intertidal sands	Areas showing water on photographs
Intertidal banks	Extensive intertidal areas off the northern margin of the eastern estuary margin barrier islands
River bench	Narrow strip of intertidal areas off the eastern margin marsh
Beaches and and supratidal non-vegetated sandy areas	White or light grey featureless areas (usually in the vicinity of water)
Dunes with sparse vegetation	Single plants and bushes separated by not more than 3-5 m of sand from each other; the outer limit to sandy areas was the last vegetated point; to all other areas it was the beginning of that area
Well-vegetated dunes	Vegetation forms continuous carpet
Dunes with trees	Areas with more than one tree, where the trees are separated by a distance of not more than the average diameter of the tree crowns; the outer limit of this area was the line connecting the crowns of the outermost trees
Sand marshes	Topographically low-lying areas, wet at spring high tides and morphologically related to marshes, dominated by mixtures of mud, silt and fine-grained sandy sediments; light grey on images
Marshes	Mud dominated marsh areas; dark grey on images
Channels	Channel areas always filled with water
Areas associated with channels	'Wet' areas in the vicinity of channels and channel margins, not permanently filled with water
Paths	Not paved
Roads	Paved
Urban areas	All constructions and buildings, including jetties

The photographs were interpreted beginning with the most recent. Mapped objects were carefully ground-truthed in the field and interpreted images corrected where necessary. The created map was subsequently used as a basis to define objects on the next older image, and so on.

The smallest mapped features have a diameter in the range of 2-4 m, or less where objects are well visible on the photos and are of diagnostic significance, for instance some tidal channels. The great majority of digitised objects are larger.

All areas were mapped several times in order to test the error involved in the mapping process. This error was found to be largest for very small areas of just a few hundred square meters in size (up to 10-15%), but usually much smaller, with the size of re-mapped areas being within about 1-3% of each other. Consequently the calculations in this study use the average size of mapped areas. An example of two of the digitised maps showing both the western and eastern margins of the estuary is shown in Figure 3.

The only mapped intertidal areas are the intertidal bank and the river bench (or terrace) of the eastern margin, because of their importance as a link between the margin and the Guadiana River. Although the limit of the river bench can be clearly seen on most photographs as a steep slope, the limits of bank and bench towards the river are considered to be estimates.

The quantifications of overlap between different types of areas on temporal successions of photographs were calculated using split functions in MapInfo5.5. Furthermore, an estimate of the possible error in these calculations was made using buffer functions in MapInfo. It was assumed that the limits of areas might (due to the error in overlap between photographs) vary as much as 2.5 m on either side from their position, and overlaps were calculated for maximum and minimum extensions of areas. The error in overlap was strongly related to the size of areas. While it could be as high as 60-70% for the smallest areas, it was more in the region of 5% for large areas. For instance, a calculated overlap of 500 m^2 might be as small as 150 m^2 or as large as 850 m^2, while an overlap in the order of 50,000 m^2 could be as small as 47,500 m^2 and as large as 52,500 m^2. It should be pointed out that these errors are not cumulative throughout the analysed time series, as the overlap of photos of all ages was compared and optimised prior to digitising.

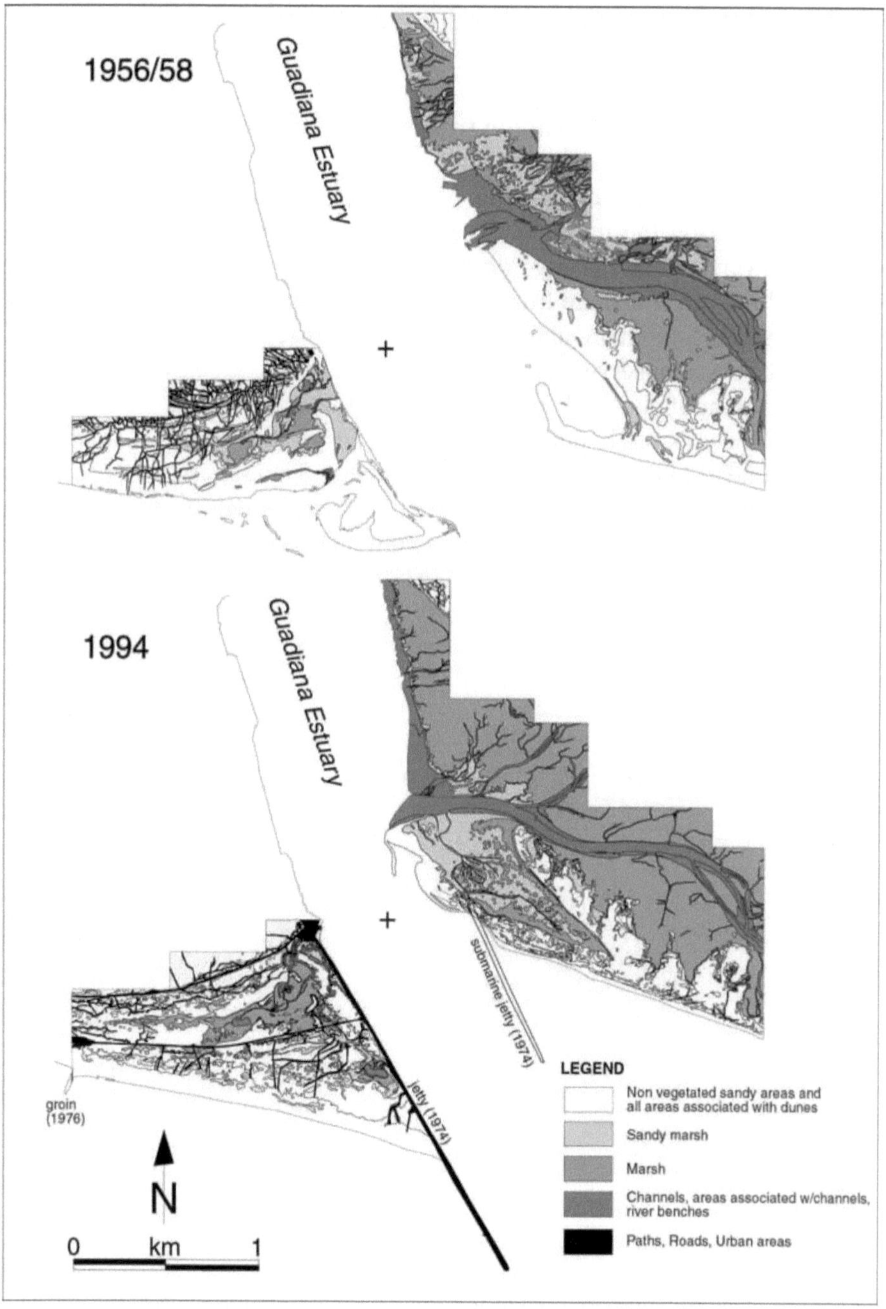

Figure 3. The Guadiana estuary mouth in 1956/58 and 1994 showing the accumulation of sand against the western margin jetty, and the progradation of the eastern margin barrier island and marsh systems into the main estuarine channel. The cross locates the same geographical position on both maps.

3.2 Relation between land-cover types

3.2.1 Differentiating random from significant changes

The method presented in this paper uses a modified Markov chain analysis, based on the methods developed by De Raaf *et al.* (1965), Miall (1973), Powers and Easterling (1982) and Harper (1984). It tests whether an observed land-cover change from type *a* to type *b* is significant or not. While the original Markov chain analysis is based upon used the number of vertical facies transitions observed in a determined vertical succession, this paper uses the amount of changed area in time to quantify transitions.

Figure 4 compares hypothetical land-cover types *a*, *b*, and *c* at an older point in time *i*, and a younger point in time *j*. The superposition of both images shows transitions that occurred between land-cover types. If the land-cover type was found to be the same on both images, it was assumed that the land-cover type either changed to itself or did not change.

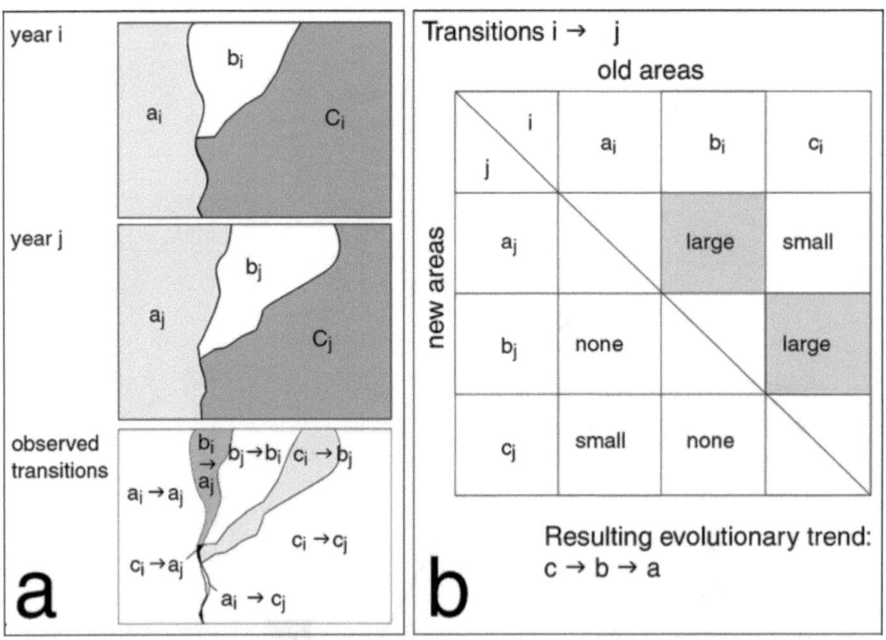

Figure 4. Simplified example for land-cover transitions from hypothetic time i to j, and the simplified matrix showing corresponding statistically relevant transitions.

The transitions from time *i* to time *j* can be listed in a matrix (Fig. 4). Some transitions are more important than others. The method picks out these important changes, and determines whether an occurring change was a coincidence (for instance, as related to rectification errors or misinterpretation, or related to random/chance processes) or not.

In order to do this, the data are first organized in a matrix $M_{[i,\ j]}\ (a,b)$, where the cells of the matrix correspond to the amount of area in m^2 that changed from land-cover *a* to land-cover *b* from point in time *i* to point in time *j*.

A matrix *C (a,b)* calculates whether a change of land-cover in *M* is coincidental:

$$C\ (a,b) = s_b\ /\ (T - s_a) \qquad (1)$$

Where s_b is the sum of land-cover area changes in column *b*, *T* all observed area changes in *M*, and s_a the sum of area changes in row *a*.

A second matrix *P (a,b)* calculates the probability of a land-cover change from *a* to *b* in matrix *M*:

$$P(a,b) = M\ (a,b)\ /\ s_a \qquad (2)$$

A final matrix *D (a,b)* compares the probability that a land-cover area transition is larger with the probability that this transition was a coincidence:

$$D\ (a,b) = P\ (a,b,) - C\ (a,b) \qquad (3)$$

All positive values in (3) are land-cover changes likely not to be coincidental within the analysed area.

In the hypothetic example given on Figure 4, only changes from b_i to a_j, and c_i to b_j were relevant. The result suggests an evolutionary trend from *c* to *b* to *a*.

Changes from a particular land-cover type to itself are not included in these calculations. Often the amount of area changing is relatively small in comparison to the area keeping its land-cover characteristics. If changes from areas to themselves are considered, this can distort the results of the matrix, and important changes may not be detected.

The total size of the analysed area has a large influence in whether a change will be identified as relevant or not. This makes sense if the environmental context of the area is taken into account. For instance, while the switch of a channel to marsh might be important if only the area in the vicinity of the channel is considered, it becomes relatively irrelevant when looking at an entire delta. Nevertheless, all results should be carefully checked, and it is recommended to compute the relevance of changes in the area of interest using different sized sub-areas. Furthermore, this type of iteration might help to determine if different types of processes influence different parts of the analysed area.

3.2.2 Determining changes in time

Once significant changes have been separated from random ones, changes over longer periods of time can be analysed using the same method. All changes from each cell *M (a,b)* are summarized through all calculated time intervals, into a new matrix *M'(a,b)*. From this new matrix *M'*, a matrix *D'* is calculated as above to determine the significance of changes over the entire span of time.

The importance of this iteration lies in separating short-term changes, often associated with particular events (e.g. the construction of a road) from long-term trends. The addition of changes over time usually strongly reduces the number of significant land-cover transitions in a dataset. Furthermore, some changes are slower than others, and their significance may only become relevant when looking at longer time intervals (for instance the growth of trees on dunes compared to other vegetation).

4. EXAMPLES OF THE LAND-COVER CHANGE ANALYSIS METHOD

4.1 The western Guadiana Estuary mouth margin

The following is a step-by-step description of the analytic procedure undertaken for the western margin of the Guadiana Estuary mouth. This area is well suited to demonstrate the method as it has well-defined natural limits, its evolution over the past 100 years is well known, and environmental and anthropogenic parameters influencing the area during this time have been described in detail (Gonzalez *et al.,* 1999, 2001).

Each of the ten aerial photographs showing the area between 1945 and 1999 was digitised. Nine matrices containing land-cover changes that occurred from one time frame to the next were analysed to determine whether observed land-cover transitions were larger than coincidental. From a possible 132 transitions in 12 different types of land-cover (not counting transitions from a land-cover type to itself; see listing of identified land-cover areas on the western margin on Fig. 5), 72 turned out to be significant at one or the other of the analysed transitions.

These changes are all relevant, but many of them are related to specific events, for instance a large amount of sand eroded from a beach after a storm. Of the 72 observed significant changes, only 12 were found to be significant at all times (i.e. significant in each time-step).

In order to reduce the total number of transitions and to determine which changes were relevant over the entire period of time, all area changes from one land-cover to another between 1945 and 1999 were added up into one matrix (Fig. 5a). Figure 5b shows which of these changes were larger than coincidental, i.e. had a probability of more than 0 as determined by equation (3). This computation reduced the number of relevant transitions to 36, of which 13 were just significant (probability between 0 and 0.05), and 23 highly significant (probability larger than 0.05). The cut-off value 0.05 between 'significant' and 'highly significant' changes was chosen arbitrarily, and is useful to make a qualitative distinction between types of significant transitions.

To better illustrate and understand the underlying relationships between significantly interacting land-cover types, they can be represented schematically (Fig. 6). The figure shows a graphic summary of long-term significant transitions between land-cover types occurring on the western Guadiana Estuary margin in form of a flow chart. The chart shows that in this area transitions occur essentially within two separate 'domains' (Fig. 6).

a

	UR(45)	RO(45)	PA(45)	SM(45)	MA(45)	CH(45)	CH2(45)	BE(45)	DR(45)	DN(45)	DO(45)	SEA(45)
UR(99)		2330.40	2323.12	1012.53	650.30	12.76	219.11	13883.58	4038.99	1789.54	23.43	79512.60
RO(99)	6014.70		9030.80	1218.04	1115.73	0.00	0.00	7023.10	16772.90	17385.13	608.71	5393.00
PA(99)	388.84	2809.76		12653.86	5018.20	2029.28	312.84	359875.00	171083.00	165849.00	70011.00	2433.10
SM(99)	876.30	164.37	28076.00		120124.60	12297.70	33758.00	143249.00	55609.00	36116.00	0.00	73192.70
MA(99)	273.90	377.83	27782.00	198094.00		29949.40	46221.86	42186.00	22243.70	76647.20	0.00	33902.00
CH(99)	149.20	0.00	308.83	10186.90	27641.80		17277.74	4448.83	2869.38	5290.48	0.00	9582.00
CH2(99)	28.42	0.00	206.16	25193.60	29988.50	29765.80		8748.70	10594.65	2209.35	0.00	9221.60
BE(99)	9609.30	247.52	61843.00	82553.90	20513.40	3736.75	10886.10		128220.00	129541.40	1604.33	557061.00
DR(99)	2160.58	3614.00	164033.00	35590.90	9836.20	828.83	723.80	504340.00		215288.00	12372.90	664.10
DN(99)	744.07	5914.90	235725.00	94593.90	71811.00	281.47	4233.45	123722.00	558348.00		75475.00	0.00
DO(99)	0.00	134.81	88434.00	211.48	2310.69	0.00	0.00	8541.00	31557.50	120435.90		0.00
SEA(99)	1862.34	0.00	0.00	18320.00	0.00	1599.00	962.50	131091.80	221.10	0.00	0.00	

b

	UR(45)	RO(45)	PA(45)	SM(45)	MA(45)	CH(45)	CH2(45)	BE(45)	DR(45)	DN(45)	DO(45)	SEA(45)
UR(99)		0.019	-0.089	-0.076	-0.045	-0.014	-0.018	-0.110	-0.141	-0.121	-0.028	0.612
RO(99)	0.059		0.029	-0.066	-0.034	-0.014	-0.020	-0.131	0.051	0.131	-0.019	-0.054
PA(99)	-0.004	0.000		-0.082	-0.052	-0.013	-0.023	0.117	0.010	0.051	0.055	-0.155
SM(99)	-0.002	-0.002	-0.063		0.182	0.008	0.044	0.023	-0.083	-0.077	-0.030	-0.003
MA(99)	-0.003	-0.002	-0.060	0.322		0.047	0.074	-0.171	-0.146	0.012	-0.030	-0.077
CH(99)	-0.002	-0.002	-0.106	0.045	0.303		0.201	-0.183	-0.142	-0.069	-0.028	-0.014
CH2(99)	-0.003	-0.002	-0.109	0.130	0.203	0.242		-0.167	-0.088	-0.119	-0.028	-0.059
BE(99)	0.004	-0.003	-0.070	-0.020	-0.041	-0.013	-0.013		-0.087	-0.036	-0.032	0.388
DR(99)	-0.002	0.000	0.041	-0.064	-0.050	-0.016	-0.023	0.245		0.053	-0.020	-0.162
DN(99)	-0.004	0.001	0.064	-0.025	-0.002	-0.017	-0.021	-0.193	0.254		0.028	-0.171
DO(99)	-0.004	-0.002	0.237	-0.087	-0.044	-0.014	-0.021	-0.214	-0.059	0.336		-0.142
SEA(99)	0.008	-0.002	-0.111	0.032	-0.052	-0.004	-0.014	0.609	-0.179	-0.139	-0.029	

LEGEND

The numbers (45) and (99) refer to the year the aerial photograph was taken.

UR Urban areas, buildings
RO Roads, paved tracks
PA Footpaths
SM Sand marsh
MA Marsh
CH Tidal channels
CH2 Areas associated to channels
BE Supratidal, non-vegetated sands
DR Dunes with sparse vegetation
DN Densely vegetated dunes
DO Dunes with trees
IT Sub- and Intertidal sands

Significant transition (p = 0-0.05) — Highly significant transition (p > 0.05)

Figure 5. An example for the calculation of observed and significant land-cover changes: a) All observed land-cover transitions between 1945 and 1999 on the western estuary margin in m^2; b) calculated significant and highly significant transitions based on a), occurring more often than by mere coincidence.

The first domain is associated with dunes and related areas, the second one to marshes and lagoons. An additional factor is the human component, which, in terms of significant changes, only affects the dune domain (Fig. 6). The only significant link between the dune and marsh domains is from beaches and supra-tidal non-vegetated sands to sand marsh. The analysis of the photographs is not able to make a distinction between topographic lows and highs. It can be assumed that low-lying sandy areas eventually become marshes, and elevations become dunes. The fact that there is no other link between the two domains indicates that there is no significant migration of dunes in the analysed area once morphological highs and lows are established.

The dune domain shows a straightforward trend from non-vegetated supratidal sands, to dunes with sparse vegetation, to well-vegetated dunes, to dunes with trees (Fig. 6). A significant reversal of the trend is only found between well-vegetated dunes and dunes with sparse vegetation. This fact may be linked to the destruction of vegetation close to paths and roads. Within the marsh domain, links between areas are more complex and dynamic (Fig. 6). Here, all areas show significant links to each other.

Figure 7 shows the cumulative area transitions between some of the land-cover areas in the marsh domain during the past sixty years. In this type of graphic representation one of the two compared land-cover areas is chosen to be dominant, usually the area showing overall increase at the expense of the other. For each analysed time segment, transitions to the subordinate area are subtracted from transitions to the dominant one, and added to the previous total. Therefore, if more transitions to the subordinate area occurred at any particular time interval, the amount of cumulative area transition will decrease (for instance transitions from channels (CH) to marsh (MA) between 1958 and 1972 on Fig. 7).

One of the ways to use this type of graph is to interpret shifts in the dynamics of an area. For instance, Figure 7 shows that transitions from sand marsh (SM) to marsh (MA) always occur in a single direction, showing an increase of the area of the marsh. This indicates a continuous infilling of the lagoon within the western estuary margin with fine-grained material under low energy conditions.

Trends are not as uniform when looking at shifts from channels (CH) and areas related to channels (CH2) to marsh (MA). Before the construction of the jetty in 1974 both showed a dominant trend of marsh erosion by channels, indicating relatively high-energy conditions within the channel system. After the jetty was built the flow of water inside the lagoon was strongly reduced, resulting in a narrowing, silting up and abandonment of channels, reflected on the graph by a transition to marsh. (Supposedly no water should flow through the jetty; however, these results and field observations show that water does indeed penetrate between the boulders that make up the jetty).

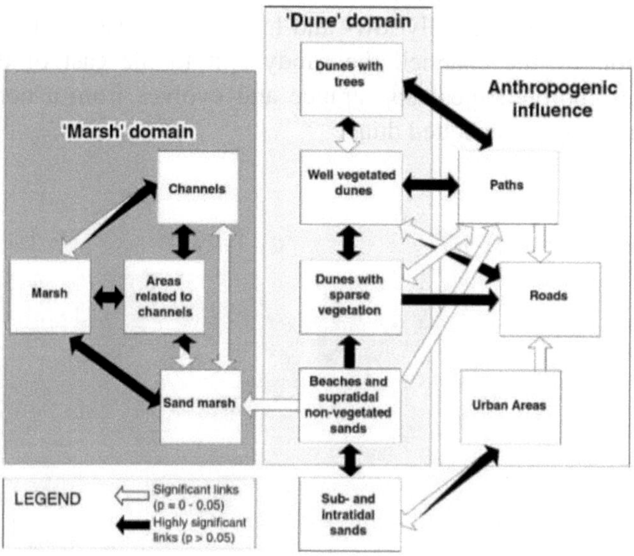

Figure 6. Graphic visualization of significant land-cover interaction on the western estuary margin between 1945 and 1999. The direction of arrows indicates the evolutionary trend. Note that in most cases this trend can go both ways, although normally one direction is quantitatively dominant.

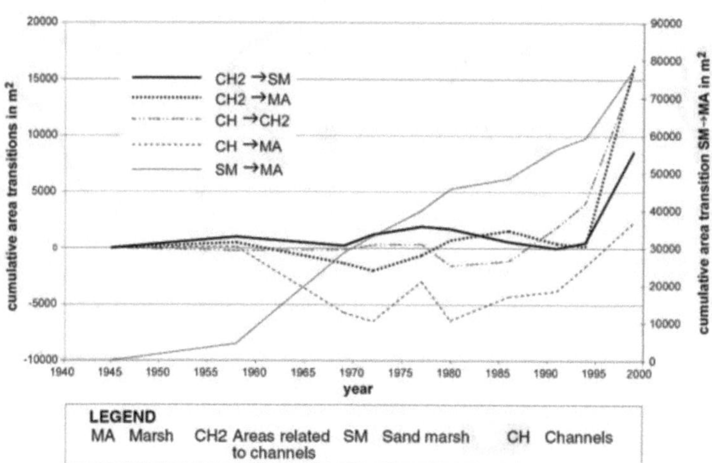

Figure 7. Quantifying some of the significant land-cover transitions over time. The curves are obtained by cumulatively adding up the difference between transitions from A to B and transitions that occur from B to A.

Figure 8 illustrates, using the example of a supplier channel, how this trend can be identified on the aerial photographs. The figure also demonstrates the lack of significant interaction between the dune and marsh domains (i.e. morphologically low- and high-lying areas): although there is some erosion by the channel, the sandy spit to the east of the channel preserves its morphological appearance and evolves from a non-vegetated sandy high to a well vegetated dune.

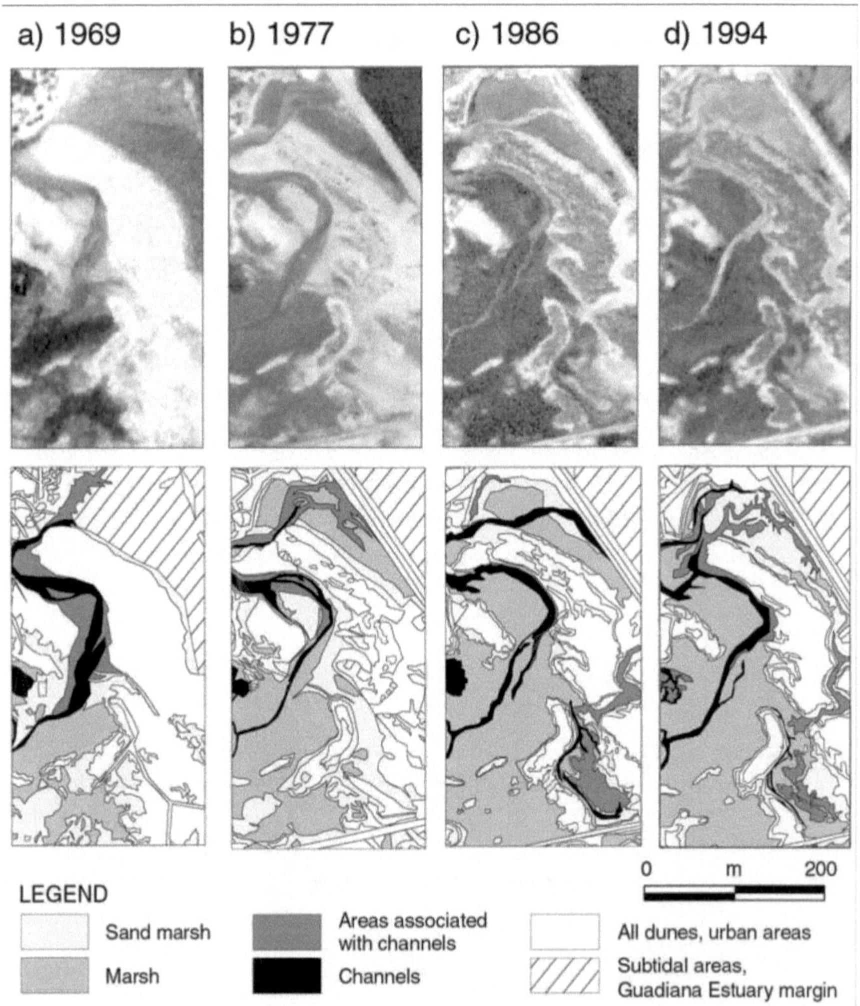

Figure 8. Example for changes in the marsh domain on the western estuary margin between 1969 and 1994. Above are the original images, below the interpretations. All images show the same geographic area. Note the section of the jetty delimiting the area on the upper right hand corner of all images post 1977.

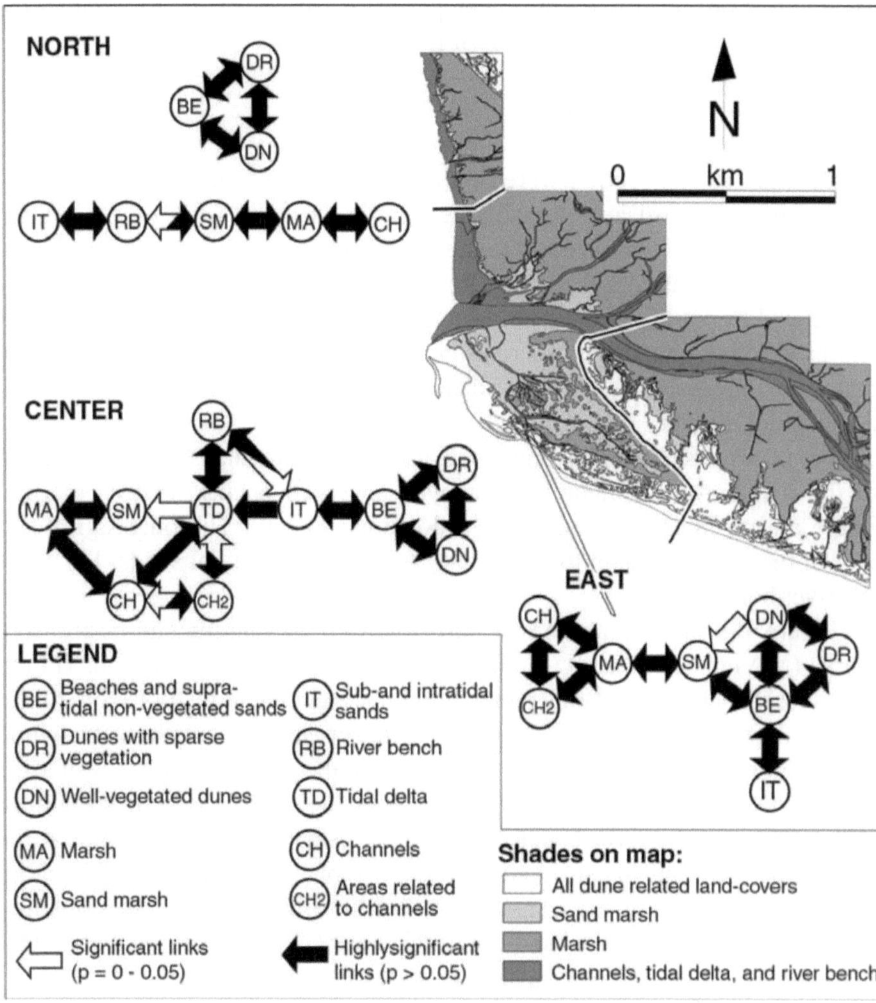

Figure 9. Significant land-cover interactions in three areas on the eastern estuary margin between 1956 and 1994. The map is based on the image from 1994.

4.2 The eastern Guadiana Estuary mouth margin

Because of the size and diversity of the area considered in the eastern estuary margin, it was divided into 3 sections, each of which was analysed separately (Fig. 9): 1) a northern section comprising the eastern estuary margin, where the sub- and intratidal river bench leads into a marsh bordered by old dune belts and beaches; 2) the central section, characterized by the area where the estuarine mouth borders the Atlantic coastline; and 3) the eastern section, formed by the older part of a barrier island adjacent to the estuary, and the marsh to the north of it.

4.2.1 Northern section

Beaches in this section were probably active until the late 19th century, before the westward accretion of the barrier island in front of the system closed off the direct connection to the ocean, and the infilling of what is now the back-barrier marsh (*cf.* Fig. 2b; Morales, 1997). This area is mainly influenced by the flow of tidal currents through the main estuarine channel, and by floods occurring in the river channel. Field observations have shown that waves produced by ship traffic can significantly contribute to erosion of the channel margin.

Two domains can be distinguished, areas linked to dunes and areas linked to the marsh, very similar to what was observed on the western margin. There are no links between the two domains (Fig. 9). The only significant link of the marsh domain to the estuary (sub- and intra-tidal sands; IT) is via the river bench (RB).

Figure 10a shows that the two statistically significant transitions are from marsh (MA) to sand marsh (SM), and from sub- and intra-tidal sands (IT) to the river bench (RB). Both show highly significant transitions of 2,000-2,500 $m^2\ yr^{-1}$ up to 1969, interpreted as an erosion of marsh areas, while simultaneously the river bench increases at the expense of the estuarine channel. A reversal of this trend occurs after the construction of the jetties, with erosion of the river bench by the main estuarine channel, while at the same time the sand marsh is silted up and replaced by marsh. After about 1980 the magnitude of all transitions decreases significantly. The transition of river bench (RB) to sub- and intratidal sands (IT) between 1980 and 1994 of about 700 $m^2 yr^{-1}$ indicates a slow erosion of the river margin.

There are links between all land-cover areas in the dune domain indicating that this domain is highly dynamic (Fig. 9). However, the quantification of links shows that although these changes are significant, the changes in area are very small and of only local relevance.

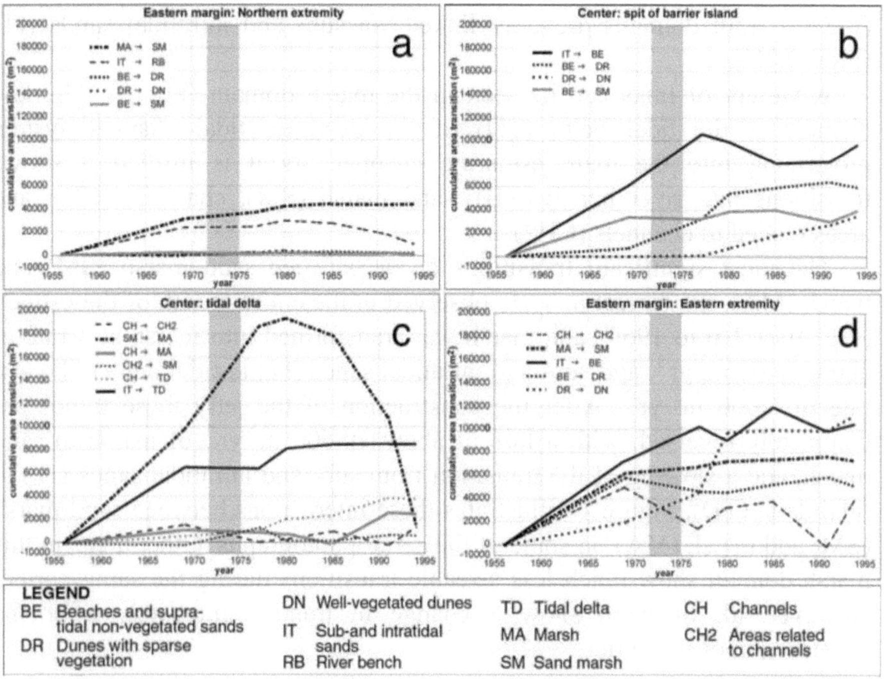

Figure 10. Quantification of land-cover transitions in different areas of the eastern Guadiana estuary margin (for location of areas see Fig. 8). a) Northern extremity of eastern margin; b) spit of barrier island in central area, mainly dune related transitions; c) tidal delta in central area, mainly marsh related transitions; d) eastern extremity of eastern margin. The gray bar indicates the period during which the jetties were built.

4.2.2 Central section

This area is fronted towards the ocean by a long spit, first formed between 1938 and 1956, as the easternmost extremity of the barrier island delimiting the Guadiana delta west of the main estuarine channel (Figs, 1,2). A marsh containing a system of dendritic tidal channels has developed in the back of the spit. The analysed area includes the terminal part of this channel system draining into the main channel of the Guadiana Estuary, and an ebb-tidal delta associated with this channel system (Fig. 9; cf. Figs. 2b,3). This section is influenced by a large number of environmental factors, such as the flow of tidal currents through the main estuarine channel, floods occurring in the river channel, and the wave regime and direction (Morales, 1995, 1997). The section is the only one directly adjacent to the submerged eastern jetty.

The analysis of land-cover transitions shows a similar dissociation between the dune and marsh domains as the other discussed sections

(Fig. 9). Both domains are only linked via sub- and intratidal sands (IT) (Figs. 9,10b,c).

A variety of links can be seen in the marsh domain (Fig. 9). The key element is the tidal delta (TD) of the dendritic tidal channel system prograding into the main estuarine channel. As it is infilled it shows transitions into river bench (RB), sand marsh (SM), channels (CH), and areas related to channels (CH2).

The most significant transitions occur between sand marsh (SM) and marsh (MA), the largest in scale observed in the entire mouth of the estuary (Fig. 10c). Up to 1969 sand marsh was transformed into marsh at a rate of about 7,700 $m^2 yr^{-1}$, and sub- and intratidal sands (IT) to tidal delta (TD) at a rate of 5,000 $m^2 yr^{-1}$. After the construction of the jetty these trends are completely reversed, with a loss of about 10,600 $m^2 yr^{-1}$ of marsh to sand marsh, and a slowing of the transition from sub- and intratidal sands (IT) to tidal delta (TD). Figure 11 illustrates these interactions between sand marsh (SM) and marsh (MA) on some of the aerial photographs. Other areas in this marsh domain show much less sizeable transitions during the same period. However, all of them show a change in trend immediately after the construction of the jetties.

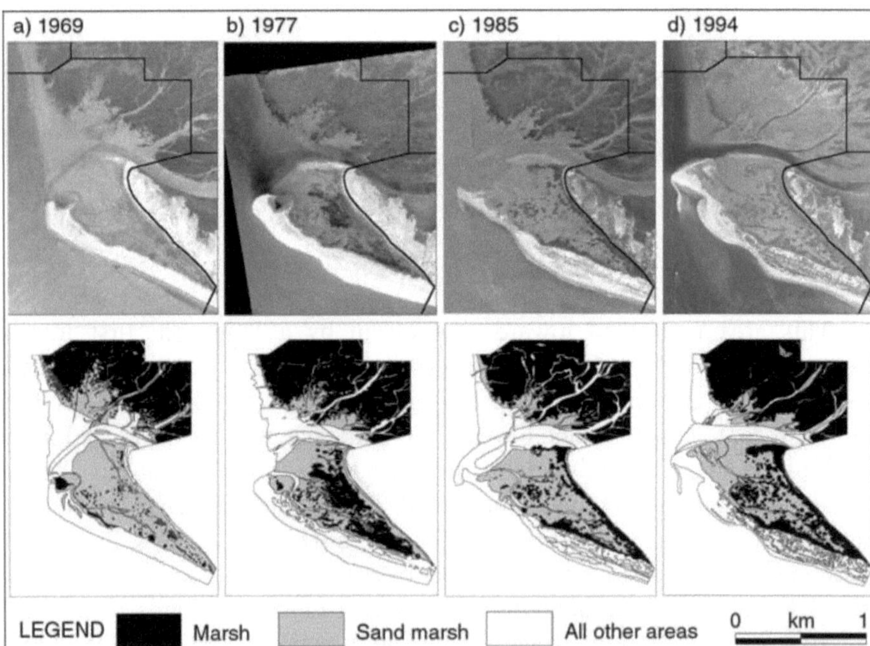

Figure 11. Example for transitions from sand marsh to marsh and vice versa in the central section of the eastern estuary margin between 1969 and 1994. Above are the original images, below the interpretations. All images show the same geographic area.

The dune domain of the central section shows significant transitions from sub- and intratidal sands (IT) to supratidal non-vegetated sands (BE) of around 4,500-5,500 $m^2\ yr^{-1}$ reversed to a loss of about 2,200-3,500 $m^2\ yr^{-1}$ until 1985 (Fig. 10b). Subsequently, supratidal non-vegetated sands begin accumulating again. None of the other land-cover areas in the dune domain show a similar change of trends after the construction of the jetties (Fig. 10b).

4.2.3 Eastern section

This section is the eastward continuation of the central section (cf. Fig. 1). The barrier island delimiting the system towards the ocean is mainly influenced by the wave regime. The only factors influencing the backbarrier marsh are tidal currents coming through the dendritic tidal channel system.

This section shows a relatively simple pattern of links (Fig. 9). The marsh domain is sheltered from the open ocean by the barrier island (Fig. 9). Consequently, and interestingly, the only link of the marsh to the dunes is from vegetated dunes (both DR, and DN) to sand marsh (SM). This is the only direct interaction found between dune and marsh domains in the entire area.

Most observed trends show stagnation, or even reversal after the construction of the jetties (Fig. 10d). Transitions in this section are much less uniform than in the others. A lack of defined trends in transitions between sub- and intratidal sands (IT) and supratidal non-vegetated sands (BE) indicates strong variations in the coastline between periods of accretion and erosion (Fig. 10d). Similarly, no definite trend was found in transitions between channels (CH) and areas related to channels (CH2). This may indicate a reduction in the energy of the environment in this area, similar to what can be observed on the western estuary margin.

5. DISCUSSION

The method presented in this study analyses changes in land-cover area first by identifying significant changes, and subsequently by quantifying these changes.

The significance of land-cover shifts is examined using a variation of the Markov chain analysis. Essentially, the method proposed here replaces the vertical spatial dimension of the traditional Markov chain analysis by time, as vertical aerial photographs can be thought of as superimposing each other in time. Even though there is no unequivocal proof for the premise of this study that lateral facies changes are not random, most results seem to both confirm this premise and make intuitive sense in the field.

In some cases the abstract concept of superimposing land-cover areas in time in fact corresponds to typical vertical facies successions as identified in the sedimentary record. For instance, Figure 6 shows that in the marsh domain channels can develop into areas related to channels and finally to marsh, which is comparable with typical fining-upward distributary channel fill successions (e.g. Bhattacharya and Walker, 1992). Equally, the succession in the dune domain shown on Figure 6, from sub- and intratidal sands to beaches and supratidal non-vegetated sands to dunes with sparse vegetations, corresponds to shallowing-upward prograding successions commonly found in barrier-island and estuarine systems (e.g. Reinson, 1992).

An interesting aspect of the model is the fact that, in contrast to traditional Markov chain analysis, it also includes erosional features (which would leave no sedimentary record). Furthermore, it includes anthropogenic objects, which are not facies in the traditional sense (but the remnants of which may in fact be found in recent vertical sedimentary successions, for instance in archaeological excavations).

It should be pointed out that the quantification of changes from one land-cover area to another can be used independently from the determination of whether such changes are significant. The usefulness in determining their significance lies mostly in narrowing down the selection of land-cover areas that will be compared to each other, since the analysis of an area can involve a large amount of possible changes. For instance, if 10 different types of land-cover areas were identified, 100 types of changes are mathematically possible (if changes from one area to itself are included).

It is also important to differentiate between changes that are relevant in the short term (i.e. from one photograph to the next), and changes relevant for the entire observation period. In this context, the choice of the right time scale seems to be crucial to obtain relevant results. The larger the area of interest, the longer the period covered by the data set should be. This is obviously limited by the availability of aerial photographs covering the studied area, which in most cases does not go further back in time than 40-50 years. This study concentrated on changes relevant over the entire analysed period. It may well be that in other locations some changes only relevant over part of the analysed time show important insights into the dynamic behaviour of an area.

One of the goals of this study using data from the Guadiana Estuary mouth was to compare the results provided by the method with what is known about the recent history of this area in order to validate the proposed method in practice. The data indicate that there is very little, if any, interaction between the dunes and the marsh in the area. This is in accordance with evidence from other regional data, indicating that the dune systems active here can, once established, sometimes remain in place for decades, or even centuries (e.g. Barnhardt *et al.*, 2002).

The identified links between individual land-cover types were found to accurately reflect field observations, and to possess a high degree of inherent logic, for instance the development from non-vegetated sands (BE) to dunes with trees (DO) on the western margin of the estuary (Fig. 6). The quantification of changes from one land-cover type to another in time, with particular emphasis on alterations in the rate of these changes (Figs. 7 and 10), showed that although many of these changes fall within the range of error of area change calculations, particularly in the northern section of the eastern margin of the Guadiana, many others are significant in magnitude.

Of particular interest is the alteration of trends as related to changes in the physical environment, such as the construction of the jetties. Many changes in area transition trends do not occur immediately after the jetty construction, but with a lag of a few years (Fig. 10). It is possible that some of these changes also reflect alterations in sediment supply from the river basin, increasingly reduced by the construction of dams during the sixties and seventies (Gonzalez *et al.*, 1999, 2001). Furthermore, the construction of the jetties could have caused knock-on effects, for instance an increase of ship traffic in the estuary due to the construction of the jetties, and thus an increase in steep, short-period waves, which can cause erosion of fine-grained sediments on river margins (Paolo Ciavola, *pers. comm.*). In this context it is interesting to note that the western and eastern marsh areas show a strong difference in behaviour. While the western margin shows an increasing, or at least uninterrupted transition from sand marsh (SM) to marsh (MA) (Fig. 7), the eastern margin shows a reversal of trends after construction of the jetties (Fig. 10c). This might be explained by the fact that the western margin marshes are in fact sheltered from waves by the jetty.

The method gives two possibilities to analyse the relative dynamics of the environment. One of them is the amount of significant links between areas, for instance, where all dune areas are significantly linked to each other, as it is the case in the central section of the eastern margin. This indicates a highly dynamic, quickly-changing environment. The quantification of area changes shows that, between 1969 and 1977, more than 40% of the total area of the barrier island spit changed land-cover type. This is due to the growth of vegetation, the growth of dunes on the spit, and to the quick relocation of the spit as it adapts to constantly changing current and wave conditions.

The second possibility is to gather information about the relative dynamic behaviour of the environment and to analyse which of the transitional trends is dominant at any given point in time. If, for instance, a trend from sand marsh to marsh is reversed, this possibly indicates an increase in energy. A good example for this is the inversion of transitions from channels (CH) to areas related to channels (CH2) on the eastern portion of the eastern margin (Fig. 10d).

6. CONCLUSIONS

The method presented here implies that alterations in coastal zone land-cover are forced by external factors. More often than not a series of external changes will affect an area at once, making an interpretation of results extremely difficult. However, while results consequently don't necessarily always point in the direction of the culprit for an alteration of trends, they help to better understand the past and present behaviour of a system.

It should be noted that all the Markov chain analysis determines is whether changes occur more frequently than random. Many results are very persuasive in their apparent logic, and there is a distinct danger of over-interpreting data. Furthermore, there are many indirect links and influences between areas that are not at all considered by the method, for instance sewage flowing from urban areas into marsh etc. For a thorough interpretation of any area of interest all existing supplemental information has to be considered. The method here described should only be part of any analysis.

Even though we cannot present conclusive prove our premise, that lateral facies or land-cover changes are not random and do in fact obey rules imposed by environmental parameters, we found no evidence to prove the contrary. More research is needed to provide such proof. As it stands, we suggest that the altered Markov chain analysis is used carefully and as a guideline for which changes might be more significant than others.

However, the quantification of changes of land-cover from one type to another in time can be used independently from this premise. The spatial and temporal analysis of land-cover shifts in the Guadiana estuary clearly showed that the method indeed allowed a better understanding of timing and magnitude of the environmental impact of anthropogenic changes in the Guadiana estuary mouth. It was found that some land-cover area changes do significantly react to changes in environment, for instance the alteration in the rate of change from marsh (MA) to sand marsh (SM) and *vice versa* after the construction of the jetties in the Guadiana estuary mouth. As such the presented method can be a valuable tool to better understand the impact and magnitude of environmental changes in other areas.

7. ACKNOWLEDGEMENTS

We wish to thank Darius Bartlett (University College Cork), Gonzalo Malvarez (University of Ulster), and Jasper Knight (University of Ulster) for their insightful reviews of the manuscript. Financial support for this study was provided by the European project SWAMIEE (FMRX-CT97-0111), as well as the national projects RIMAR (PRAXIS 2/2.1/MAR/1743/95), and CRIDA (PLE/8/00).

REFERENCES

Barnhardt, W.A., Gonzalez, R., Kelley, J.T., Neal, W., Pilkey, O.H., Monteiro, J.H. and Dias, J.M.A. 2002. Geologic evidence for the incorporation of flood tidal deltas at Tavira Island, southern Portugal. *Journal of Coastal Research, Special Issue,* in press.

Bartlett, D. 1999. Working on the Frontiers of Science: Applying GIS to the Coastal Zone. In: Wright, D. and Bartlett, D. (eds) *Marine and Coastal Geographical Information Systems*. Taylor & Francis, London, 11-24.

Bhattacharya, J. and Walker, R.G. 1991. River- and wave-dominated depositional systems in the Upper Cretaceous Dunvegan Formation, northwestern Alberta. *Bulletin of Canadian Petroleum Geology,* 39, 165-191.

CEEPYC 1979. Plan de estudio de la dinámica litoral de la Provincia de Huelva. *Inform Dirección General de Puertos y Costas*, Madrid, 37 pp.

Costa, C. 1994. *Wind- Wave Climatology of the Portuguese Coast.* Final Report of Sub-Project A. NATO PO-WAVES Report, 6/94-A, 80 pp.

Cuena, G.J., 1991. *Proyecto de Regeneracion de las Playas de Isla Cristina.* Servicio de Costas, MOPT, 100 pp.

De Raaf, J.F.M., Reading, H.G., and Walker, R.G. 1965. Cyclic sedimentation in the Lower Westphalian of North Devon, England. *Sedimentology,* 4, 1-52.

Dias, J.M.A. 1988. Aspectos Geológicos do Litoral Algarvio. *Geonovas*, 10, 113-128.

Dolan, R., Fenster, M.S. and Holme, S.J. 1991. Temporal Analysis of Shoreline Recession and Accretion. *Journal of Coastal Research*, 7, 723-744.

Dyer, K.R., 1973. *Estuaries. A Physical Introduction.* Wiley, Chichester. 195 pp.

Gonzalez, R., Dias, J.M.A. and Ferreira, Ó. 1999. Evolução recente da margem oeste do Rio Guadiana. EUROCOAST – Portugal: *Os Estuarios de Portugal e os Planos de Bacia Hidrográfica*, 147-158.

Gonzalez, R., Dias, J.M.A. and Ferreira, Ó. 2000. Altering the natural balance of sedimentation and its consequences: Recent evolution of the Guadiana Delta (SW Iberian Peninsula). *3rd Symposium on the Iberian Atlantic Margin*, 125-126.

Gonzalez, R., Dias, J.M.A. and Ferreira, Ó. 2001. Recent rapid evolution of the Guadiana Estuary (Southern Portugal/Spain). *Journal of Coastal Research, Special Issue,* 34, 516-527.

Harper, C.H. 1984. Improved Method of Facies Sequences Analysis. In: Walker, R.G. (ed) *Facies Models*. Geoscience Canada, Reprint Series, 1, 11-13.

Instituto Hidrográfico 1998. *Tidal Charts 1999.* Volume I.

Kemp, K.K. 1997. Fields as a framework for integrating GIS and environmental process models. Part 1: Representing spatial continuity. *Transactions in GIS*, 1, 219-234.

Li, R. and Saxena, N. 1993. Development of an integrated marine geographic information system. *Marine Geodesy*, 16, 293-307.

Loureiro, J.J.M., Nunes, M.N. and MACHADO, M. 1986. Bacia hidrográfica do Rio Guadiana. In: *Monografías Hidrológicas dos Principais Cursos de Água de Portugal Continental*. M.P.A.T., S.E.A.R.N. Direcção- Geral dos Recursos e Aproveitamentos Hidráulicos, 341-407.

Lucas, A. 1999. Representation of Variability in Marine Environmental Data. In: Wright, D. and Bartlett, D. (eds) *Marine and Coastal Geographical Information Systems*. Taylor & Francis, London, 53-74.

Miall, A.D. 1973. Markov chain analysis applied to an ancient alluvial plain succession. *Sedimentology*, 20, 347-364.

Morales, J.A. 1995. *Sedimentologia del Estuario del Rio Guadiana*. Unpublished PhD thesis, Huelva University, 322 pp.

Morales, J.A. 1997. Evolution and facies architecture of the mesotidal Guadiana River delta (S.W. Spain – Portugal). *Marine Geology*, 138, 127-148.

Ortega, J. A. and Garzón, G. 1997. Inundaciones históricas en el río Guadiana: sus implicationes climaticas. *Cuaternario Iberico*, 365-167.

Powers, D.W. and Easterling, R.G. 1982. Improved methodology for using embedded Markov chains to describe cyclical sediments. *Journal of Sedimentary Petrology*, 52/3, 913-923.

Reinson, G.E. 1992. Transgressive barrier island and estuarine systems. In: Walker, R.G. and James, N.P. (eds) *Facies Models*. Geological Association of Canada, 179-194.

Rocha, C., Galvão, H. and Barbosa, A. 2002. Role of transient silicon limitation in the development of cyanobacteria blooms in the Guadiana estuary, south-western Iberia. *Marine Ecology Progress Series*, 228, 35-45.

Walther, J. 1894. *Einleitung in die Geologie als historische Wissenschaft. Bd. 3: Lithogenesis der Gegenwart.* Fischer Verlag, Jena, 535-1055.

Weinholtz, M. De Bivar 1978. *Rio Guadiana, Elementos para o estudo da evolução da sua desembocadura.* Direcção Geral de Portos, 11 pp.

Chapter 5

COMPARISON OF THE HYDRODYNAMIC CHARACTER OF THREE TIDAL INLET SYSTEMS

Elizabeth A. Pendleton[1] and Duncan M. FitzGerald[2]

[1]*U.S. Geological Survey, Coastal and Marine Geology Program, 384 Woods Hole Road, Woods Hole, MA 02543-1598 USA (ependleton@usgs.gov)*
[2]*Boston University, Department of Earth Sciences, 685 Commonwealth Avenue, Boston, MA 02215, USA (dunc@bu.edu)*

1. INTRODUCTION

New Inlet formed on 2 January 1987 when a northeasterly storm passed over the southern portion of Cape Cod, Massachusetts, USA (Fig. 1). Breaching of Nauset Spit during this event presented an excellent opportunity to describe and understand how tidal inlets evolve hydrodynamically and morphologically after their formation (FitzGerald and Montello, 1991, 1993; Weidman and Ebert, 1993). In this study an analysis of the evolution of a newly formed, frictionally-dominated bay is conducted and results are compared to two well-documented inlet systems.

Tidal asymmetries (duration and velocity) of bays have long been recognized as important factors in controlling the direction and magnitude of net sediment transport (Postma 1961, 1967). Classical perspective of tidal duration asymmetry (unequal rise and fall of the tide) has shown that bays with shorter flood than ebb durations lead to flood-dominance by producing shorter, stronger flood currents (flood-dominated; e.g. Nauset Inlet, Massachusetts) and subsequent sediment infilling of the bay. Conversely, shorter ebb durations lead to stronger ebb currents (ebb dominated; e.g. North Inlet, South Carolina), sediment flushing, and long-term inlet stability

D.M. FitzGerald and J. Knight (eds.),
High Resolution Morphodynamics and Sedimentary Evolution of Estuaries, 83-100.

(Boon and Byrne, 1981). Although this seems like a logical conclusion, it is insufficient to assume that duration and velocity asymmetries within an entire inlet and backbarrier system are always consistent and in the same direction. Because shallow bays are frictionally dominated, they produce a variety of nonlinearities associated with friction and time varying channel depths and bay widths (Friedrichs *et al.*, 1993). Bays that vary in geometry through the tidal cycle due to flooding and draining of intertidal areas are thought to produce flood-dominated shallow areas, and ebb-dominated channels (Aubrey and Speer, 1983; FitzGerald and Nummedal, 1983; Friedrichs *et al.*, 1992). Thus, variations in bay geometry can produce small-scale (tens of meters) variations in velocity dominance. On longer time scales (decades), bays may also change hydrodynamic response by shifting from flood-dominance to ebb-dominance or *vice versa* as a result of morphologic changes (Aubrey and Speer 1983; Friedrichs *et al.*, 1992).

Non-linearities in shallow tidal bay hydrodynamics are produced by the propagation of the tidal wave from the deep ocean into shallow coastal waters. As the tidal wave moves into shallow water, the spectral components of the wave change from one dominated strictly by the deep-water constituents to one controlled by primary constituents and their associated harmonics (overtides). Overtides are a result of non-linear terms (quadratic friction, advection of momentum, and changes in bay geometry) in the shallow-water integrated equations of motion, and they are produced by the distortion of the tidal wave as it propagates into shallow water. The shallow water equation of motion can be written as (e.g. Speer and Aubrey, 1985);

$$b\frac{\partial \zeta}{\partial t} + \frac{\partial (b_c h_c u)}{\partial x} = 0 \tag{1}$$

for continuity, and

$$\frac{\partial u}{\partial t} + u\frac{\partial u}{\partial x} + g\frac{\partial \zeta}{\partial x} + \frac{c_d u|u|}{h_c} = 0 \tag{2}$$

for momentum, where t is time, x is the distance into the embayment, b is cross-sectional width of the embayment, b_c is the cross-sectional width of the channel, h_c is mean channel depth, u is depth-averaged cross-sectional velocity within the channel, ζ is the water surface elevation, g is acceleration due to gravity, and c_d is the drag coefficient. Because channel width (b_c) and depth (h_c), bay width (b) and surface elevation (ζ), quadratic friction, and advective momentum all change with time, the shallow water 1-D equation of motion is complicated. To bypass the analytical difficulty of solving the continuity equation with non-linear terms, methods have been developed in which amplitude and phase values of tidal constituents and simplified bay geometries are used to aid understanding.

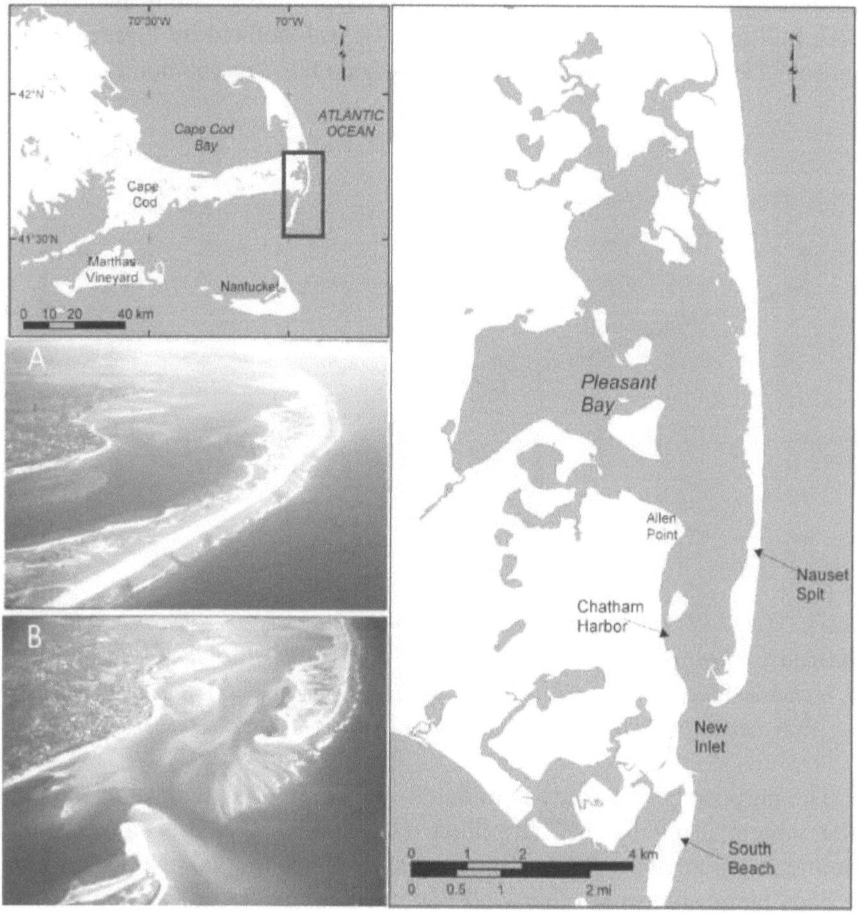

Figure. 1 Location of New Inlet, Chatham Harbor, and Pleasant Bay along the outer coast of Cape Cod. Chatham Harbor is the portion of the backbarrier south of Allen Point and North of the New Inlet entrance. A) Before (1985) and B) after (1987) photographs of the location of inlet formation.

In most areas along the East Coast of the United States, M2 (a lunar component) is the primary tide-producing constituent, which produces the M4, M6, M8 and M10 overtides. The rate of growth of M4 as the tide wave propagates into shallow water relates directly to the flood- versus ebb-dominance of the bay. The ratio of the amplitude of M4 and M2 (M4/M2) has been used in previous studies to determine the velocity asymmetry of a bay, where high M4/M2 ratios suggest flood dominance, and low M4/M2 ratios indicate ebb-dominance (Nummedal and Humphries, 1978; Aubrey and Speer, 1983). Similarly, the phase difference (*Δθ*) in M2 and M4 has been used to further evaluate the overall tidal asymmetry in a bay. An additional parameter γ, which is the difference in bay width and channel depth multiplied by tidal amplitude (Fig. 2), was defined by Friedrichs and Madsen (1992) and can also be used investigate flood or ebb dominance.

$$\gamma = a(\alpha - \beta)$$

$$\alpha = 5/3h_o,\ \beta = \Delta b/ab_o$$

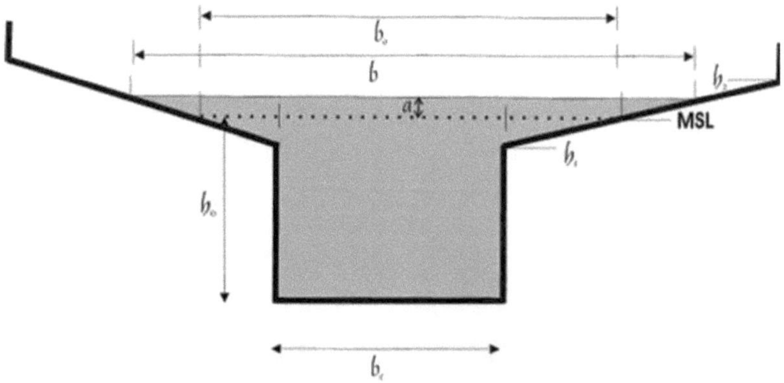

Figure. 2 Idealized channel showing variation in embayment width and channel depth during a tidal cycle (after Friedrichs and Madsen, 1992); where a is tidal amplitude relative to MSL, b_c is channel width, h_o is the channel depth, and b is total embayment width including tidal flats. Other variables relate to time-varying embayment width and channel depth.

The purpose of this study is to determine the hydrodynamics of the New Inlet system using M4/M2 amplitude ratios, relative phase values, and a nonlinear parameter (γ) as proxies. Our results are compared to other hydrodynamic studies at other well-documented inlet systems (Nauset Inlet, MA, and North Inlet, SC), as well as previous work within New Inlet, MA (Fig. 3).

1.1 Study area

The barrier island and spit system that extends from Coast Guard Beach to Monomoy Point on the southern outer coast of Cape Cod is one of New England's longest barrier chains, approximately 30 km in length (Fig. 1). The barriers that make up the chain from north to south are: Nauset Spit, South Beach, North and South Monomoy Island. They serve to protect Chatham Harbor and Pleasant Bay from the open Atlantic Ocean, providing an environment of high productivity and biodiversity (Ramsey, 1998). Because of high wave energy, storm frequency, and a large longshore sediment supply, this barrier system has exhibited drastic morphological changes over short geologic time scales (decades) (Giese, 1988, 1990).

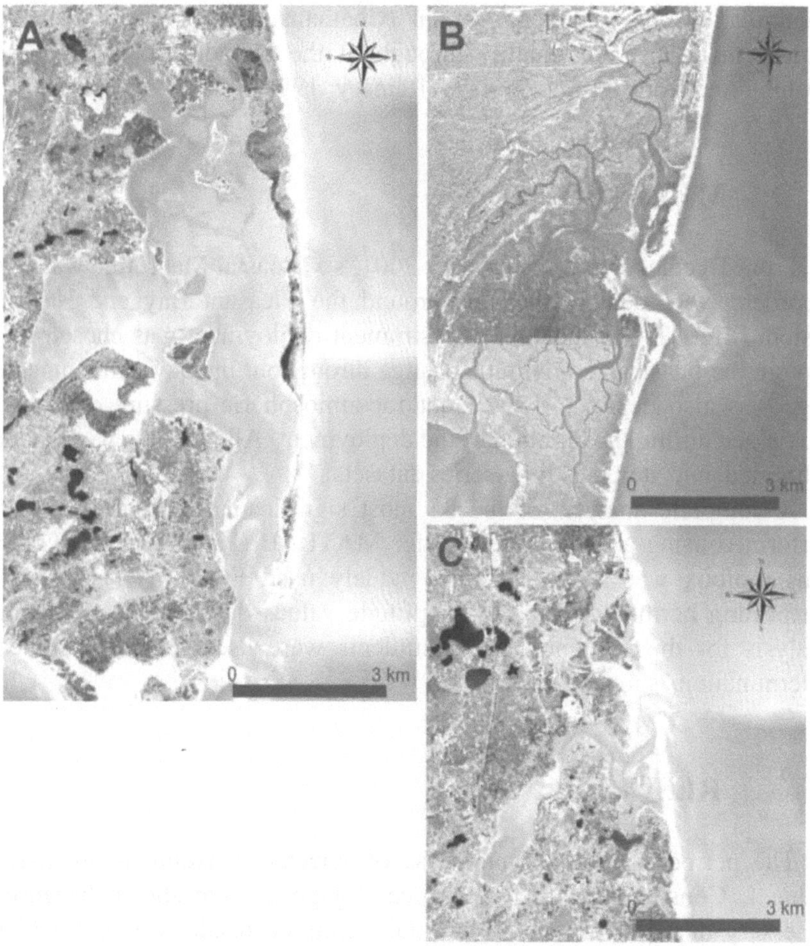

Figure. 3 Three tidal inlets having different hydrodynamic regimes. (A) New Inlet - assumed flood-dominance, (B) North Inlet, SC - ebb-dominated, (C) Nauset Inlet, MA - flood-dominated. Scale bar is 3 km in all cases.

The dominant waves approach the southern outer coast of Cape Cod from the east-northeast, which produces a southerly longshore transport system estimated to move between 3.6 x 10^5 m^3 yr^{-1} (Weishar *et al.*, 1989) to 5.0 x 10^5 m^3 yr^{-1} (Liu *et al.*, 1993) of sediment eroded from glacial outwash cliffs to the north. Average deep-water wave height and period for this area are 1.5 m every 8 s, respectively (Brooks and Brandon, 1995).

The formation of New Inlet on 2 January 1987 changed the hydrodynamics of the bay. Tidal range inside the bay increased by 0.3-0.4 m (from 1.1 to 1.4 m at the Chatham fish pier) producing stronger tidal current velocities (> 10% within Chatham Harbor) and a 12-16% increase in tidal prism (Ramsey, 1998). Since the inlet opened, there has been a net accumulation of sediment within Chatham Harbor and an overall improvement in water quality throughout the bay due to an increase in flushing rate from 1.1 to 0.98 days (Ramsey, 1998).

2. METHODS

From December 2000- February 2001, six Coastal Macrotide water level recorders were deployed in and around the Pleasant Bay and New Inlet system (Fig. 4). The location of instrument deployment was chosen to best capture the phase and amplitude change throughout the bay. An atmospheric gage was also deployed to account for atmospheric pressure changes that took place during the duration of the deployment. All tide data were reduced, calibrated, and stored as time-series datasets.

A complete analysis of the 35 most significant tidal constituents was performed using a toolbox created for MATLAB (Pawlowicz *et al.*, 2002). This toolbox was adapted from a widely used FORTRAN program for performing harmonic analysis. Amplitude values determined in constituent analysis for the M2 and M4 constituents were used for the nonlinearity determinations shown here.

3. RESULTS

The period, amplitude and phase of selected constituents are listed in Table I. Constituent data for the ocean gage are not shown because the necessary duration (2 weeks) to obtain accurate results was not achieved. Amplitude and phase values for the ocean were corrected using another investigation in the area (Friedrichs *et al.*, 1993), because hydrodynamics for the open ocean do not change in response to inlet formation.

Since M2 (the primary lunar constituent) has the most significant effect on tidal amplitude, a graph of amplitude decay and phase lag can be produced from M2 values (Fig. 5). Overall, New Inlet is backed by a shallow frictionally-dominated bay as determined by the degree of amplitude decay and phase lag of the propagating tide wave. Ocean M2 values were corrected using amplitudes determined by Friedrichs *et al.* (1993). The role of friction in dampening the tidal wave and increasing the harmonic overtides in shallow bays is illustrated in figure 6. The data show the decay of major constituents and the growth of the M4 harmonic as the tidal wave propagates into the inlet.

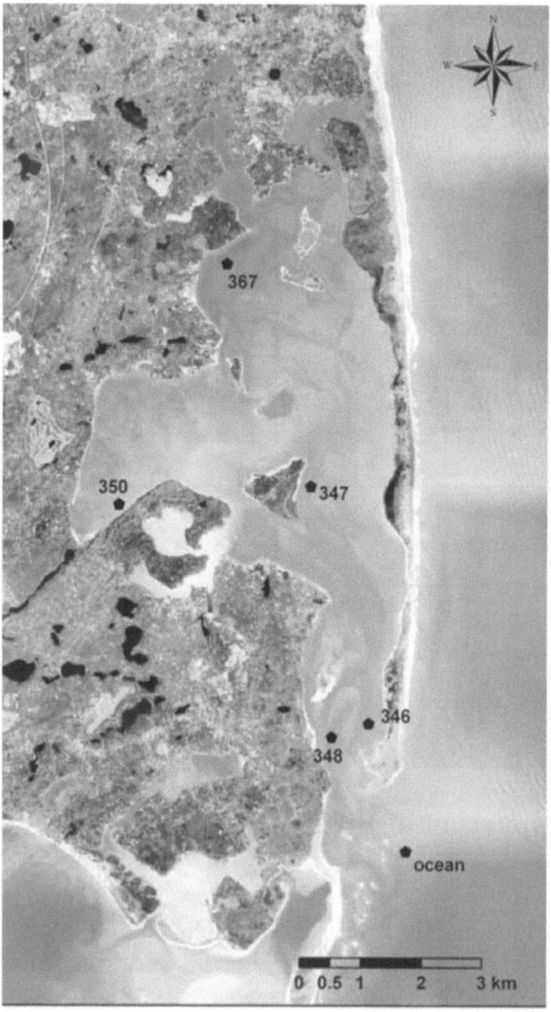

Figure. 4 Location of water level recorders deployed within Pleasant Bay.

Table I Major tidal constituents for tide gages within Pleasant Bay. Gage locations shown in figure 4.

	Gage 348				Gage 346		
	Period (hrs)	Amplitude (m)	Phase (degrees)		Period (hrs)	Amplitude (m)	Phase (degrees)
M2	12.42	0.7201	349.472	M2	12.42	0.6732	355.089
S2	12	0.0736	26.028	S2	12	0.0798	45.489
N2	12.66	0.1731	327.888	N2	12.66	0.1714	326.349
K1	23.93	0.1229	164.303	K1	23.93	0.1068	176.834
M4	6.21	0.0365	270.532	M4	6.21	0.0445	276.737
	Gage 347				Gage 350		
	Period (hrs)	Amplitude (m)	Phase (degrees)		Period (hrs)	Amplitude (m)	Phase (degrees)
M2	12.42	0.5655	23.964	M2	12.42	0.5503	32.127
S2	12	0.0549	63.257	S2	12	0.053	74.339
N2	12.66	0.1313	5.63	N2	12.66	0.1324	14.23
K1	23.93	0.1115	189.234	K1	23.93	0.1111	193.882
M4	6.21	0.0444	344.867	M4	6.21	0.0827	4.34
	Gage 367						
	Period (hrs)	Amplitude (m)	Phase (degrees)				
M2	12.42	0.5308	38.685				
S2	12	0.0483	80.044				
N2	12.66	0.1306	20.214				
K1	23.93	0.109	196.394				
M4	6.21	0.1049	17.345				

3.1 Comparison of Inlets

The nonlinearity measurements of the New Inlet system were compared to two hydrodynamically well-documented tidal inlets, North Inlet, Massachusetts, which is an ebb-dominated system (stronger peak ebb velocities) (Nummedal and Humphries, 1978), and Nauset Inlet, Massachusetts, which is a flood-dominated system (Aubrey and Speer, 1983) (Fig. 3). Morphologically North Inlet consists of deep ebb-dominated tidal channels flanked by wide intertidal marshes. During a tidal cycle, the area of the bay changes significantly due to flooding and draining of the marsh. Conversely, the Nauset system is a shallow bay with poorly defined channels, and the width of the bay changes little throughout a tidal cycle (Fig. 3). The M4/M2 amplitude ratios, relative phase values, and the calculation of nonlinear parameter γ demonstrate that New Inlet exhibits characteristics of both hydrodynamic end-member inlets, such that Chatham Harbor is ebb-dominated and Pleasant Bay is flood-dominated (Fig. 7).

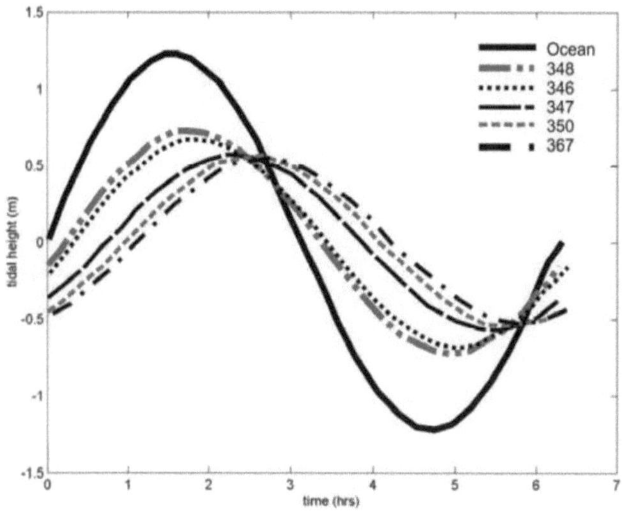

Fig. 5 Amplitude decay and phase lag of M2 tide wave from the ocean into New Inlet. See figure 4 for gage locations.

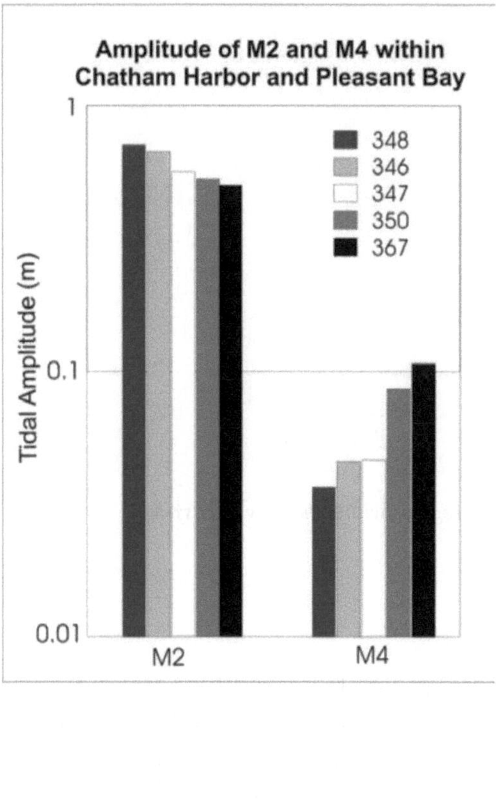

Fig. 6 Decay and growth of selected tidal constituents. Notice that the amplitude of M4 increases as the tide wave propagates into the inlet.

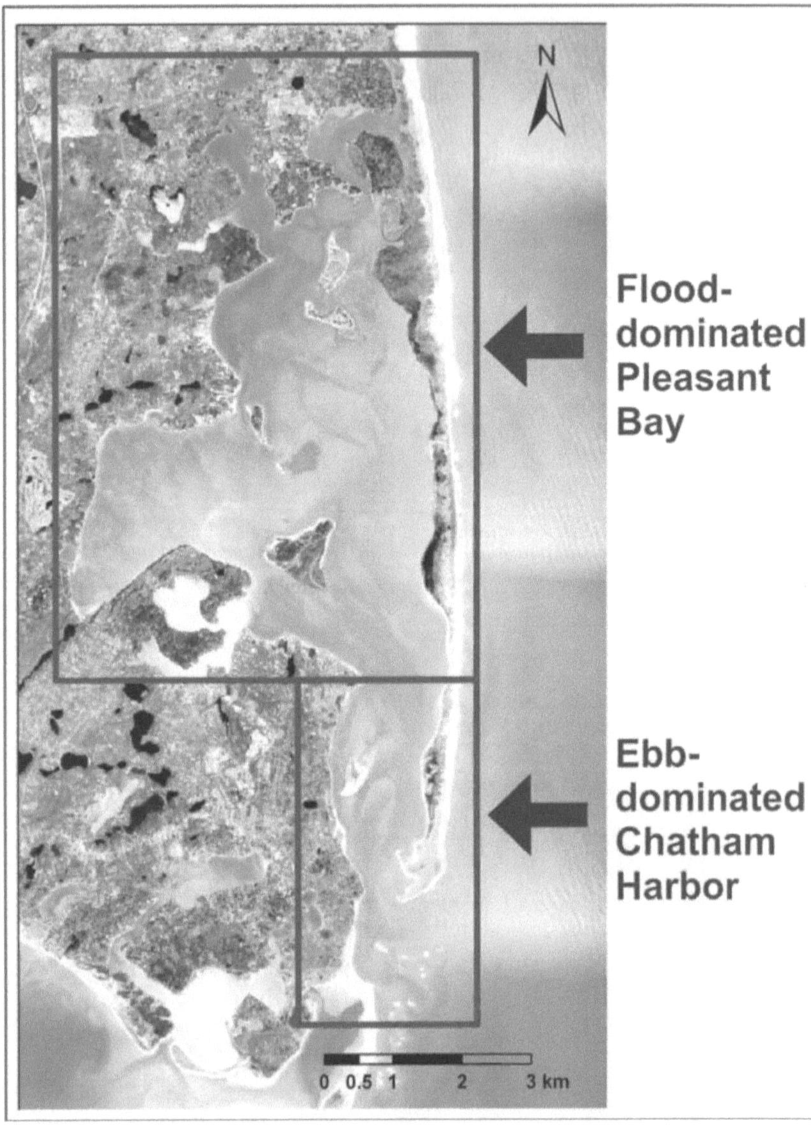

Figure. 7 Hydrodynamic regions within the New Inlet system.

3.2 M4/M2

North Inlet, South Carolina, is a well-documented example of an ebb-dominated system, and has a low growth rate of M4 because the tidal channels are narrow and deep and friction is relatively low (Fig. 3). Conversely, Nauset Inlet, is flood-dominated and has a high growth rate of

M4 as the tide wave propagates through the shallow bay. Aubrey and Speer (1983) attributed relatively high M4/M2 amplitude ratios in Nauset Inlet to a strong tidal asymmetry (flood-dominated). They distinguished the Nauset system from North Inlet, which also exhibits tidal asymmetry, but with shorter ebb duration (ebb-dominated) and much smaller M4/M2 ratios (Nummedal and Humphries, 1978) (Table II).

Table II M4/M2 values for three inlets

South Channel, Nauset, MA	**x/L**	**M4/M2**
Aubrey and Speer, 1985	1	0.007
Aubrey and Speer, 1985	0.86	0.083
Aubrey and Speer, 1985	0.73	0.119
Aubrey and Speer, 1985	0.47	0.142
Aubrey and Speer, 1985	0.32	0.142
Aubrey and Speer, 1985	0.12	0.207
North Inlet, SC		
NOS, 1985	1	0.007
Nummedal and Humphries, 1978	0.91	0.043
Nummedal and Humphries, 1978	0.85	0.062
Eiser and Kjerfve, 1975	0.65	0.053
NOS, 1985	0.46	0.058
Eiser and Kjerfve, 1975	0.18	0.074
New Inlet, MA		
Friedrichs *et al.*, 1992	1	0.025
Aubrey, unpublished (1992)	0.98	0.008
This study	0.96	0.051
This study	0.91	0.066
Friedrichs *et al.*, 1992	0.8	0.052
Friedrichs *et al.*, 1992	0.68	0.074
This study	0.55	0.079
This study	0.36	0.15
Friedrichs *et al.*, 1992	0.3	0.159
This study	0.22	0.198
Friedrichs *et al.*, 1992	0	0.219

Comparison of the M4/M2 amplitude ratios for Pleasant Bay to M4/M2 ratios for North Inlet and Nauset Inlet shows that the values for the southern portion of the bay (Chatham Harbor) are similar to that of North Inlet at high x/L (distance into embayment/embayment length) values (Fig. 8).

Conversely, north of Allen Point in wider and shallower Pleasant Bay, M4/M2 ratios become much more similar to those in Nauset Inlet.

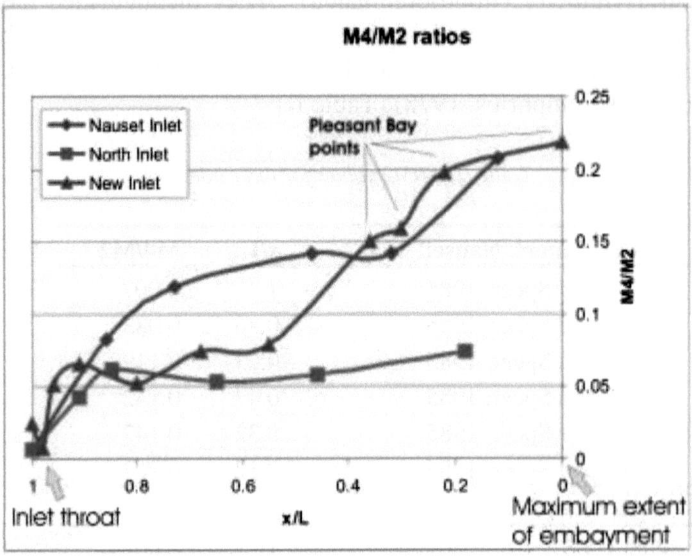

Figure 8. M4/M2 amplitude ratios for Nauset Inlet, MA, North Inlet, SC, and New Inlet, MA. Extremely flood-dominated systems like Nauset Inlet have high M4/M2 ratios; notice that New Inlet has low values near the inlet throat and high values in Pleasant Bay.

3.3 Relative Phase Difference (Δθ)

A relative phase difference calculation indicates the overall tidal asymmetry of the bay.

$$\Delta\theta = 2(\theta_{M2} - \theta_{M4}) \qquad (3)$$

where θ_{M2} is the phase (in degrees) of M2, and θ_{M4} is the phase (in degrees) of M4. For $\Delta\theta$ (relative phase difference) between 270 and 90 degrees, the embayment is overall flood-dominated, and when $\Delta\theta$ is between 90 and 270 the system is overall ebb-dominated. Maximum flood or ebb dominance occurs at 0° and 180° respectively (Speer, 1985; Speer and Aubrey, 1985; Fry, 1987). Relative phase difference calculations suggest that North Inlet is ebb-dominated and Nauset and New Inlet are flood-dominated (Fig. 9).

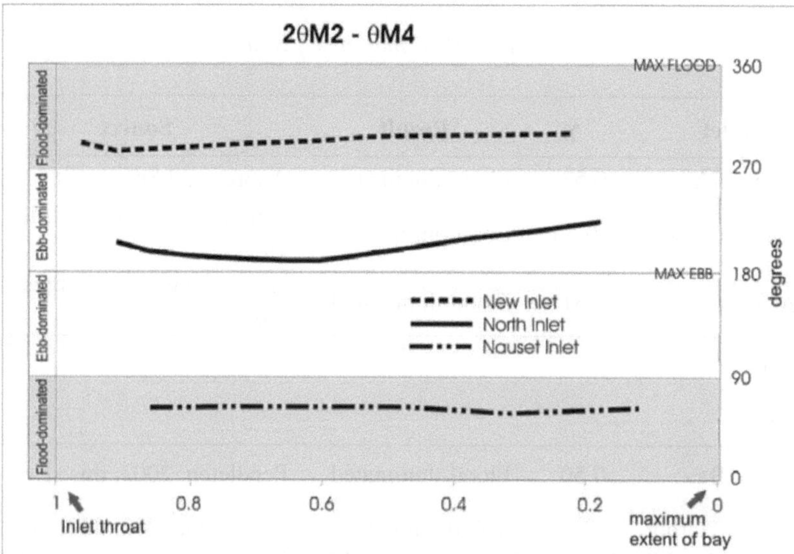

Figure. 9 Relative phase differences for Nauset, MA, North Inlet, SC, and New Inlet, MA. Relative phase values from 90-270° (shaded) are ebb dominated, and values from 270-90° are flood-dominated, with maximum dominance at 180° and 0°.

3.4 Nonlinear Parameter γ

A final nonlinear parameter γ was determined for each region of the bay to investigate the hydrodynamics. This parameter, γ as first defined by Friedrichs and Madsen (1992) is a measure of how flood or ebb dominated a system is and is based on variations in bay geometries. The parameter is calculated as

$$\gamma = a(\alpha - \beta) \tag{4}$$

where a is tidal amplitude, α is measure of the variation in channel depth within the bay, and β is a measure of the change in bay width during a tidal cycle (Fig. 2). When γ is negative, changes in embayment width are more important than variations in channel depth, and the bay is ebb-dominated. Conversely, when γ is positive, variation in channel depth are more significant than changes in embayment width, so the bay is flood-dominated. These variations in bay geometry associated with tidal asymmetries are consistent with findings of FitzGerald and Nummedal (1983). New post-breaching γ values were calculated for Pleasant Bay and Chatham Harbor using recent bathymetric and aerial photo data and compared to those calculated for North Inlet, Nauset Inlet and New Inlet (Friedrichs and Madsen, 1992) (Table III).

Table III Calculated γ values

Inlet	γ	Result	Source
Nauset, MA	0.53	Flood-dominated	Aubrey and Speer, 1985
North Inlet, MA	-0.53	Ebb-dominated	Nummedal and Humphries, 1978
New Inlet, MA	0.51	Flood-dominated	Aubrey, unpublished data, 1988
Pleasant Bay	0.50	Flood-dominated	Pendleton, 2002; this study
Chatham Harbor	-0.41	Ebb-dominated	Pendleton, 2002; this study

4. DISCUSSION

New Inlet system can be seperated into two separate hydrodynamic regions: flood-dominated Pleasant Bay and ebb-dominated Chatham Harbor (Fig. 7). The hydrodynamic character of each region is a result of the antecedent geomorphology of the bay, inlet formation, and the ensuing changes in hydraulic regime.

M4/M2 ratios suggest that the New Inlet system based on 2001 tidal data has both flood and ebb dominated regions; such that narrower Chatham Harbor is similar to North Inlet and dominated by ebb flow, and wider, more extensive Pleasant Bay is similar to Nauset Inlet and is flood-dominated (Fig. 7). Original calculations of γ for Nauset Inlet and North Inlet were consistent with M4/M2 and relative phase results, where Nauset Inlet is flood-dominated and North Inlet is ebb-dominated (Friedrichs and Madsen, 1992). However, Aubrey data from 1988 for New Inlet suggest that it is flood-dominated (Table III). If New Inlet is partitioned into two regions and γ values are recalculated with more recent data (2001), it is seen that New Inlet has a flood-dominated upper bay and an ebb-dominated lower bay (Table III). This is consistent with M4/M2 data for New Inlet, but this characteristic is not revealed in the relative phase data. This can be explained because relative phase calculations are a measure of the overall dominance of the bay, therefore New Inlet system is overall flood-dominated based on $\Delta\theta$ because Pleasant Bay is flood-dominated and accounts for 80% of the tidal prism.

When comparing the morphology of New Inlet to Nauset Inlet and North Inlet, it is apparent that New Inlet is more complex (Stauble, 2001). North Inlet is a well-established stable inlet that has reached a dynamic equilibrium configuration, whereby sediment is flushed from the system by ebb-dominated tidal channels. Morphologically the backbarrier of North Inlet lacks large open water areas, and is characterized by deep channels that are flanked by extensive intertidal marsh. Conversely, Nauset Inlet opens into a wide, shallow bay, which diffuses flow across a large area. Less mature, flood dominated systems such as Nauset Inlet will tend to fill through time by landward sediment transport until the inlet closes or the amount of open water area has been sufficiently reduced to reverse the hydrodynamic response (Friedrichs *et al.*, 1992). In these systems the backbarrier is gradually converted to intertidal flats and/or marsh incised by tidal creeks. New Inlet is sedimentologically and hydrodynamically young and much of its morphology is inherited from previous episodes of inlet formation (Giese, 1988), flood-tidal delta evolution, and southerly spit accretion. These processes coupled with inherited morphology have created a very complex hydrodynamic configuration.

By 1998, shoal growth had slowed, tidal channels reached stable configurations, and overall sediment transport approached a dynamic equilibrium within the backbarrier of New Inlet (Pendleton, 2002). The increase in intertidal area from 11% (pre-breach) to 25% (2001, post-breach) effectively reduced the amount of open water area over which Chatham Harbor could experience peak ebb velocities because peak ebb velocities occur near low tide. Peak ebb flow in Chatham Harbor is now confined to the channels, whereas peak flood velocities (which occur near high tide) are spread out over the entire bay (Boothroyd, 1985). Open water area differences within Chatham Harbor are responsible for the hydrodynamic shift from flood-dominance (1988-98) to ebb-dominance (1998-present).

Pleasant Bay experienced little morphologic change as a result of inlet formation, and thus has remained flood-dominated due to its substantial width and lack of well-developed channels (Pendleton, 2002). Additionally, the size of Pleasant Bay has a strong influence on the overall inlet hydrodynamics. Because of this, New Inlet is considered flood-dominated as suggested by relative phase differences. Even though the bay is overall flood-dominated, this does not necessarily correspond to sediment filling of the bay through time as might be expected if the system were completely flood-dominated (e.g. Nauset Inlet) (Boon and Byrne, 1981).

Because Chatham Harbor is ebb-dominated and Pleasant Bay is flood-dominated (Fig. 9), there is a contradiction in the tendency of the bay to fill with sediment or maintain itself through self-flushing. Chatham Harbor maintains itself by keeping deep ebb-dominated channels swept free of sediment deposited by flood currents, and establishing a net seaward transport of sediment. Conversely, Pleasant Bay lacks channels and is flood-dominated, therefore having a tendency to fill with sediment over time. However, Pleasant Bay is located far from the new sediment source, (longshore transport system along Nauset Spit), and thus will only be transporting sediment landward within Pleasant Bay. Presently, the backbarrier of New Inlet has achieved a dynamic equilibrium and the system will continue to evolve toward ebb-dominance until the next breaching event.

5. CONCLUSIONS

The M4/M2 ratios of Chatham Harbor (south of Allen Point) are similar to ebb-dominated North Inlet, SC. However, as the tidal wave moves into Pleasant Bay M4/M2 ratios rapidly increase and become similar to M4/M2 ratios of flood-dominated Nauset Inlet, MA. From M4/M2 ratios, the embayment can be divided into two hydrodynamically and morphologically distinct regions: ebb-dominated Chatham Harbor and flood-dominated Pleasant Bay (Fig. 7). The overall flood-dominated signature determined in the $\Delta\theta$ measurement is explained because Pleasant Bay accounts for nearly 80% of the bay tidal prism, and thus determines the overall hydrodynamics of the bay.

Numerical modeling has confirmed that all of the major channels within Chatham Harbor are ebb-dominated, and all of Pleasant Bay is flood-dominated, with a transitional area in between these two regions (Pendleton, 2002). Shallow intertidal areas are flood dominated in all parts of the bay because they are exposed during peak ebb near low tide, but are covered during peak flood near high tide (Boothroyd, 1985). This observation is consistent with velocity asymmetry predictions by Friedrichs *et al.* (1992) who suggested that small-scale differences in tidal dominance do exist, such as an ebb-dominated channel adjacent to a flood-dominated shoal. Additionally, these small-scale (tens of meters) dominance differences are independent of the overall flood/ebb dominance of the system as shown in the $\Delta\theta$ results. Results presented here suggest that there can also be medium-scale (hundreds of meters to kilometers) dominance disparities, where a narrow portion of the bay (Chatham Harbor) is ebb-dominated whereas the wide shallow portions are flood-dominated (Pleasant Bay). The area of the bay which contains most of the tidal prism will determine the overall flood or ebb dominance. However, the region closest to the inlet entrance will control the overall sediment filling or self-flushing characteristics of the bay.

REFERENCES

Auubrey, D.G. and Speer, P.E. 1983. *Sediment transport in a tidal inlet.* Technical Report WHOI-83-20, Woods Hole Oceanographic Institution, Woods Hole, MA, 110pp.

Aubrey, D.G. and Speer, P.E. 1985. A Study of non-linear tidal propagation in shallow inlet/estuarine systems. Part I. Observations. *Estuarine Coastal and Shelf Science*, 21, 185-201.

Boon, J.D. and Byrne, R.J. 1981. On basin hypsometry and morphodynamic response of coastal inlet systems. *Marine Geology*, 40, 27-48.

Boothroyd, J.C. 1985. Tidal inlets and tidal deltas. In: Davis, R.A. (ed) *Coastal Sedimentary Environments*. Springer-Verlag, New York, 445-532.

Brooks, R.M. and Brandon, W.A. 1995. *Hindcast wave information for the U.S. Atlantic coast: Update 1976–1993 with hurricanes.* WIS Report 33, U.S. Army Engineer Waterways Experiment Station, Vicksburg, MS.

Eiser, W.C. and Kjerfve, B. 1975. Marsh topography and hypsometric characteristics of a South Carolina salt marsh basin. *Estuarine Coastal and Shelf Science*, 23, 595-605

FitzGerald, D.M. and Nummerdal, D. 1983. Response characteristics of an ebb-dominated tidal inlet channel. *Journal of Sedimentary Petrology*, 53, 833-845.

FitzGerald, D.M. and Montello, T.M. 1991. *A preliminary study of changes in bedform distribution and shoal morphology, sedimentation trends, and inlet hydraulics at New Inlet, Cape Cod, Massachusetts.* Report for the U.S. Army Corps of Engineers Coastal Engineering Research Center Waterways Experiment Station., 68pp.

FitzGerald, D.M. and Montello, T.M. 1993. Backbarrier and inlet sediment response to the breaching of Nauset Spit and formation of New Inlet, Cape Cod, Massachusetts. In: Aubrey, D.G. and Giese, G.S. (eds) *Formation and Evolution of Multiple Tidal Inlet Systems.* American Geophysical Union, Washington D.C., 44, 158-185.

Friedrichs, C.T., Lynch, D.R. and Aubrey, D.G. 1992. Velocity asymmetries in frictionally-dominated tidal embayments: longitudinal and lateral variability. In: Prandle, D. (ed) *Dynamics and Exchanges in Estuaries and the Coastal Zone, Coastal and Estuarine Studies*. Springer-Verlag, New York, 277-312.

Friedrichs, C.T., Aubrey, D.G., Giese, G.S. and Speer, P.E. 1993. Hydrodynamical modeling of a multiple-inlet estuary/barrier system: insight in to tidal inlet formation and stability. In: Aubrey, D.G. and Giese, G.S. (eds) *Formation and Evolution of Multiple Tidal Inlet Systems*. American Geophysical Union, 44, 95-112.

Friedrichs, C.T. and Madsen, O.S. 1992. Nonlinear diffusion of the tidal signal in frictionally dominated embayments. *Journal of Geophysical Research*, 97 (C4), 5637-5650.

Fry, V.A. 1987. *Tidal velocity asymmetry and bedload sediment transport in shallow embayments.* MS thesis, WHOI Technical Report No. 87-51, Woods Hole, MA, 55pp.

Giese, G.S. 1988. Cyclic behavior of the tidal inlet at Nauset Beach, Chatham, MA. In: Aubrey, D.G. and Weishar, L. (eds) *Hydrodynamics and Sediment Dynamics of Tidal Inlets.* Springer-Verlag, New York, 269-283.

Giese, G.S. 1990. The story behind the New Inlet at Chatham. *Nor'easter, magazine of Northeast Sea Grant Programs*, 2, 28-33.

Liu, J.T., Stauble, D.K., Giese, G.S. and Aubrey, D.G. 1993. Morphodynamic evolution of a newly formed tidal inlet. In: Aubrey, D.G. and Giese, G.S. (eds) *Formation and Evolution of Multiple Tidal Inlet Systems*, American Geophysical Union, 44, 62-94.

Metcalf and Eddy Inc. 1975. *The hydrogeology of the mainland parts of the Town of Chatham, MA*. Metcalf and Eddy Inc., 18pp.

National Ocean Service (NOS). 1985. *Index of tide stations, United States of America and miscellaneous other locations*. NOAA/National Ocean Serves, Rockville, Md.

Nummedal, D. and Humphries, S.M. 1978. *Hydraulics and Dynamics of North Inlet, SC, 1975-1976*. GITI Report 16. U.S. Army Engineer Research and Development Center, Vicksburg, MS. 214pp.

Pawlowicz, R., Beardsley, B. and Lentz, S. 2002. Classical tidal harmonic analysis including error estimates in MATLAB using T_TIDE. *Computers and Geosciences*, 28, 929-937.

Pendleton, E.A. 2002. *Morphologic and hydrodynamic evolution of a shallow tidal inlet, Chatham, MA*. Unpublished MS thesis, Boston University, 97pp.

Postma, H. 1961. Transport and accumulation of suspended matter in the Dutch Wadden Sea. *Netherlands Journal of Sea Research*, 1, 148-190.

Postma, H. 1967. Sediment transport and sedimentation in the estuarine environment. In: Lauff, G.H. (ed) *Estuaries*. American Association for the Advancement of Science, 16, 651-668

Ramsey, J.S. 1998. *Hydrodynamic and tidal flushing study of Pleasant Bay Estuary, MA*. Final Report for the Pleasant Bay Steering Committee. Aubrey Consulting, Inc. Caaumet, Massachusetts.

Speer, P.E. 1985. *Tidal distortion in shallow estuaries*. Unpublished PhD thesis, WHOI-MIT Joint Program in Oceanography, Woods Hole, MA, 210pp.

Speer, P.E. and Aubrey, D.G. 1985. A study of non-linear tidal propagation in shallow inlet/estuarine systems. Part II: Theory. *Estuarine Coastal, and Shelf Science*, 21, 207-224.

Stauble, D.K. 2001. *Impacts of navigation channel maintenance dredging on the coastal processes of Chatham, Massachusetts*. ERDC/CHL TR-01-26, U.S. Army Engineer Research and Development Center, Vicksburg, MS, 95pp.

Weidman, C.R. and Ebert, J.R. 1993. Cyclic spit morphology in a developing inlet system. In: Aubrey, D.G. and Giese, G.S. (eds) *Formation and Evolution of Multiple Tidal Inlet Systems*. American Geophysical Union, Washington D.C., 44, 158-185.

Weishar, L., Stauble, D. and Gingerich, R. 1989. *A study of the effects of the new breach at Chatham, MA*. Reconnaissance report, Coastal Engineers Research Center, Army Corps of Engineers, Vicksburg, MS, 164pp.

Chapter 6

SUSPENDED SEDIMENT FLUXES IN THE MIDDLE REACH OF THE BAHIA BLANCA ESTUARY, ARGENTINA

Gerardo M. E. Perillo[1,2], Jorge O. Pierini[1,3], Daniel E. Pérez[1,4], M. Cintia Piccolo[1,5]

[1]Instituto Argentino de Oceanografía, CC 804, 8000 Bahía Blanca, Argentina
[2]Departamento de Geología, Universidad Nacional del Sur, San Juan 670, 8000 Bahía Blanca, Argentina
[3]CIC - Comisión de Investigaciones Científicas de la Provincia de Buenos Aires
[4]Departamento de Ingeniería, Universidad Nacional del Sur, Av. Alem 1253, 8000 Bahía Blanca, Argentina
[5]Departamento de Geografía, Universidad Nacional del Sur, 12 de Octubre y San Juan, 8000 Bahía Blanca, Argentina

1. INTRODUCTION

Puerto Galván (Fig. 1) is one of the five harbors that form the Bahía Blanca Harbor System, the largest and deepest of Argentina. The harbors are all located along the Canal Principal of the Bahía Blanca Estuary, a mesotidal, coastal plain environment (Perillo, 1995) formed by a series of major NW-SE trending channels separating extensive tidal flats, low salt marshes and islands. The geomorphology and physical characteristics of the estuary are described in detail elsewhere (Perillo and Piccolo, 1999) including a recent review of its major environmental features (Perillo *et al.*, 2000).

The main economic activity of the harbor system is related to the export of agriculture products, especially grain, but in recent years there has been a marked increase in the export of oil, gas and, most commonly, petrochemical

D.M. FitzGerald and J. Knight (eds.),
High Resolution Morphodynamics and Sedimentary Evolution of Estuaries, 101-114.

derivatives. In 1989-90 a major dredging operation augmented the nominal depth of the navigation channel and harbor sites to 13.8 m (45 ft) which induced an immediate increase in the economic activity of the harbors as larger vessels were allowed to enter into the system.

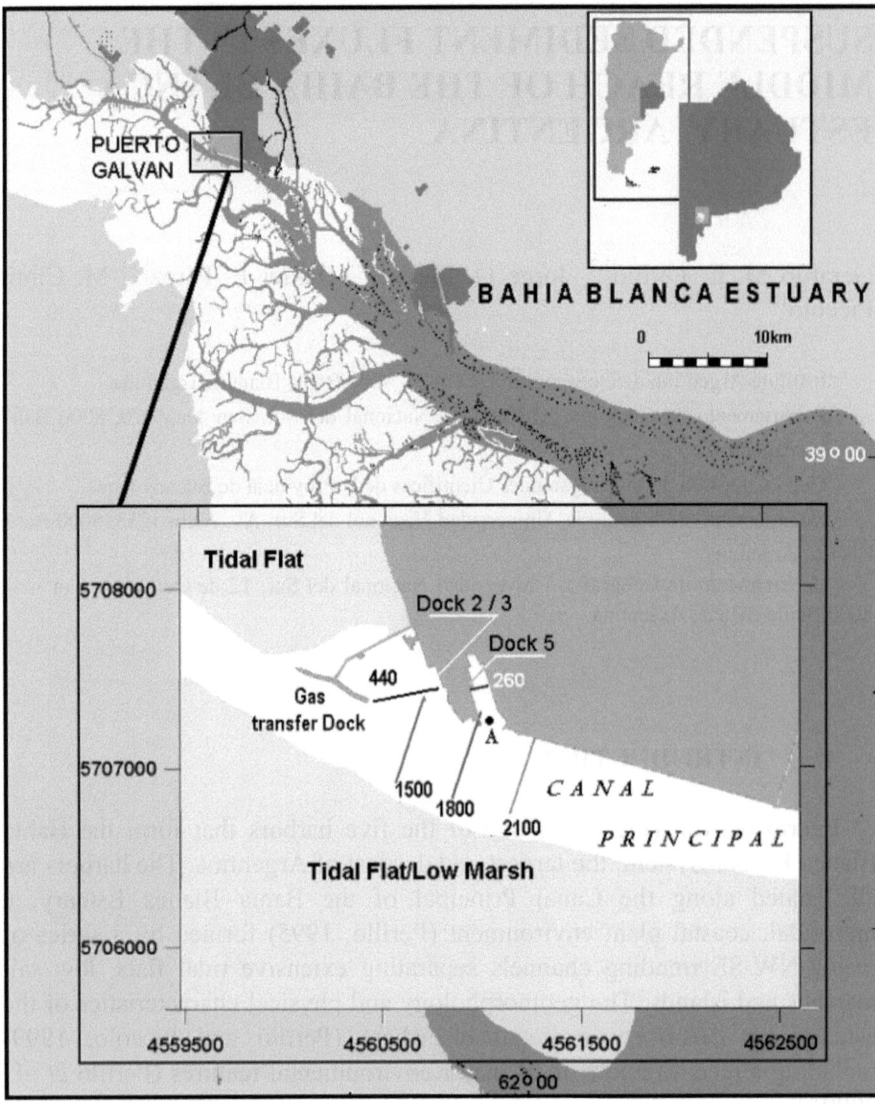

Figure 1. Location of the study area and positions of the cross-sections surveyed. A *marks the location of the station where the ADCP was calibrated.*

Although the Canal Principal at Puerto Galván has not registered significant sedimentation, some minor dredging to maintain some parts is required periodically. However, the docks require frequent dredging as sedimentation of fine materials is very active. The aim of the present article is to describe the tidal current circulation and fluxes of suspended sediment within a specific reach of the Canal Principal at Puerto Galván and its interaction with the circulation and transport in and out the docks.

1.1 Study site

Puerto Galván is located on the northern margin of the Canal Principal of the estuary (Fig. 1). It is constituted by two docks (5 and 2/3) located to the east and west, respectively, of a wharf. All harbor operations take place along the wharf and the docks have maximum operational depths of 9-12 m along it.

Dock 5 (to the east of the wharf) is closed on three sides and open only to the Canal Principal. This dock is approximately 600 m in length and 120 m wide at the mouth with maximum dredged depths of the order of 9 m; however, the operating maximum width is about 80 m. The eastern shore of the dock was originally a tidal flat that was reclaimed during the 1989-90 dredging. Two major petrochemical plants were built there, but they have no access to the dock facility. Although no detailed study of the siltation rate of this dock was made, the values are quite large and require at least one or two maintenance dredgings per year. Bottom sediments are silty clays with no traces of sand.

Even though Dock 2/3 is also about 600 m in length and 120 m wide at the mouth, it continues further inland in the Galván tidal channel. Maximum dredged depth at this site is 12 m. The western border is formed by a low tidal flat that is fully covered by at least 0.5 m of water even during neap tides. The only 'barrier' is an open bridge (Fig. 1) towards the natural gas transfer dock located on the Canal Principal. Bottom sediments are mostly silty clays with less than 5% sand.

On the other hand, the Canal Principal at the study site has been dredged to a nominal depth of 15 m and includes the turning area for vessels (just to the east of line 1500) providing, therefore, over 500 m width at low tide. Dredging of the Canal Principal was done up to 200 m inland of the gas transfer dock. Since then, there has been no need for maintenance as the dynamics of this reach of the estuary is very strong and prevents siltation over the hard bottom, which is formed by a Pliocene semi-consolidated silty sand formation (Perillo and Piccolo, 1999).

Bahía Blanca Estuary is a mesotidal system governed by a semidiurnal, quasi-stationary tidal regime. Mean tidal range along the 68 km of the Canal

Principal varies from 2 m at the mouth to 3.5 m at the head (Perillo and Piccolo, 1991). At Puerto Galván mean tidal range is 3.3 m, varying from 3.8 m to 2.75 m for spring and neap conditions, respectively. Wind influences water circulation and tidal conditions significantly. Forty percent of the time wind blows from the northwest and north and 20% from the southeast. These directions are parallel to the Canal Principal inducing both set up and set down of the tide plus advance and retardation of the time of occurrence of both high and low tide. This results in an added effect upon the normal asymmetry of the tidal wave. Observed set ups have reached up to 1.5 m over the forecasted astronomical tide (Perillo and Piccolo, 1991).

2. METHODOLOGY

In order to define the water and suspended sediment fluxes in the middle reach of the Bahía Blanca Estuary, two field cruises were made on November 27 and December 27 2000. Field operation consisted of the determination of current velocities and suspended sediment concentration continuously along predefined lines during one tidal cycle and calibration at a fixed station. Data were gathered with a 600 kHz Broad Band RD Instruments Acoustic Doppler Current Profiler (ADCP) owned and operated by ISDK (The Netherlands). Simultaneous meteorological and tidal data were obtained from stations located within 2 km of the study site.

The ADCP was located looking downward on the side of the research vessel. The equipment obtained vertical profiles of the three components (longitudinal, transversal and vertical) of the velocity and the acoustic backscatter integrated in vertical cells of 0.5 m every 2s. Due to the inherent limitations of the instrument, data are not received within 3 m from the surface and from 1 m to 3 m from the bottom. This is especially true for the backscatter data. Therefore, all the measurements analyzed here correspond to the core of the channel. In many cases velocity data were within acceptable ranges up to 1 m above the bed, but this seldom occurred for suspended sediment concentration.

In each cruise, field calibration of the ADCP was done using a MiniCTD InterOcean which also included an OBS-3 for suspended sediment concentration (C) and an acoustic current meter FSI. The vessel was positioned with a differential GPS which was directly connected to the ADCP. Roll and heave were also electronically controlled.

Four cross-sections (Fig. 1) were defined such that they covered all possible interactions of the study area with other parts of the Canal Principal and the docks. A fifth cross-section was located in the middle point to control the flows within the channel. The study was carried out for over 13 h in each cruise to ensure that a full tidal cycle was covered. Measurements started one hour prior to high tide and ended one hour after the next high water.

A typical operation consisted of running the vessel along the predesigned line (a navigation software allowed the continuous monitoring of the vessel) within the accepted draught depth for the boat. All five lines were made sequentially in the same order and following the same path during the measurement periods. At the end of one round of lines, the boat was moored to a buoy located near the mouth of the Dock 5 (Station A in Fig. 1) and a calibration of the ADCP was performed. The calibration consisted in running the ADCP continuously while vertical profiles with the CTD-OBS and the current meter were made. Water samples were also taken for laboratory calibration of the OBS.

Normally, one round of cross-sections took about 1 h whereas a calibration required on the order of 15-20 min, depending on water depth at the time of operation. Therefore, each cross-section was repeated about once every lunar hour which is the maximum allowable for residual flux studies (Kjerfve, 1979).

With the information obtained during each calibration, a special software made and operated by ISDK was employed to transform the acoustic backscatter data into suspended sediment concentration (C). These values in combination with longitudinal component of the flow (U) for each individual ADCP profile and cell, the flux of suspended sediment (F) was calculated by

$$F(z,t) = U(z,t)\ C(z,t)$$

where z is the depth of the cell and t is the time at which the measurement was made.

3. RESULTS AND DISCUSSION

Data on the longitudinal component of the velocity (U, positive in the ebb direction) and suspended sediment concentration (C) were gathered during one tidal cycle in two cruises separated by a month during late spring 2000 at the middle reach of the Bahía Blanca Estuary. Values of C for the study area during both cruises vary from 130 mg/l to near 1600 mg/l (Table I). Maximum concentration was observed at Line 1800 with 1585.4 mg/l. In both cases the larger values occurred at high water and minimal values during low tide.

Table I. Maximum and minimum suspended sediment concentration (mg/l) at each cross-section and for each cycle.

	27/11/01		27/12/01	
Line	Min C	Max C	Min C	Max C
260	203.5	1069.2	143.6	984.4
440	143.6	984.4	203.5	1069.2
1500	300.8	1031.3	164.9	1104.5
1800	304.7	1585.4	134.7	826.8
2100	365.8	1103.4	130.8	1019.1

Maximum velocities occurred during ebb at the three cross-sections located on the Canal Principal with values of 1.5 m/s, while maximum flood currents at the same sections were between 0.79 to 0.86 m/s. Maximum ebb currents at section 260 never were higher than 0.83 m/s and flood ones were always below 0.4 m/s. In all cases, the strong asymmetry of the tidal current curves indicated in previous studies which define the Canal Principal as ebb-dominated (Perillo and Piccolo, 1999) was confirmed.

For the present case, we are going to discuss only the distributions of longitudinal currents (U) and fluxes (F) of C. As examples, one graph for U and F for each cross-section is presented in Figures 2 to 6. Previous studies using vertical profiles of U and C at a station located in the deepest portion of line 440 (Fig. 2) indicated a marked difference in the circulation of the dock (Perillo *et al.*, 2001a). During the present study, the boat approached the cross-section from the Canal Principal crossing on top of the extreme of the tidal flat which appears as a depth reduction in Figure 2. Perillo *et al.* (2001b) mapping of the C distribution at three different depths (3, 6 and 9 m) indicated that during ebb a plume appears at the extreme of the flat and moves along the northern third of the Canal Principal. Although the sediment plume has a general trend along the Canal Principal, the currents introduce it into the docks. During flood, the sediment from the flat is put into suspension but in lower concentrations than during ebb.

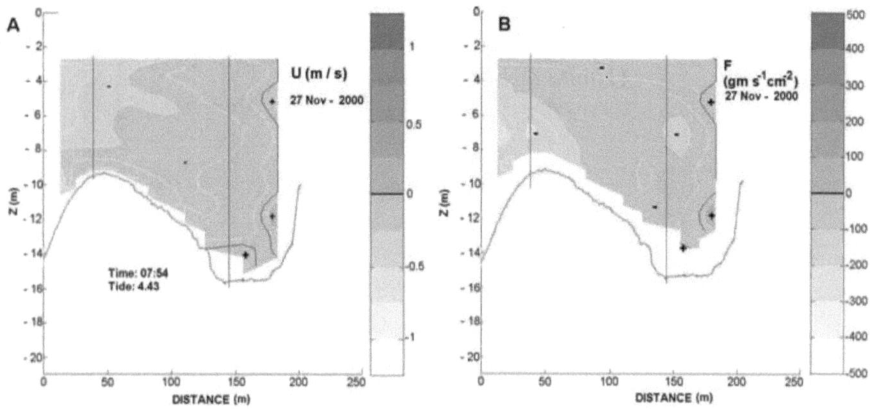

Figure 2. Cross-section looking into the Dock 2/3 (Line 440) of the (a) longitudinal component of the current (U), positive in the ebb direction, and (b) suspended sediment fluxes (F) at the end of the flood during the first cruise. Note the outflow at the bottom of the channel and at the side of the wharf (right portion of the figure). The lower depths at the left indicates the tip of the tidal flat crossed with the boat.

As is shown in Figure 2, current directions reverse along the wharf and bottom of the channel. Particularly during the flood, although it was observed during the ebb too, an eddy forms at Dock 2/3 probably due to flow separation at the tip of the wharf.

Data on cross-section 260 (Fig. 3) were very difficult to obtain as the width of the dock, especially during low tide condition, was considerably reduced. This allowed relatively few vertical profiles. Thus adequate appreciation of the circulation may be acceptable only from about half tide and higher. Figure 3 is an example of the type of circulation pattern observed in which there are upper and lower layers of water and sediment outflowing from the dock while inflow is concentrated in a middle depth layer. Reviewing the transversal component of the flow, it becomes clear that the water that entered into the dock over the lateral tidal flat is flowing towards the channel inducing a compensating outflow.

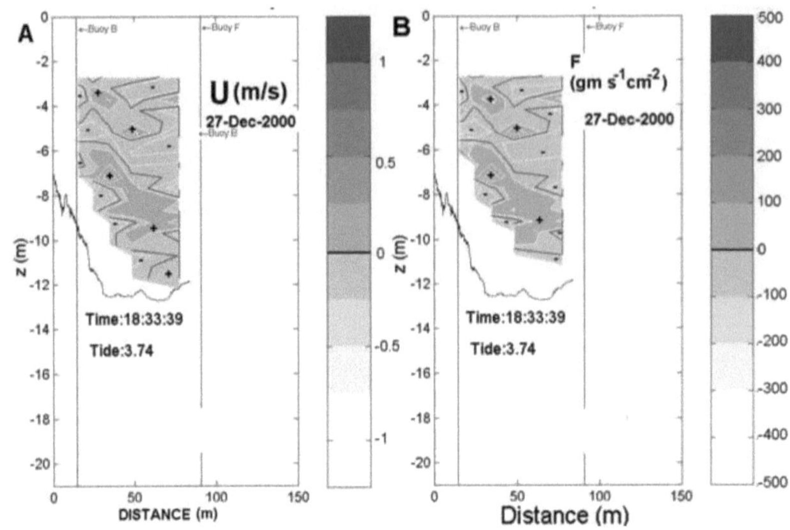

Figure 3. Cross-section looking out of Dock 5 (Line 260) of the (a) longitudinal component of the current (U), positive in the ebb direction, and (b) suspended sediment fluxes (F) near the end of the flood during the second cruise. Note the complex current and flux pattern having both positive (ebb) and negative (flood) directions. The wharf is located to the left of the graphs.

The circulation in the Canal Principal was analyzed based on lines made at three places 300 m apart. In lines 1800 and 2100, the general geomorphology is very similar: the channel has a general 'U' shape produced by dredging with the bottom relatively flat at about 13 m below Datum Plane. On the northern side the channel flank is well defined with an inclination of 0.5° and water is fully contained within the channel. However, on the southern side, an extensive tidal flat is present and, when the tide is over 2.5 m, water covers it producing a lateral circulation that is very difficult to monitor by means of the methodology used in the present study. In the case of line 1500, both flanks are open to tidal flats.

Cross-sectional distribution of the longitudinal component of the velocity varies similarly for all three lines in the Canal Principal. Changes in direction occur near the bottom and along the lower sides before low and high-water slack (Fig. 4a). During ebb, maximum currents concentrate in the upper middle portion of the Canal Principal diminishing with depth and towards the lateral (Fig. 5a). However, maximum flood currents tend to concentrate along the southern portion of the channel (Fig. 6a). This is coincident with the various studies of circulation made at this area with 2 or 3 individual stations located in various cross sections of the study reach (e.g. Piccolo and Perillo, 1990; Pérez and Perillo, 1998), that indicated a typical vertically-homogeneous estuary but only for the residual current distribution. Residual currents are mostly seaward on the northern side whereas in the southern side they are headward.

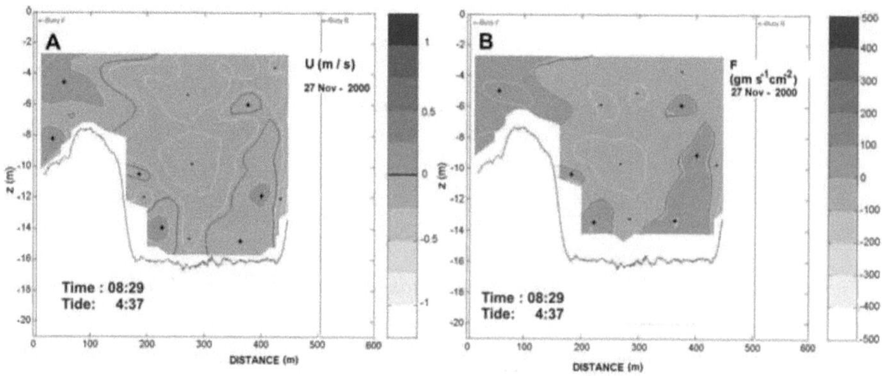

Figure 4. Cross-section at Line 1500 of the (a) longitudinal component of the current (U), positive in the ebb direction, and (b) suspended sediment fluxes (F) at high tide during the first cruise. The lower depths at the left indicates the tip of the tidal flat crossed with the boat.

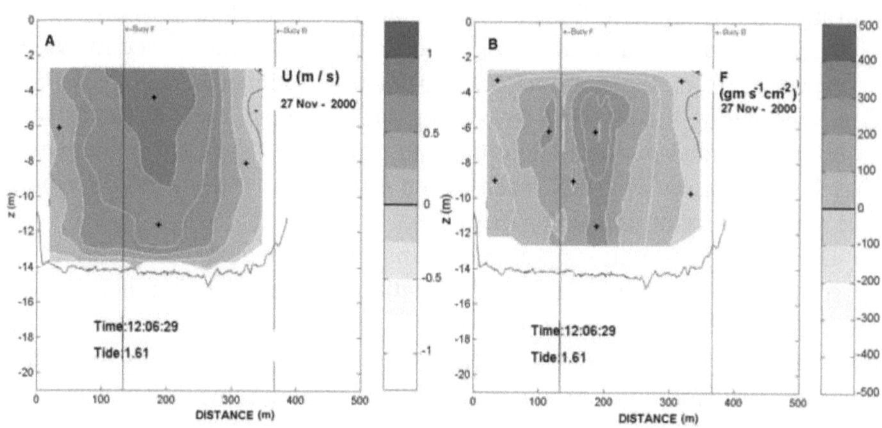

Figure 5. Cross-section at Line 1800 of the (a) longitudinal component of the current (U), positive in the ebb direction, and (b) suspended sediment fluxes (F) near low tide during the first cruise. Note the high values of F at the center of the cross - section resembling a plume of resuspended sediments from the channel bottom.

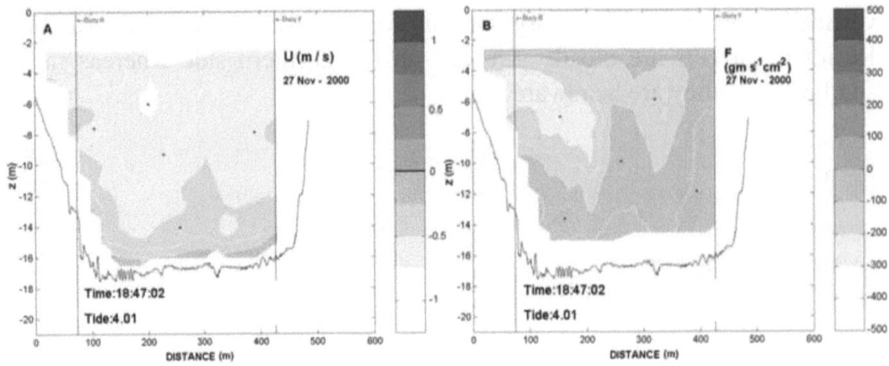

Figure 6. Cross-section at Line 2100 of the (a) longitudinal component of the current (U), positive in the ebb direction, and (b) suspended sediment fluxes (F) near the end of flood during the first cruise. Note the plume of suspended sediment coming from the southern tidal flats in the F pattern.

Regarding the fluxes of suspended sediment, in general, they follow the trends defined by the longitudinal current component. Nevertheless, a number of interesting features can be deduced from the cross-section distributions, which are valid for all three lines along the Canal Principal. During ebb, maximum fluxes concentrate at the middle portions of the channel, but as the currents are stronger in these conditions, sediments are resuspended from the bottom (Fig. 5b). These sediments may have been deposited during the previous low-water slack. The fact that sediments seem to be resuspended in large quantities and transported seaward immediately, may be the reason why this reach of the estuary has little or no deposition and requires very little maintenance.

During flood conditions, vertical plumes were not observed in any of the cross-section and, in general, the distribution of F is rather homogeneous in most cases. However, in several profiles we detected horizontal plumes coming from the southern tidal flats (Fig. 6b). These plumes are either concentrated along the southern flank and being moved landward and restricted to that portion of the channel, or they spread horizontally near the surface. In the latter, the whole upper layer with high sediment concentration is transported headward.

Surprisingly, lateral plumes all appear during flood, when we were expecting them with the ebb currents. As indicated, ebb currents are stronger and also considering that the flats are eroding, the expectancy was that more material was coming off from them when larger dynamic conditions were prevalent. Although we cannot demonstrate it from our data, we hypothesize that as the tide enters and concentrates along the southern channels, it puts into suspension sediment from the channel side. Perillo *et al.* (2001b) also point out a similar increase in C along the northern flank of line 2100 but they assigned it to erosion of a mud layer that is deposited due to the presence of a pier. Following the same reasoning, the material that is being eroded on the southern flank may be the one deposited during the previous low-water slack. The net result of these processes is that all material deposited during slack water is later removed by the strong currents precluding any sediment accumulation. Our data do not allow us to define where this material is deposited but, due to the general erosional conditions of the whole estuary, the sediments are most probably exported from it.

Meteorological conditions, especially wind, play a major role in the dynamics of the estuary. In the first cruise wind was mostly a breeze (< 18 km/h) and no direct effect upon circulation was detected. However, during the second cruise, wind velocity was relatively strong reaching mean values of 34 km/h and gusts of over 60 km/h, especially in the afternoon. Besides being strong, wind direction changed significantly during the cruise on December 27. From the beginning of the tidal cycle to about 1100 h, wind blew from the NW. At that time, wind direction changed abruptly to the SW with increasing velocities until 1900 h when it rotated again to the SE and maintaining high velocities.

Wind effect was then observed in the suspended sediment concentration near the surface and even at mid-depth. When the wind was from the NW (having the general trend of the channel), the larger values of C were located in a longitudinal pattern in the middle and also close to the southern flank. But with the rotation to the SW (coincident with the lower tidal levels) the larger values moved towards the northern flank. When the SW wind prevailed, the near surface turbidity maximum was near the northern end of line 2100, but as the SE wind became effective, the maximum moved first to the northern end of line 1800 and then to the northern end of line 1500. Also, in this later stage coincident with high tide, more sediment was coming out of the southern tidal flat probably due to the formation of short waves that attacked the flat surface and produced resuspension.

Figures 7 and 8 describe the variation along the tidal cycle of the suspended sediment fluxes integrated over the cross-section for the three lines located along the Canal Principal for each cruise, respectively. In both cases, the asymmetry of the F curves are well-defined. Ebb fluxes are almost double in magnitude but much shorter in duration than flood fluxes. Also, both graphs show a relatively long period around low water slack in which fluxes were very small.

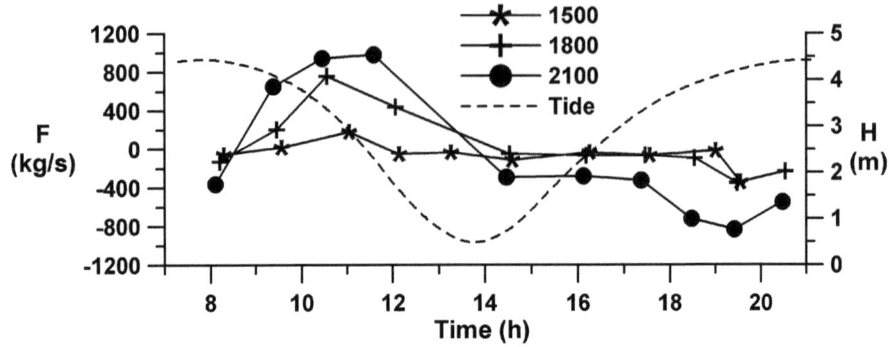

Figure 7. Cross-sectional integrated suspended sediment fluxes (F) (kg/s) during the tidal cycle for the three cross-sections at the Canal Principal for the first cruise (27/11/00) compared with the corresponding tidal curve. Positive values are ebb-oriented.

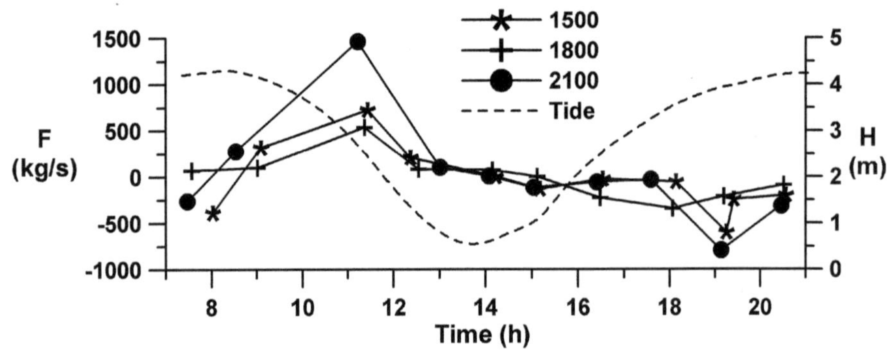

Figure 8. Cross-sectional integrated suspended sediment fluxes (F) (kg/s) during the tidal cycle for the three cross-sections at the Canal Principal for the second cruise (27/12/00) compared with the corresponding tidal curve. Positive values are ebb-oriented.

Thus most of the sediment transport occurred in less than 6 h and, also, sediments in the lower portion of the water column were more prone to be deposited. A marked difference between cross-sections is also observed. Line 2100 has the highest values in all conditions, in many cases double that of the other fluxes, and Line 1500 presented very low fluxes. The maximum velocities and fluxes that occur in Line 2100 may be attributed to the structure of the Canal Principal. During ebb at Line 2100, the channel reduces its width after moving from the vessel turning area thus the flows strongly accelerate, whereas Line 1500 is located at the turning area having the widest portion of the channel, then flows are reduced.

4. CONCLUSIONS

Water and suspended sediment fluxes at Puerto Galván have been studied by continuous vertical profiling at specific cross-sections covering both the Canal Principal and the entrance to the docks. The study comprised two spring tidal cycles separated by one month having different climatic conditions. In the first cruise, wind was calm; however, in the second, wind was relatively strong and changed direction during all the study period.

Although in general in both cases the suspended sediment fluxes provided similar results, wind influenced the input of suspended sediments from the adjacent tidal flats, especially during high tide and ebb. Wind waves generated during high tide resuspended fine sediments that are then transported to the channel during ebb.

The general circulation in the Canal Principal confirmed previous studies that the estuary behaves as a vertical homogenous one, with residual outflow along the northern margin and residual inflow along the southern margin. However, the fact that we could not cover the whole cross-section (for operational and equipment reasons) may have affected the estimation of the residual circulation.

On the other hand, we were able to observe in detail the reversing circulation that occurs at the mouth of Dock 2/3 clearly demonstrating that flow separation is important at the tip of the wharf. Also, the tidal flat forming the western border of the dock is a major source of suspended sediment that in part is driven into the two docks, and the rest is flushed out like a core current in the middle of the Canal Principal.

5. ACKNOWLEDGEMENTS

The authors wish to thank the President, Lic Jorge Scoccia, and the Directory of the Consorcio de Gestión del Puerto de Bahía Blanca (CGPBB) for financial support and the authorization to use their data for the present article as well to its technical staff, Ing. Osvaldo Abitante and Agr. Miguel Schengelberger for their support. We also thank Mr Jurgen Nieuwenhoven and Mr Eduardo Calot from HAM Dredging for logistic support during the study and Mr Lennard van der Hulst from ISDK for obtaining the ADCP data and pre-processing them. The comments and suggestions by Prof Keith Dyer and Dr Jasper Knight have greatly improved this paper.

REFERENCES

Kjerfve, B. 1979. Measurement and analysis of water current, temperature, salinity and density. In: Dyer, K.R. (ed) *Estuarine Hydrography and Sedimentation*, Cambridge University Press, 186-226.

Pérez, D.E. and Perillo, G.M.E. 1998. Residual fluxes of mass, salt, temperature and suspended sediments through a section of Bahía Blanca Estuary. *Geoacta*, 23, 56-65.

Perillo, G.M.E. 1995. Definition and geomorphologic classifications of estuaries. In: Perillo, G.M.E. (ed) *Geomorphology and Sedimentology of Estuaries*. Elsevier Science BV, Amsterdam, Development in Sedimentology Vol. 53, 17-47.

Perillo, G.M.E. and Piccolo, M.C. 1991. Tidal response in the Bahía Blanca Estuary. *Journal of Coastal Research*, 7, 437-449.

Perillo, G.M.E. and Piccolo, M.C. 1999. Geomorphologic and physical characteristics of the Bahía Blanca Estuary. Argentina. In: Perillo, G.M.E., Piccolo, M.C. and Pino Quivira, M. (eds) *Estuaries of South America: their geomorphology and dynamics*. Springer-Verlag, Berlin, Environmental Science Series, 195-216.

Perillo, G.M.E., Piccolo, M.C., Parodi, E. and Freije, R.H. 2000. The Bahía Blanca Estuary, Argentina. In: Seeliger, U. and Kjerfve, B. (eds) *Coastal Marine Ecosystems of Latin America*. Springer-Verlag, Berlin, Environmental Science Series, 205-217.

Perillo, G.M.E., Pierini, J.O, Pérez, D.E. and Gómez, E.A. 2001a. Suspended sediment circulation in semienclosed docks, Puerto Galván, Argentina. *Terra et Aqua*, 83, 13-20.

Perillo, G.M.E., Pierini, J., Gómez, E.A. Pérez, D.E. Cuadrado, D.G., Piccolo, M.C. and Ginsberg, S.S. 2001b. *Evaluación del desplazamiento de una capa de fango fluido generada por el dragado a inyección*. IADO, Informe Técnico OF-CGPBB-02, 45 pp.

Piccolo, M.C. and Perillo, G.M.E. 1990. Physical characteristics of the Bahía Blanca estuary (Argentina). *Estuarine, Coastal and Shelf Science*, 31, 303-317.

Chapter 7

TEMPORAL VARIABILITY IN SALINITY, TEMPERATURE AND SUSPENDED SEDIMENTS IN A GULF OF MAINE ESTUARY (GREAT BAY ESTUARY, NEW HAMPSHIRE)

Larry G. Ward[1] and Frank L. Bub[2]

[1]*Department of Earth Sciences and Jackson Estuarine Laboratory, University of New Hampshire, 85 Adams Point Road, Durham, NH 03824, USA (lgward@cisunix.unh.edu)*

[2]*Modeling Division, Naval Oceanographic Office, 1002 Balch Blvd, Stennis Space Center, MS 39522, USA (bubf@navo.navy.mil)*

1. INTRODUCTION

Determining temporal and spatial variations of suspended sediments and other water column physical properties (e.g. temperature, salinity, turbidity) in estuarine systems require high-resolution observations over several scales of space and time (Uncles *et al.*, 1988; Dyer, 2000; Grabemann and Krause, 2001; Schmidt and Luther, 2002). Although obtaining these types of measurements can be difficult due to time, equipment and monetary constraints, they are important for developing a fundamental scientific understanding of many estuarine processes, such as primary and secondary productivity, the transport and fate of contaminants, nutrient cycling, or sedimentation (Pritchard and Schubel, 1981; Ward *et al.*, 1984; Fisher *et al.*, 1988; Bilgili *et al.*, 1996; Allen *et al.*, 1998; Lee and Cundy, 2001; Sanford *et al.*, 2001; Johnston *et al.*, 2002; Verity, 2002). Accordingly, numerous studies have been conducted over the last several decades that seek to describe and quantify basic estuarine physics and sedimentological processes (see Kennedy, 1984; Nichols and Biggs, 1985; Eisma, 1993, and Dyer, 2000 for reviews). For instance, it has been long understood that the combination

D.M. FitzGerald and J. Knight (eds.),
High Resolution Morphodynamics and Sedimentary Evolution of Estuaries, 151-142.

and balance of freshwater input from rivers and tidal energy controls or strongly influences net non-tidal circulation (density driven), water column stratification, and sedimentation (Pritchard, 1952; Schubel and Biggs, 1969; Biggs, 1970; Schubel, 1972; Allen *et al.*, 1980; Biggs and Cronin, 1981; Ward and Twilley, 1986; Dyer, 2000; Sanford *et al.*, 2001; Schmidt and Luther, 2002). Wind energy has also been shown to have a major impact on estuaries via wind-driven circulation, mixing of the water column, and local wave-forced resuspension of bottom sediments (Anderson, 1970; Eliot, 1978; Ward *et al.*, 1984; Ward, 1985; Blumberg and Goodrich, 1990; Sanford *et al.*, 1991; Sanford, 1994; Dyer, 2000). Furthermore, high winds (and heavy precipitation) during intense storms can dominate estuarine processes (Hayes, 1978; Hirschberg and Schubel, 1979; Althausen and Kjerfve, 1992). However, the importance of wind events and wave resuspension will vary depending on the morphology of the system (i.e. depth, length of fetch, substrate type).

Despite the significant amount of progress that has been made concerning estuarine sedimentological processes, gaps still exist in our understanding. This is largely due to the complex interactions of controlling processes and the variability that exists between estuaries, as well as within estuaries (see Wolfe, 1986 for a review). Furthermore, the relative importance of all the major controls of estuarine sedimentation strongly depends on climatic setting (and change), tidal characteristics, the geomorphology of the system, and ultimately sea level trends (Schubel and Hirshberg, 1978; Stone *et al.*, 1978; Wolfe and Kjerfve, 1986).

The objectives of the study presented here are twofold. First, to evaluate the temporal and spatial variability of suspended sediments, water clarity, and physical structure of the water column in a temperate Gulf of Maine estuary. And second, to assess the impact of two aperiodic forcings (river discharge and wind waves) on these parameters. Great Bay Estuary (GBE), New Hampshire, USA (Fig. 1) was chosen for study due to its morphology, range of controlling processes and size. The system is characterized by deep, narrow channels where tidal current velocities and turbulence are very strong, and wide, shallow flats where tidal current velocities diminish and the importance of wind-forced waves increases (Fig. 2). Also, several rivers periodically provide sufficient fresh water to increase turbidity and stratify the water column. In addition, GBE's moderate size (~ 25 km in length from Portsmouth Harbor to Great Bay) allows observations to be made over the entire system in relatively short time periods (several hours).

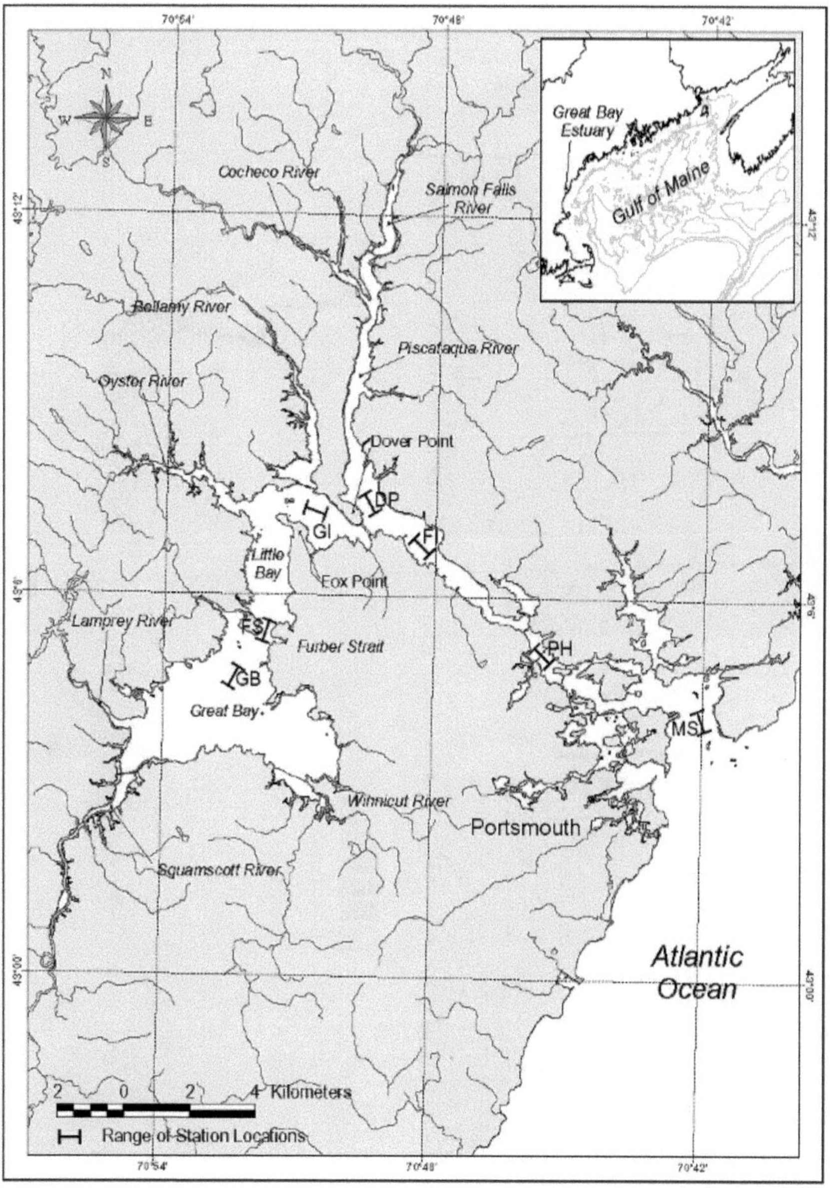

Figure. 1. Location map of study area with sampling stations. Station abbreviations are shown in bold; MS (mouth), PH (Portsmouth Harbor), FI (Frankfort Island), DP (Dover Point), GI (Goat Island), FS (Furber Strait), GB (Great Bay).

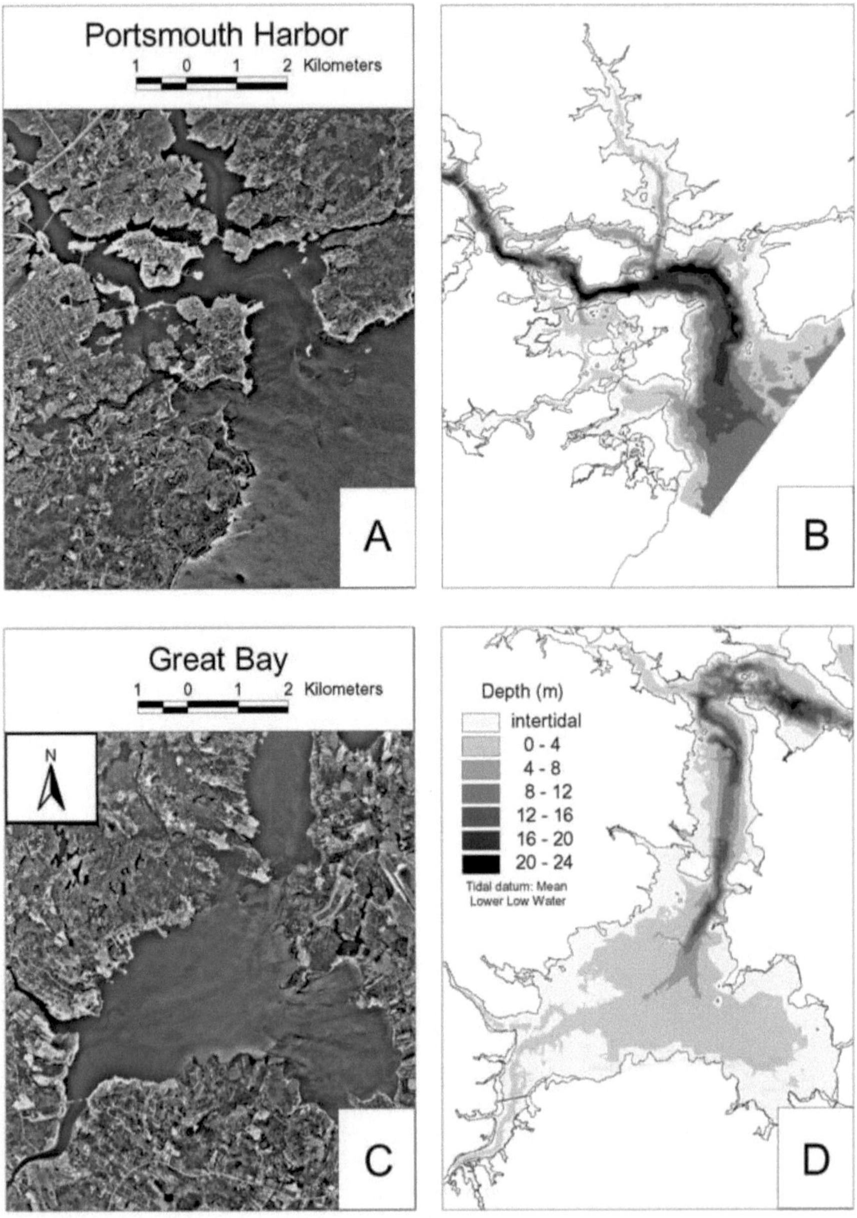

Figure. 2. Aerial photographs and bathymetric maps of Portsmouth Harbor (A and B) and Great Bay (C and D).

2. GREAT BAY ESTUARINE SYSTEM

2.1 Previous studies

Previous studies in GBE have shown that temperature, salinity, turbidity and suspended sediment concentrations have large temporal and spatial variations. For instance, detailed sampling along the axis of GBE over a thirty-month period from 1976 to 1978 showed that high concentrations of suspended sediments occurred in the middle and upper estuary when salinities were reduced due to river discharge (Loder *et al.*, 1983). Although these discharge events could occur over any time of the year, many were associated with a spring freshet. However, high suspended sediment concentrations also occurred during other periods when salinities did not indicate significant freshwater inputs from the rivers. It is likely that during many of these periods elevated suspended sediment concentrations were caused by wind-forced wave resuspension of muddy bottom sediments, especially in the shallow middle and upper estuary areas. Monitoring of suspended sediments and other water column parameters in upper Great Bay and several tributaries from 1988 to recently have shown similar results, although suspended sediment concentrations lacked clear seasonal patterns (see Langan and Jones, 1999; Jones, 2000 for reviews). Anderson (1983) in his review of processes influencing fine-grained sedimentation on intertidal flats demonstrated that wind-generated waves were very effective in resuspending bottom sediments and increasing suspended particulate concentrations in intertidal flats along the flanks of the estuary. This was especially true for the upper reaches of GBE (Great Bay or Little Bay). For example, wind wave resuspension on an intertidal mudflat near Furber Strait (Figs. 1, 2) increased suspended sediment concentrations several fold (Anderson, 1970, 1972). However, the resuspended material settled out very rapidly, within a day, perhaps within several hours (Anderson, 1976). Nevertheless, Anderson (1972) hypothesized that increases in suspended particulate concentrations at the confluence of the Bellamy River with Little Bay were due to wave resuspension on the shallow mudflats found inside the Bellamy (Fig. 1). A number of investigations also have shown GBE is strongly affected by tidal currents, which break down stratification and enhance mixing (Brown and Arellano, 1980; Swift *et al.*, 1996), resuspend and transport fine-grained sediments (Anderson, 1970), and force active bedload transport in the lower estuary (Bigili *et al.*, 1996). Although resuspension clearly plays a critical role in controlling suspended sediment distribution in GBE, it is not understood if these effects are local or collectively influence the entire system.

2.2 Physical Characteristics

GBE is physically complex due to the existence of seven riverine tributaries, several major constrictions, and three large, shallow bays (Little Bay, Great Bay and lower Bellamy River) (Figs. 1, 2). However, this degree of morphologic complexity is not unusual for Gulf of Maine estuarine systems. The drainage basin for GBE and its tributaries covers ~ 2409 km^2 (Reichard and Celikkol, 1978), while the estuary itself has a surface area of ~ 44 km^2 and a perimeter (shoreline) of ~ 160 km. The mean tidal range at the mouth of the estuary is 2.6 m (3.0 m for spring tides), decreasing to 1.9 m (2.2 m for spring) at Dover Point, and increasing to 2.1 m (2.4 m for spring) at the mouth of the Squamscott River at the southern extent of Great Bay (NOAA, 1992). Tidal currents in GBE are very strong with maximum velocities in Little Bay and Great Bay on the order of ~ 0.5 m s^{-1}. Maximum currents in the lower Piscataqua River are typically in the range of 0.5-2.0 m s^{-1}, depending on channel cross-sectional areas (Swift and Brown, 1983). The average high tide volume of the estuary is ~ 230 x 10^6 m^3 and the mean tidal prism is ~ 64 x 10^6 m^3. The total estimated average freshwater input of GBE for all the tributaries is ~32 m^3s^{-1} (Short, 1992), which is only 1-2% of the tidal prism. However, the freshwater input can increase dramatically during high discharge periods. Although the Lamprey has the largest discharge of all the rivers accounting for over 25% of the total flow (Short, 1992), all of the gauged rivers in GBE (Lamprey, Salmon Falls, Cocheco, Squamscott and Oyster) track each other reasonably well (Ward and Bub, 2000). Two of the rivers flowing into GBE (Winnicut and Bellamy) are not gauged. However, based on the size of their watersheds, the Winnicut's discharge is probably very small in comparison to the other rivers and the Bellamy's discharge is probably similar to Oyster River (Short, 1992). Therefore, the Lamprey was used to characterize flow conditions in the estuary during this study. Lowest flows in the Lamprey typically occur in late summer and early fall, while highest discharges usually occur in late winter to spring during the spring freshet (Table I). However, there are significant departures from average conditions when viewing discharge records on a yearly basis (Fig. 3). New Hampshire has a temperate climate with distinctive seasons. Winter conditions can be harsh with significant ice in the marshes and in the upper estuary (Meese *et al.*, 1987). Based on nine years of climate records at the Isles of Shoals located ~ 11 km offshore of GBE, the mean air temperatures ranges from –1.6°C in February to 19.2°C in August (Ward *et al.*, 2001). Average precipitation for New Hampshire averages 107.8 cm yr^{-1} (1895-2002) with little seasonal pattern (NCDC, 2003).

Table I. Discharge characteristics of gauged rivers flowing into the Great Bay Estuary (see Figure. 1). Data from USGS (2003).

Daily Average Discharge for Gauged Rivers		
River	**Discharge (m^3s^{-1})**	**Period**
Lamprey River	8.0	1934-1999
Salmon Falls River	5.4	1969-1999
Cocheco River	4.7	1995-1999
Squamscott River	3.1	1996-1999
Oyster River	0.6	1935-1999

Monthly Average Discharge for the Lamprey River for 1934-1999⁺			
October	3.7 m^3s^{-1}	April	19.5 m^3s^{-1}
November	7.5 m^3s^{-1}	May	9.9 m^3s^{-1}
December	9.4 m^3s^{-1}	June	5.4 m^3s^{-1}
January	8.2 m^3s^{-1}	July	2.7 m^3s^{-1}
February	8.8 m^3s^{-1}	August	2.0 m^3s^{-1}
March	17.2 m^3s^{-1}	September	2.0 m^3s^{-1}

⁺Upper 10% of the flow during the year > 18.6 m^3s^{-1}

⁺50% of the flow during the year exceeds 4.8 m^3s^{-1}

⁺Lowest 10 % of the flow during the year is < 0.7 m^3s^{-1}

3. METHODS

Twelve cruises were conducted from February 22, 1997 to April 4, 2000. Almost half of the cruises (five) were done in 1998 (covering all four seasons), while the remaining observations were made in the late winter and early spring of 1997, 1999 and 2000. All sampling cruises were aboard the University of New Hampshire Research Vessel *Gulf Challenger*. During each cruise water column characteristics were measured at five to ten stations along a channel transect from the lower estuary (Portsmouth Harbor) to the upper estuary in Great Bay (Fig. 1). Station locations were determined with an onboard differential GPS with an accuracy of ~ 15 m. The physical characteristics of the water column were measured with a Sea-Bird SBE-25 Sealogger CTD and associated integrated beam transmissometer (WET Labs C-Star). The transmissometer projected a collimated pulsed 660 nm beam over a 25 cm path length. Light not in synch with the pulse (i.e. ambient) was excluded. The CTD system measured depth, salinity, temperature, and beam transmittance near continuously from the surface to normally 1-3 m off the bottom (to avoid damage to the sensors).

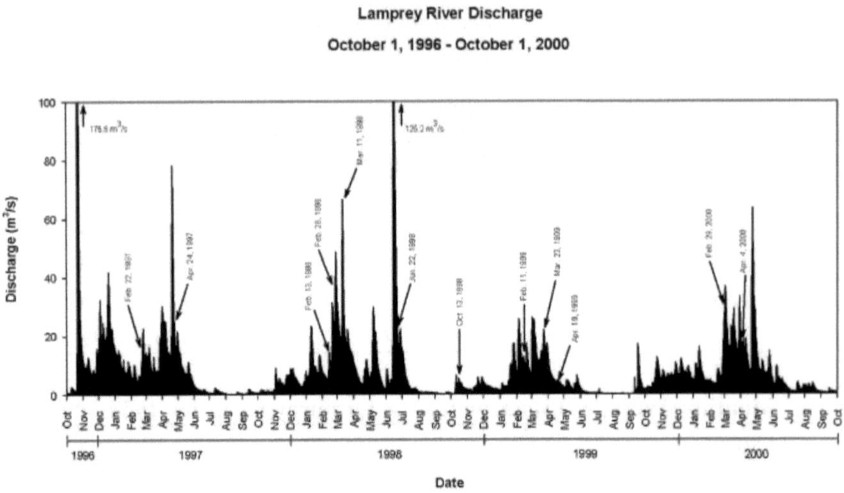

Figure. 3. Daily discharge record for the Lamprey River, New Hampshire from October 1996 to October 2000. The date of each sampling cruise is shown by the downward pointing arrows.

Initial editing of the Sea-Bird data involved separating and then averaging the observations in either 0.5 m or 1.0 m bins from the surface to the bottom. Subsequently, the profiles for each cruise with the exception of the Dover Point station served as the bases for contouring temperature, salinity and beam transmittance along the axis of the estuary. The Dover Point station was excluded because it is not in the main channel (thalweg) of GBE, being located slightly into the upper Piscataqua system. In addition to the contouring, average temperature, salinity and % beam transmittance were determined for each station (Fig. 1) by averaging over the upper and lower one to three meters of the vertical profiles (Table II). Three-meter averages were normally used unless the total water depth at a station was less than 6 m, which frequently was the case at Dover Point and in Great Bay. Using the averages dampened some extreme values resulting from thin surface layers, but was felt to be the most representative for determining trends.

At each station water samples were collected for suspended sediment analyses at the surface with either a clean plastic bucket or a 5-l Niskin bottle. All samples were normally stored on ice or at ~5°C and in the dark after collection until filtered. All samples were filtered within 1 to 2 days after collection (if not the samples were discarded). The exception to this was the samples collected on February 28, 1998. Due to a handling error these water samples were stored at 5°C for a week before processing. Only the near surface suspended sediment were considered during this study in order to evaluate general ambient water conditions instead of local resuspension. The suspended sediment concentrations were determined by vacuum filtration with glass fiber filters (Pall Gelman Type A/E) with a nominal pore size of 1 micron (modified after Banse *et al.*, 1963, Strickland and Parsons, 1968). The organic fraction of each sample was estimated by loss-on-ignition (LOI) by heating the filters to 450°C for 4 hours (modified after Ball, 1964).

The transmissometer was calibrated with clear (e.g. distilled) water representing 100% light or beam transmittance. Beam transmittance values were converted to estimated suspended sediment concentrations by using a regression analysis between average % beam transmittance for the upper water column (1 to 3 m) from each station and the suspended sediment concentration determined from a single surface water sample at that station. The best fit for the regression analysis was for all the stations and cruises when suspended sediment concentrations were available using a logarithmic (natural log) relationship ($R^2 = 0.81$):

$$Y = -8.6221 \, \mathrm{Ln}(X) + 38.98 \qquad (1)$$

Because of inherent problems with calibrating transmissometers to suspended sediment concentrations (Wells and Kim, 1991), all observations are reported here as % beam transmittance, along with the estimated suspended sediment concentrations (computed from the regression analysis shown in equation 1). Although the calibration of the transmissometer was overall very good for most of the range of the suspended sediments observed, the lowest values appeared to overestimated in some cases, while the highest values appeared underestimated.

Freshwater discharge for the Lamprey River was obtained from the United States Geological Survey (USGS, 2003). Wind velocity and direction were obtained from the National Weather Service station at the Isles of Shoals (IOS), New Hampshire (NDBC, 2003), which is located approximately 11 km offshore of Portsmouth Harbor. For this study average wind speeds were computed from the hourly means.

Table II. Water column characteristics for six stations along the main channel of Great Bay Estuary (see Fig. 1 for station locations). Surface (surf) and bottom (bott) values are averages for the upper or lower 1-3 m of the vertical profiles (depending on station depth). Physical conditions for each cruise are given at the bottom of the table. Discharge is the mean of the daily averages for the Lamprey River for five days preceding each cruise. Average wind velocity is the mean of the hourly averages (regardless of direction) for ~ 36 to 44 hours preceding each cruise.

		Cruise 1 Feb 22 1997		Cruise 2 Apr 24 1997		Cruise 3 Feb 13 1998		Cruise 4 Feb 28 1998	
Station name and parameter		Surf	Bott	Surf	Bott	Surf	Bott	Surf	Bott
MS	Salinity (psu)	30.9	30.9	26.4	29.0	28.9	31.4	20.4	29.9
	Temp (°C)	4.7	4.6	7.3	6.8	2.8	2.9	3.1	3.0
	Beam Trans (%)	70.2	81.1	70.1	73.0	71.2	76.7	55.2	57.1
PH	Salinity (psu)	28.4	29.1	24.6	27.1	---	---	---	---
	Temp (°C)	4.3	4.3	7.4	7.1	---	---	---	---
	Beam Trans (%)	72.0	75.9	62.4	69.3	---	---	---	---
FI	Salinity (psu)	---	---	---	---	29.0	29.4	15.3	16.4
	Temp (°C)	---	---	---	---	2.8	2.8	3.0	3.0
	Beam Trans (%)	---	---	---	---	76.6	77.5	44.9	46.0
GI	Salinity (psu)	23.5	24.4	17.1	22.4	26.6	27.4	13.5	15.2
	Temp (°C)	3.8	3.9	8.5	7.7	2.7	2.8	3.0	3.0
	Beam Trans (%)	56.1	60.6	55.3	61.0	70.9	73.2	40.2	42.2
FS	Salinity (psu)	22.3	22.2	10.4	11.8	24.0	24.3	11.4	12.3
	Temp (°C)	3.9	3.8	9.7	9.3	2.6	2.6	3.0	2.9
	Beam Trans (%)	44.2	44.4	44.1	42.4	64.4	66.7	38.3	34.0
GB	Salinity (psu)	21.8	21.9	9.7	11.2	23.9	23.9	11.5	11.7
	Temp (°C)	3.9	3.9	9.7	9.4	2.6	2.6	3.0	2.9
	Beam Trans (%)	38.6	39.1	37.0	39.6	64.8	65.1	36.4	35.0

Cruise 1: Average discharge = 9.6 m^3s^{-1}; Average wind velocity = 11.7 m s^{-1}; Resuspension event.

Cruise 2: Average discharge = 52.0 m^3s^{-1}; Average wind velocity = 6.3 m s^{-1}; Spring freshet.

Cruise 3. Average discharge = 7.4 m^3s^{-1}; Average wind velocity = 10.4 m s^{-1}.

Cruise 4. Average discharge = 36.7 m^3s^{-1}; Average wind velocity = 4.8 m s^{-1}; Spring freshet.

Table II (continued). Definitions given on the preceding page.

		Cruise 5 Mar 11 1998		Cruise 6 June 22 1998		Cruise 7 Oct 13 1998		Cruise 8 Feb 11 1999	
Station name and parameter		Surf	Bott	Surf	Bott	Surf	Bott	Surf	Bott
MS	Salinity (psu)	20.5	30.7	---	---	29.2	29.9	27.0	28.0
	Temp (°C)	3.6	3.3	---	---	11.4	11.4	2.6	2.8
	Beam Trans (%)	53.2	70.3	---	---	69.3	71.2	68.1	69.8
PH	Salinity (psu)	---	---	18.4	18.4	28.0	29.0	22.0	21.9
	Temp (°C)	---	---	15.7	15.7	11.7	11.5	2.0	2.0
	Beam Trans (%)	---	---	58.0	57.9	66.7	68.9	63.9	64.1
FI	Salinity (psu)	25.4	26.5	---	---	26.2	26.5	17.7	18.2
	Temp (°C)	3.6	3.5	---	---	12.2	12.1	1.6	1.6
	Beam Trans (%)	61.1	62.4	---	---	62.1	62.6	52.6	53.4
GI	Salinity (psu)	16.8	20.9	12.9	21.9	25.0	25.8	16.3	26.0
	Temp (°C)	3.9	3.7	17.3	14.3	12.4	12.3	1.4	2.4
	Beam Trans (%)	40.2	45.7	56.7	64.5	60.2	62.7	42.8	66.8
FS	Salinity (psu)	12.1	12.3	6.5	8.8	25.7	25.9	13.0	14.8
	Temp (°C)	4.1	4.2	19.6	18.6	12.7	12.7	1.5	1.2
	Beam Trans (%)	27.6	30.4	36.0	46.4	59.9	59.8	13.3	25.4
GB	Salinity (psu)	12.2	12.2	5.8	9.2	24.6	25.0	13.3	14.1
	Temp (°C)	4.2	4.2	20.2	18.1	12.9	12.9	1.5	1.8
	Beam Trans (%)	28.9	28.7	32.1	50.0	56.6	58.6	25.0	33.4

Cruise 5. Average discharge = 33.0 m^3s^{-1}; Average wind velocity = 12.0 m s^{-1}; Spring freshet/resuspension event.

Cruise 6. Average discharge = 62.4 m^3s^{-1}; Average wind velocity = 1.9 m s^{-1}; High discharge event.

Cruise 7: Average discharge = 3.4 m^3s^{-1}; Average wind velocity = 6.3 m s^{-1}.

Cruise 8: Average discharge = 16.8 m^3s^{-1}; Average wind velocity = 9.2 m s^{-1}.

Table II (continued). Definitions given on the preceding page.

Station name and parameter		Cruise 9 Mar 23 1999		Cruise 10 Apr 19 1999		Cruise 11 Feb 29 2000		Cruise 12 Apr 4 2000	
		Surf	Bott	Surf	Bott	Surf	Bott	Surf	Bott
MS	Salinity (psu)	23.1	30.9	30.1	30.7	26.6	27.8	30.7	31.8
	Temp (°C)	4.0	3.0	5.8	5.1	2.8	2.9	5.5	5.1
	Beam Trans (%)	30.4	20.2	55.5	55.6	53.9	53.4	61.6	64.8
PH	Salinity (psu)	24.7	26.2	28.7	29.6	22.8	22.7	30.4	31.2
	Temp (°C)	3.7	3.5	6.3	5.9	2.6	2.6	5.5	5.3
	Beam Trans (%)	35.7	34.7	44.8	50.4	61.5	61.9	62.0	64.5
FI	Salinity (psu)	19.8	20.8	28.2	28.3	20.0	20.7	27.7	29.5
	Temp (°C)	4.1	4.0	6.7	6.6	2.5	2.5	6.0	5.6
	Beam Trans (%)	25.5	27.2	43.9	43.9	57.9	60.1	57.9	61.7
GI	Salinity (psu)	18.6	19.1	25.3	25.4	20.2	26.7	23.2	27.1
	Temp (°C)	4.2	4.2	7.9	7.9	2.4	2.6	6.9	6.1
	Beam Trans (%)	19.7	21.7	34.8	34.9	61.3	70.4	52.6	58.5
FS	Salinity (psu)	18.3	18.4	23.9	24.0	13.9	18.6	15.4	17.4
	Temp (°C)	4.3	4.3	8.4	8.4	2.1	2.2	8.5	7.9
	Beam Trans (%)	12.8	10.0	29.7	29.3	23.5	49.2	43.3	46.7
GB	Salinity (psu)	18.0	18.4	23.6	23.6	14.6	15.4	15.8	17.6
	Temp (°C)	4.5	4.3	8.6	8.6	2.1	2.1	8.5	7.8
	Beam Trans (%)	9.4	9.7	28.4	27.8	15.3	18.3	46.5	47.2

Cruise 9: Average discharge = 15.8 m^3s^{-1}. Average wind velocity = 13.1 m s^{-1}; Major resuspension event.
Cruise 10: Average discharge = 4.3 m^3s^{-1}; Average wind velocity = 7.1 m s^{-1}.
Cruise 11: Average discharge = 17.6 m^3s^{-1}; Average wind velocity = 9.5 m s^{-1}; Spring freshet.
Cruise 12: Average discharge = 21.6 m^3s^{-1}; Average wind velocity = 7.2 m s^{-1}.

4. RESULTS AND DISCUSSION

4.1 Water temperature

Average surface and bottom water temperatures at the major sampling stations located from the lower to the upper estuary showed strong temporal and spatial variations (Table II). The largest seasonal variations occurred at the station in the upper estuary in Great Bay (GB), which ranged from 1.5 to 20.2°C, due to the morphologic characteristics of this portion of the system. Great Bay is a large, shallow embayment with extensive subtidal and intertidal flats (Fig. 2). Consequently, the large surface area relative to volume in the upper estuary, coupled with frequent mixing, results in an increased rate of atmospheric heating or cooling. The lower estuary (MS and PH stations, Fig. 1) has a smaller surface area relative to volume so atmospheric heating and cooling is less efficient resulting in a smaller range of temperatures (2.6 to 11.4°C). Additionally, tidal processes help keep the lower estuary more similar to oceanic conditions. Loder *et al.* (1983) observed a slightly larger annual range of water temperatures in the GBE from 1976 to 1978 than those observed during this study varying from ~ -2 to ~ 25°C (with the largest extremes in Great Bay as well). However, Loder *et al.* sampled at near monthly intervals over the entire year including the coldest and warmest months.

Longitudinal temperature gradients from the head to the mouth of the estuary were frequently created as a result of the differential heating between the upper and lower estuary. These gradients varied in magnitude and reversed in direction over the year (Fig. 4). Typically, in winter (e.g. February cruises), average water temperatures were slightly colder in Great Bay (by 0.1 to 1.1°C) than at the mouth of the estuary (Table II). However, by early spring (April) water temperatures were higher in Great Bay than in the lower estuary. The maximum horizontal gradient measured during this study occurred on June 22, 1998 when the water temperatures were ~ 4.5°C warmer in Great Bay than at the mouth of the estuary. In contrast to the variations in water temperatures over the length of the estuary, changes in temperature with depth in the main channel throughout GBE were small, usually less than one degree centigrade. The exception to this occurred during the extremely high discharge event on June 22, 1998 (Fig. 3) when a vertical gradient of ~ 3°C occurred in the middle estuary near Dover Point (Fig. 5).

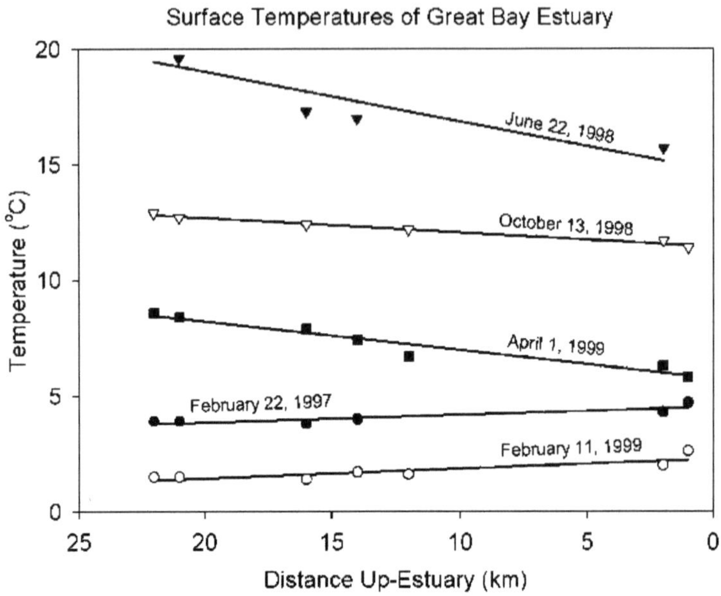

Figure. 4. Average surface water temperatures in GBE for five cruises. The station locations extend from the mouth (MS) to the upper estuary in Great Bay (GB). See Figure 1 for station locations.

4.2 Salinity Distribution and Stratification

During this study, average salinities in GBE ranged from 20.4 to 31.8 psu (practical salinity units) at the mouth station (MS) in the lower estuary, from 12.9 to 27.4 psu in the middle estuary at the Goat Island station (GI), and from 5.8 to 25.0 psu in the upper estuary at the station in Great Bay (GB) (Table II). In general, the highest salinities occurred in winter and fall when freshwater discharge typically was at a minimum (Fig. 3), while lowest salinities occurred during spring freshets associated with the late winter to early spring snow melt and precipitation. However, there were notable exceptions to this trend. High riverine discharge events occurred over almost any part of the year due to the lack of distinct seasonal patterns in precipitation in the region. For instance, the largest discharge event during this study occurred in June 1998 when peak discharge from the Lamprey reached 127.4 m^3s^{-1} (Fig. 3).

Salinity varied with depth from being nearly vertically mixed (less than 1 psu change vertically) at almost all sampling locations (three cruises) to stratified over the entire estuary (one cruise) (see Table II and Figs. 5, 6).

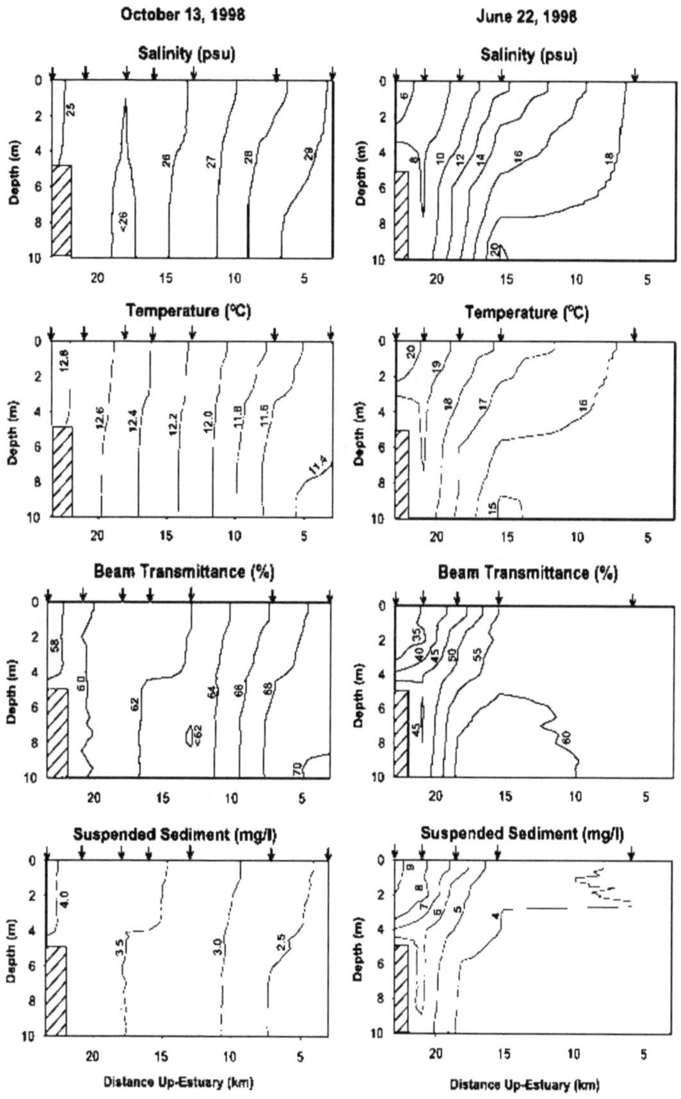

Figure. 5. Salinity, temperature, beam transmission, and suspended sediment contour diagrams for axial surveys in GBE for low discharge (October 1998) and very high discharge (June 1998) conditions. Suspended sediment concentrations were determined from beam transmittance. Distance up-estuary from the mouth of GBE is shown on the lower horizontal axis. The downward pointing arrows along the top axis indicate sampling locations.

However, most commonly, the estuary was stratified over some segments, while being vertically mixed at others. GBE has multiple riverine tributaries along its axis that provide freshwater to the system enhancing stratification. Conversely, GBE has several channel constrictions where intense turbulence and strong currents tend to mix the water column (Swift and Brown, 1983; Swift *et al.*, 1996).

The transition from being relatively well mixed during winter to stratified over segments or the entire system was demonstrated by the cruises in February and April 1997 (Fig. 6). During the February cruise salinities ranged from ~ 31 psu at the mouth to ~ 19 psu in the Great Bay, but the estuary was essentially vertically mixed throughout. The spring freshet in 1997 began around the beginning of April and peaked on April 20 (Fig. 3). Observations made on April 24 showed the average surface salinity in GBE was reduced by ~ 4.5 psu at the mouth and ~ 12.1 psu in the upper estuary from those measured during the February cruise (Table II). Also, the estuary was stratified over most of its entire length except in Great Bay (Fig. 6). Although the temporal sampling frequency during this study prohibited the determination of the duration of the stratification, it is likely the strong currents and turbulence in GBE mixed the system relatively quickly after the freshwater discharge decreased.

As expected, the most distinct stratification occurred at the sampling stations that were at junctions between major segments of the estuary or at the mouth of the estuary where very different water masses merged. For example, maximum stratifications (~ 10 to 12 psu vertical difference) observed during this study occurred on March 11 1998, June 22 1998 and February 29 2000 (Fig. 7) at the junction between the middle and lower estuary at Dover Point (DP) (Fig. 1). However, significant stratification (~ 7 to 10 psu vertical difference) also occurred in the lower estuary on several occasions following periods of heavy freshwater discharge (e.g. February 28 1998 and March 11 1998). As noted elsewhere (Dyer, 2000; Uncles 2002), the magnitude of stratification fluctuated significantly over a single tidal cycle. This effect was observed in GBE on February 28 1998 following the second highest discharge period of this study when a vertical gradient in salinity of ~ 9 psu was observed at the mouth station during the early flood tide (Fig. 7). However, the salinity observed about 4.5 hours later at a nearby station (~ 1 km) in the lower estuary had increased ~ 6 psu at the surface, reducing the vertical gradient by ~ 5 psu.

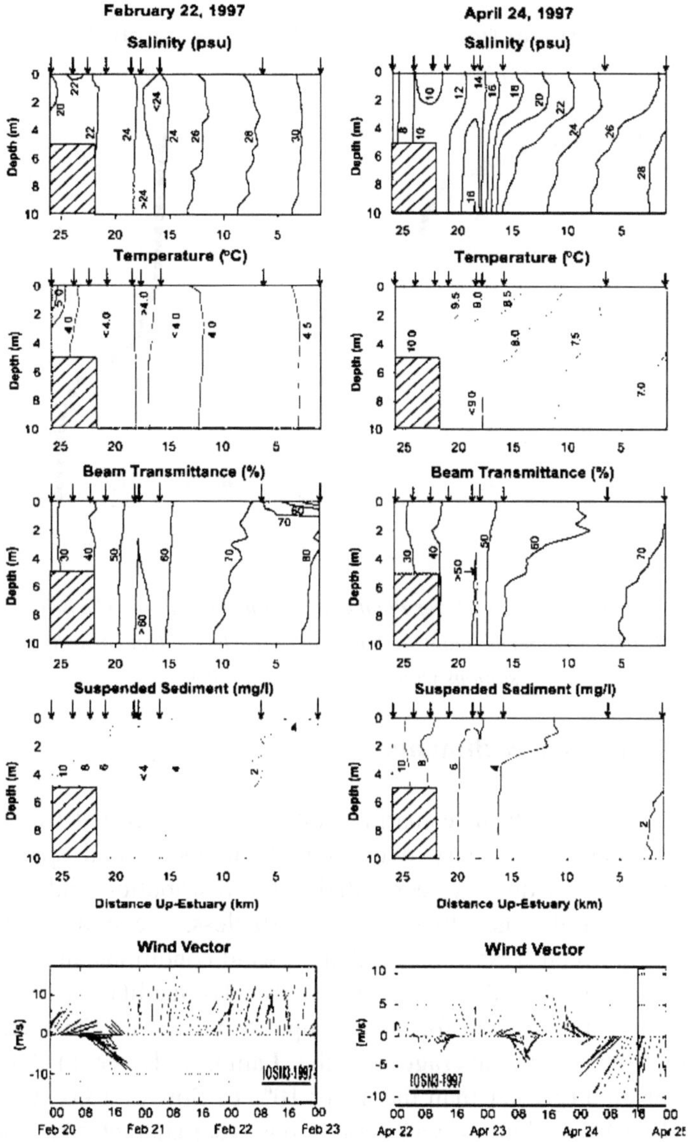

Figure. 6. Salinity, temperature, beam transmission, and suspended sediment contour diagrams for axial surveys in GBE in February and April 1997. Wind conditions before and during each cruise are shown in lower portion of the figure. Suspended sediment concentrations were determined from beam transmittance. Distance up-estuary from the mouth of GBE is shown on the lower horizontal axis. The downward pointing arrows along the top axis indicate sampling locations. The shaded area on the tide curve shows the time interval of the cruise.

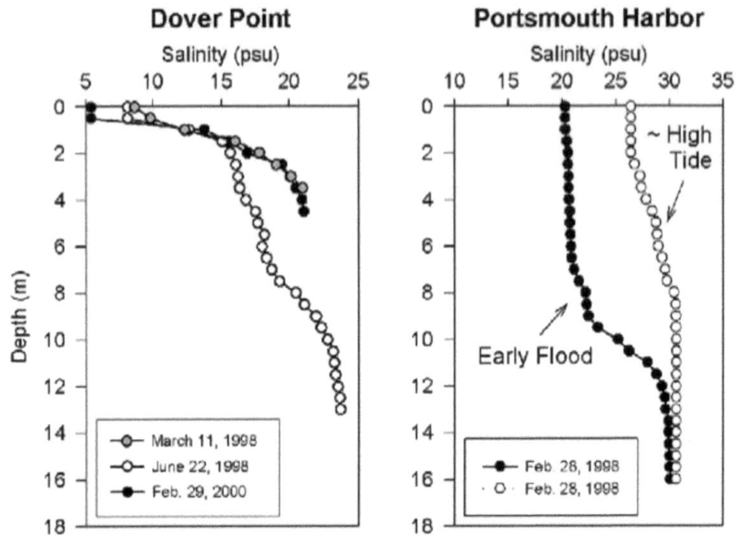

Figure. 7. Salinity profiles for Dover Point and Portsmouth Harbor. The two profiles at Portsmouth Harbor were taken about 4.5 hours apart (early flood and near high tide) from stations within ~1 km of each other.

4.3 Suspended Sediments

Suspended sediments in estuarine systems are controlled by numerous factors that interact and vary over multiple temporal and spatial scales (Eisma, 1993). Consequently, generalizing the distribution and controls of suspended materials is difficult. Nevertheless, several trends and relationships between freshwater discharge, wind conditions, and suspended sediments were observed during this study. To facilitate a comparison between these variables, the freshwater input to GBE was characterized as: low (below the yearly average for the Lamprey River of 8.0 m^3s^{-1}), moderate, or high (greater than the upper 10% of flow or 18.6 m^3s^{-1} for the Lamprey River). In addition, wind conditions were characterized simply as calm (< 6.3 m s^{-1}) or windy (> 9.2 m s^{-1}) based on the average wind speed for the ~ 36 to ~ 42 hours preceding the cruise. During this study four cruises occurred during calm conditions, while six cruises occurred when the average wind velocities were between 9.2 and 13.1 m s^{-1}. During these later periods, visual observations indicated significant wave activity and apparent resuspension of bottom sediments. These boundaries are simply meant to give general conditions and are not meant to define quantitative relationships.

Typical of many estuarine systems, the clearest water masses with the lowest suspended sediment concentrations normally occurred in the lower reaches of GBE and increased up-estuary reaching a maximum in Great Bay. However, water clarity and suspended sediment concentrations varied in response to river discharge or wind events. For instance, based on the calibration of the transmissometer observations at the mouth station (MS), estimated surface suspended sediment concentrations ranged from 2.2 to 9.5 mg l^{-1} (53.2% to 73.0% beam transmittance, Table II). However, the concentrations for two cruises during low discharge conditions were 2.2 and 2.4 mg l^{-1} (Fig. 8). During high river discharge periods the average suspended sediment concentration at the mouth station only increased slightly to 3.4 mg l^{-1}. However, during moderate to high discharge and windy periods the average increased to 4.7 mg l^{-1}.

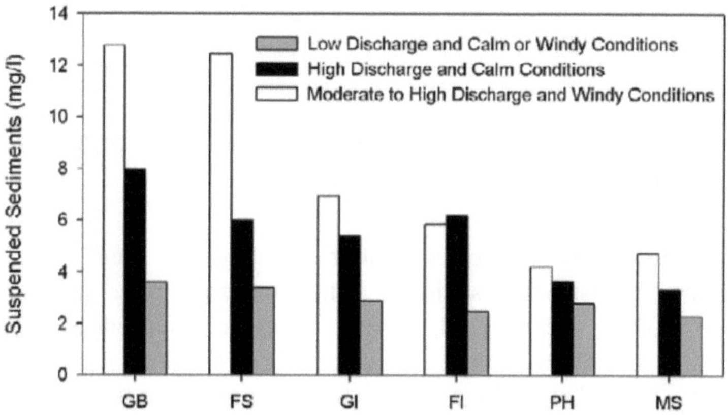

Figure. 8. Average suspended sediment concentrations for stations in the main channel of GBE extending from the mouth (MS) to the upper estuary (GB) (see Fig. 1 for locations). Suspended sediment concentrations were determined from beam transmittance. The averages are for the cruises that occurred during low flow conditions (2), high discharge conditions (4) and moderate to high discharge and windy conditions (4). Two cruises that occurred during phytoplankton blooms (April 1999 and April 2000) were excluded.

The most turbid conditions with the highest suspended sediment concentrations occurred at the Great Bay station (GB) in the upper estuary (Fig. 8). Here, suspended sediment concentrations typically were two-fold to three-fold higher than the mouth station ranging from 3.0 to 19.7 mg l^{-1} (9.4 to 64.8% beam transmittance, Table II). The suspended sediment concentrations during the two cruises during low river flow were 3.0 and 4.2 mg l^{-1}, only slightly higher than at the mouth station. However, concentrations during higher discharge periods averaged 8.0 mg l^{-1}. Consistent with results throughout the estuary, maximum suspended sediment concentrations occurred when the discharge was moderate or high and windy conditions occurred. During these periods, suspended sediment concentrations at the Great Bay station averaged 12.8 mg l^{-1} and reached 19.7 mg l^{-1}.

The station spacing during this study in the upper estuary did not provide enough detail to define whether a (typical) turbidity maximum was present as has been reported for other estuaries in the region (Kistner and Pettigrew, 2001). However, an earlier survey (July 1992), where more continuous turbidity observations were made over two tidal cycles from the mouth of the estuary into Great Bay, did not indicate the presence of a turbidity maximum, rather a more continuous increase in the upper estuary (Ward, 1995). The increases in concentrations in the upper estuary during this study more likely resulted from resuspension of bottom sediments from the shallows adjacent to the main channel and the subsequent migration of this turbid water mass into the channel during ebb flow. However, this process still needs to be examined.

4.4 River Discharge and Wind Wave Resuspension Effects on Suspended Sediments

During this study the effects of river discharge and wind wave resuspension on suspended sediment concentrations in the main channel system of GBE were similar. For example, suspended sediment concentrations during the very high discharge associated with the spring freshet in April 1997 (Fig. 3) increased from 2.3 mg l^{-1} in the lower estuary near the mouth to 7.8 mg l^{-1} in the upper estuary in Great Bay. However, these concentrations were very similar to those observed during the low river flow period in February of that same year (Fig. 6). The similarities between February and April 1997 were likely related to wind-forced wave resuspension increasing suspended sediment concentrations in Little Bay and Great Bay during the earlier cruise and riverine inputs raising concentrations during the later cruise. The winds in February approached 15 m s^{-1} from the north (Fig. 6) coinciding with maximum fetch for this area of GBE (Fig. 1). Therefore, during the February 22, 1997 cruise it is likely that wind waves caused significant resuspension of bottom sediments in the shallow flats found in this region, increasing suspended sediment concentrations in the upper estuary. During the April cruise, the winds were weaker and from the south diminishing the impact of wind wave resuspension. Consequently, the greater suspended sediment concentrations in the upper estuary were due to high river discharge associated with the spring freshet.

In contrast, the estimated suspended sediment concentrations during the very high discharge event in June 1998 did not increase dramatically compared to more moderate freshwater discharge events (compare Figs. 5 and 6). However, the winds prior to the June 22, 1998 cruise were calm eliminating any effect of wind-forced wave resuspension. In addition, the lower than expected suspended sediment concentrations on this date may be a result of the cruise being conducted several days after peak flow and on the flood tide. Earlier studies have shown suspended sediment concentrations tend to be higher at low tide than at high tide in the upper estuary (Langan and Jones, 1999; Jones, 2000).

The most turbid conditions in GBE observed during this study occurred as a result of the co-incidence of strong winds during or directly following moderate to high discharge periods. For example, the combination of these forcing factors acting in concert on March 11 1998, March 23 1999 and February 29 2000 caused the highest suspended sediment concentrations of the study period. However, this is especially true on March 23 1999. The spring freshet, which occurred in late winter and early spring 1999, was generally weaker in comparison to the other years during this study (Fig. 3). On March 23, the salinity in the estuary ranged from 22 to 31 psu at the mouth to ~ 18 psu in Great Bay. The lower estuary was stratified, but the upper estuary was well mixed, most likely due to the strong winds that occurred prior to the sampling cruise (Fig. 9). The turbidity was extremely high throughout the estuary with only 20 to 30% beam transmittance at the mouth and ~ 10% in Great Bay. At the same time the surface suspended sediment concentrations (directly measured from water samples) were the highest observed during this study increasing up-estuary from 9.2 to 22.0 mg l^{-1} (Fig. 10). The organic contents of the suspended sediments on this date were low (LOI < 16%) in the middle and upper estuary indicating most of the sediment in suspension was inorganic. Bottom sediment resuspension by wind waves during the extremely strong winds was likely the major source of suspended sediments in the upper estuary. In addition, the high concentrations of suspended sediments in the lower estuary likely resulted from the turbid water masses generated in the middle and upper estuary extending to the mouth of GBE.

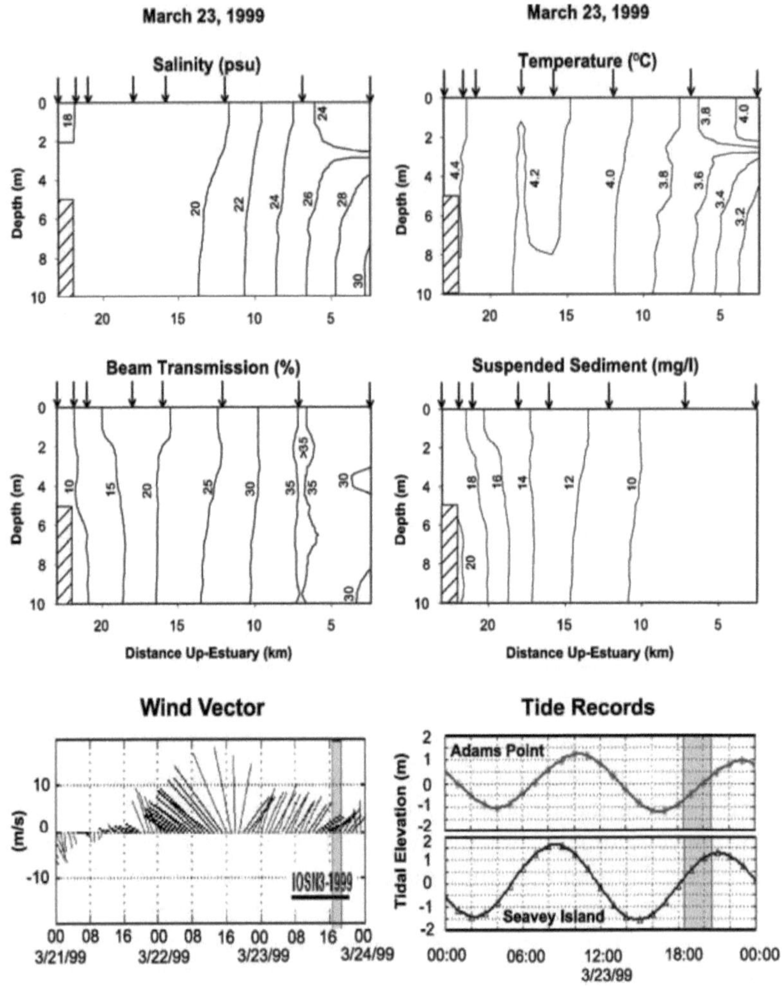

Figure. 9. Salinity, temperature, beam transmission, and suspended sediment contour diagrams for GBE in March 1999 during a major wind-forced resuspension event following the spring freshet. Wind and tidal conditions before and during the cruise are shown in lower portion of the figure. Suspended sediment concentrations were determined from beam transmittance. Distance up-estuary from the mouth of GBE is shown on the horizontal axis. The downward pointing arrows along the top axis indicate sampling locations. The shaded area on the tide curve shows the time interval of the cruise.

The effectiveness of wind waves in resuspending sediments during or after high river discharge periods is most likely related to the availability of easily eroded, unconsolidated bottom sediments. In the absence of winds it is reasonable that some proportion of the fine-grained sediments brought into the estuary by the rivers is deposited within the estuary. If wind-forced wave resuspension occurs before this material has the opportunity to dewater and consolidate, then the sediment will be more easily eroded (Sanford, 1994). Therefore, highest turbidity and suspended sediment concentrations occur when wind action coincides or soon follows discharge events. During stronger winds with longer durations, bottom sediments eroded from the upper estuary probably are distributed throughout the estuary. Although this relationship has to be examined further with consideration of the impact of tidal currents, the results of this study suggest that wind events and high river discharge had about equal impacts on turbidity and suspended sediment concentrations in the water column. However, the riverine inputs represented new suspended sediment to the system, while the wind wave resuspension reflected sediment recycling within the system.

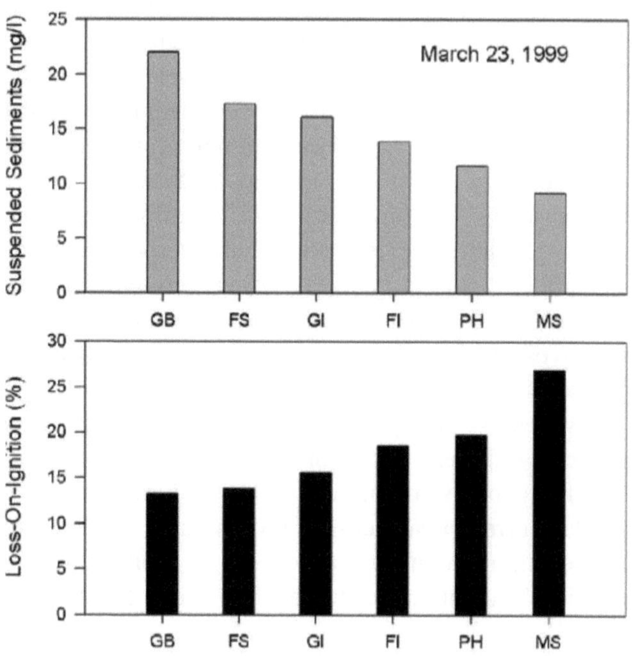

Figure. 10. Suspended sediments and particulate organic contents (based on loss-on-ignition) for surficial water samples from six stations in the main channel of GBE on March 23, 1999. Station locations are shown on figure 1.

5. CONCLUSIONS

The temporal and spatial distributions of suspended sediments and the physical structure of the water column (stratification) of Great Bay Estuary were determined by sampling during different seasonal, river discharge, and meteorological conditions over a period of four years. From these observations, insights into the responses of this mesotidal, temperate estuary to aperiodic forcings such as high river discharge or meteorological events (e.g. strong winds) were made. As expected, the freshwater discharge from riverine tributaries had a major impact on the physical structure of the estuary by creating salinity gradients (vertical and longitudinal) and stratifying the water column despite strong tidal currents and significant turbulence. However, resultant increases in suspended sediment concentrations were variable and more moderate. Based on the results of this study, river discharge and wind forced wave resuspension had similar impacts on suspended sediment distributions over the entire estuary. However, the highest turbidity and suspended sediment concentrations occurred when river discharge was higher than normal (bringing particulate material into the estuary from the rivers) and wind events caused resuspension of shallow bottom sediments distributing the material throughout the estuary. Therefore, it appears that the shoals and tidal flats in the middle and upper estuary have a strong influence on suspended sediments throughout GBE. During the frequent wind events that occur in GBE, sufficient waves are formed to resuspend sediments from these flats, creating turbid water masses. During the ebb portion of the tidal cycle, these turbid water masses most likely migrate into the deeper channels and increase suspended sediment concentrations throughout the estuary. However, the low frequency sampling interval used during this study did not allow wave or tidal current resuspension processes to be determined. The influence of these high frequency processes in controlling suspended sediment and physical structure needs to be addressed in future studies.

6. ACKNOWLEDGMENTS

This study was funded by the NOAA/UNH Cooperative Institute for Coastal and Estuarine Environmental Technology (CICEET), NOAA Grant Numbers NA77OR0357 and NA87OR0512. Most of the cruises were conducted aboard the R/V *Gulf Challenger*. We would like to thank the captain (Paul Pelletier) and the mate (Ken Houtler) for their expert seamanship and generous help. Melissa Brodeur and Jamie Adams helped with the analyses of the data and the preparation of the figures. Several graduate students participated in the field and laboratory work including John Lund, Amy Harmon, and Ken Jacobs. In addition, class members of our Nearshore Processes course (UNH Earth Sciences 796/896) joined us on a number of cruises over the four-year study period. We would like to thank both the graduate and undergraduate students in this course for their help and interest. This is University of New Hampshire Center for Marine Biology/Jackson Estuarine Laboratory contribution series number 411.

REFERENCES

Allen, G.P., Salomon, J.C., Bassoullet, P., Du Penhoat, Y. and De Grandpre, C. 1980. Effects of tides on mixing and suspended sediment transport in microtidal estuaries. *Sedimentary Geology*, 26, 69-90.

Allen, J.I., Howland, R.M.H., Bloomer, N. and Uncles, R.J. 1998. Simulating the spring phytoplankton bloom in the Humber Plume, UK. *Marine Pollution Bulletin*, 37, 295-305.

Althausen, J.D. Jr. and Kjerfve, B. 1992. Distribution of suspended sediment in a partially mixed estuary, Charleston Harbor, South Carolina, U.S.A. *Estuarine, Coastal and Shelf Science*, 35, 517-531.

Anderson, F.E. 1970. The periodic cycle of particulate matter in a shallow, temperate estuary. *Journal of Sedimentary Petrology*, 40, 1128-1135.

Anderson, F.E. 1972. Resuspension of estuarine sediments by small amplitude waves. *Journal of Sedimentary Petrology*, 42, 602-607.

Anderson, F.E. 1976. Rapid settling rates observed in sediments resuspended by boat waves over a tidal flat. *Netherlands Journal of Sea Research*, 10, 44-58.

Anderson, F.E. 1983. The northern muddy intertidal: a seasonally changing source of suspended sediments to estuarine waters – a review. *Canadian Journal of Fisheries and Aquatic Sciences*, 40, Supplement Number 1, 143-159.

Ball, D.F. 1964. Loss-on-ignition as an estimate of organic matter and organic carbon on non-calcareous soils. *Journal of Soil Science*, 15, 84-92.

Banse, K., Falls, C.P. and Hobson, L.A. 1963. A gravimetric method for determining suspended matter in seawater using Millipore filters. *Deep-Sea Research*, 10, 639-642.

Biggs, R.B. 1970. Sources and distribution of suspended sediment in northern Chesapeake Bay. *Marine Geology*, 9, 187-201.

Biggs, R.B. and Cronin, L.E. 1981. Special characteristics of estuaries. In: B.J. Neilson, B.J. and Cronin, L.E. (eds) *Estuaries and Nutrients*. Humana Press, Clifton, NJ, 3-23.

Bilgili, A., Swift, M.R. and Celikkol, B. 1996. Shoal formation in the Piscataqua River, New Hampshire. *Estuaries*, 19, 518-525.

Blumberg, A.F. and Goodrich, D.M. 1990. Modeling of wind-induced destratification in Chesapeake Bay. *Estuaries*, 13, 236-249.

Brown W.S. and Arellano, E. 1980. The application of a segmented tidal mixing model to the Great Bay Estuary, N.H. *Estuaries*, 3, 248-257.

Dyer, K.R. 2000. *Estuaries, A Physical Introduction*. 2nd

Eisma, D. 1994. *Suspended matter in aquatic environments*. Springer-Verlag, Berlin, 315 pp.

Elliot, A.J. 1978. Observations of the meteorologically induce circulation in the Potomac Estuary. *Estuarine, Coastal and Shelf Science*, 6, 285-299.

Fisher, T.R., Harding, L.W. Jr., Stanley, D.W. and Ward, L.G. 1988. Phytoplankton, nutrients, and turbidity in the Chesapeake, Delaware, and Hudson estuaries. *Estuarine, Coastal and Shelf Science*, 27, 61-93.

Grabemann, I. and Krause, G. 2001. On different time scales of suspended matter dynamics in the Weser Estuary. *Estuaries*, 24, 688-698.

Hayes, M.O. 1978. Impact of hurricanes on sedimentation in estuaries, bays and lagoons. In: Wiley, M.L. (ed) *Estuarine Interactions*. Academic Press, Inc., 323-346.

Hirschberg, D.J. and Schubel, J.R. 1979. Recent geochemical history of flood deposits in northern Chesapeake Bay. *Estuarine and Coastal Marine Science*, 9, 771-784.

Johnston, R.K., Munns, W.R. Jr., Tyler, P.L., Marajh-Whiiemore, P., Finkelstein, K., Munney, K., Short, F.T., Melville, A. and Hahn, S.P. 2002. Weighing the evidence of ecological risk from chemical contamination in the estuarine environment adjacent to the Portsmouth Naval Shipyard, Kittery, Maine, USA. *Environmental Toxicology and Chemistry*, 21, 182-194.

Jones, S.H. (ed) 2000. *A Technical Characterization of Estuarine and Coastal New Hampshire.* New Hampshire Estuaries Project, 152 Court Street, Portsmouth, NH 03801, 274 pp.

Kennedy, V.S. (ed) 1984. *The Estuary as a Filter*. Academic Press, Inc. 511 pp.

Kistner, D.A. and Pettigrew, N.R. 2001. A variable turbidity maximum in the Kennebec Estuary, Maine. *Estuaries*, 24, 680-687.

Langan, R. and Jones, S.H. 1999. *A Monitoring Plan for the Great Bay National Estuarine Research Reserve: Final Report for the Period 7/1/97 through 6/30/98.* NOS MEMD #NA770RO315-01. U.S. Department of Commerce, NOAA. Washington DC, 90 pp.

Lee, S.V. and Cundy, A.B. 2001. Heavy metal contamination and mixing processes in sediments from the Humber Estuary, eastern England. *Estuarine, Coastal and Shelf Science*, 53, 619-636.

Loder, T.C., Love, J.A., Kim, J.P. and Wheat, C.G. 1983. *Nutrient and hydrographic data for the Great Bay Estuarine System, New Hampshire – Maine, Part 11, January, 1976 – June, 1978.* University of New Hampshire Sea Grant Report Number UNH-MP-D/TR-SG-83-4. Durham, NH 03824. 149 pp.

Meese, D.A., Gow, A.J., Mayewski, P.A., Ficklin, W. and Loder, T.C. 1987. The chemical, physical and structural properties of estuarine ice in Great Bay, New Hampshire. *Estuarine, Coastal and Shelf Science*, 24, 833-840.

NDBC 2003. *NOAA National Data Buoy Center.* Available from http://www.ndbc.noaa.gov/station_history.phtml (accessed January 2003).

NCDC 2003. *NOAA National Climatic Data Center.* Available from http://lwf.ncdc.noaa.gov/oa/ncdc.html (accessed January 2003).

Nichols, M.N. and Biggs, R.B. 1985. Estuaries. In: Davis, R.A. (ed) *Coastal Sedimentary Environments*, Springer-Verlag, 77-186.

NOAA. 1992. *Tide Tables, East Coast of North and South America.* U.S. Department of Commerce, National Oceanic and Atmospheric Administration, National Ocean Service.

Pritchard, D.W. 1952. Salinity distribution and circulation in the Chesapeake estuarine system. *Journal of Marine Research*, 11, 106-123.

Pritchard, D.W. and Schubel, J.R. 1981. Physical and geological processes controlling nutrient levels in estuaries. In: Neilson, B.J. and Cronin, L.E. (eds) *Estuaries and Nutrients*, Humana Press, Clifton, NJ, 47-69.

Reichard, R.P. and Celikkol, B. 1978. Application of a finite element model to the Great Bay Estuary System, New Hampshire, U.S.A. In: Nihoul, J.C.J. (ed) *Hydrodynamics of Estuaries and Fjords*, Elsevier Scientific Publishing Company, Amsterdam, 349-372.

Sanford, L.P. 1994. Wave-forced resuspension of upper Chesapeake Bay muds. *Estuaries*, 17, 148-165.

Sanford, L.P., Panageotou, W. and Halka, J.P. 1991. Tidal resuspension of sediments in northern Chesapeake Bay. *Marine Geology*, 97, 87-103.

Sanford, L.P., Suttles, S.E. and Halka, J.P. 2001. Reconsidering the physics of the Chesapeake Bay estuarine turbidity maximum. *Estuaries*, 24, 655-669.

Schmidt, N. and Luther, M.E. 2002. ENSO impacts on salinity in Tampa Bay, Florida. *Estuaries*, 25, 976-984.

Schubel, J.R. 1972. Distribution and transportation of suspended sediment in upper Chesapeake Bay. 1972. In: Nelson, B.W. (ed) *Environmental Framework of Coastal Plain Estuaries*, The Geological Society of America Memoir 133, 151-167.

Schubel, J.R. and Biggs, R.B. 1969. Distribution of seston in upper Chesapeake Bay. *Chesapeake Science*, 10, 18-23.

Schubel, J.R. and Hirschberg, D.J. 1978. Estuarine graveyards, climatic change and the importance of the estuarine environment. In: Wiley, M.L. (ed) *Estuarine Interactions*, Academic Press, Inc., 285-303

Short, F.T. (ed) 1992. *The Ecology of the Great Bay Estuary, New Hampshire and Maine: An Estuarine Profile and Bibliography*. NOAA – Coastal Ocean Program Publication, 222 pp.

Stone, J.H., Day, J.W. Jr., Bahr, L.M. Jr. and Muller, R. 1978. The impact of possible climatic changes on estuarine ecosystems. In: Wiley, M.L. (ed) *Estuarine Interactions*, Academic Press, Inc., 305-322.

Strickland, J.D.H. and Parsons, T.R. 1968. *A Practical Handbook of Seawater Analysis*. Fisheries Research Board of Canada Bulletin 167, Ottawa, 311 pp.

Swift, M.R. and Brown, W.S. 1983. Distribution of bottom stress and tidal energy dissipation in a well-mixed estuary. *Estuarine, Coastal and Shelf Science*, 17, 297-317.

Swift, M.R., Fredriksson, D.W. and Celikkol, B. 1996. Structure and axial convergence zone from acoustic Doppler current profiler measurements. *Estuarine, Coastal and Shelf Science*, 43, 109-122.

Uncles, R.J., Stephens, J.A. and Woodrow, T.Y. 1988. Seasonal cycling of estuarine sediment and contaminant transport. *Estuaries*, 11, 108-116.

Uncles, R.J. 2002. Estuarine physical processes research: some recent studies and progress. *Estuarine, Coastal and Shelf Science*, 55, 829-856.

USGS 2003. *United States Geological Survey, Surface-Water Data for New Hampshire/Vermont*. Available from http://waterdata.usgs.gov/nh/nwis/ discharge (accessed January 1993).

Verity, P.G. 2002. A decade of change in the Skidaway River Estuary. I. Hydrography and nutrients. *Estuaries*, 25, 944-960.

Ward, L.G. 1985. The influence of wind waves and tidal currents on sediment resuspension in middle Chesapeake Bay. *Geo-Marine Letters*, 5, 71-75.

Ward, L.G. 1995. *Sedimentology of the lower Great Bay/Piscataqua River Estuary*. Department of the Navy, NCCOSC RDTE Division Report, San Diego, California, 102 pp.

Ward, L.G. and Bub, F.L. 2000. *Suspended Particulate Material and Physical Properties (Salinity, Temperature, Turbidity) of the Great Bay Estuary: Distribution and Major Controlling Processes*. NOAA/UNH Cooperative Institute for Coastal and Estuarine Environmental Technology (CICEET) Technical Report, Durham, NH, 62 pp.

Ward, L.G., Grizzle, R.E., Bub, F.L., Langan, R., Schnaittacher, G. and Dijkstra, J.A. 2001. *New Hampshire Open Ocean Aquaculture Demonstration Project: Site Description and Environmental Monitoring, Report on Activities from Fall 1997 to Winter 2000.*

NOAA/UNH Cooperative Institute for New England Mariculture and Fisheries (CINEMar) Report. Durham, NH.

Ward, L.G., Kemp, W.M. and Boynton, W.R. 1984. The influence of waves and seagrass communities on suspended particulate dynamics in an estuarine environment. *Marine Geology*, 59, 85-103.

Ward, L.G. and Twilley, R.R. 1986. Seasonal distributions of suspended particulate material and dissolved nutrients in a coastal plain estuary. *Estuaries*, 9, 156-168.

Wells, J.T. and Kim, S.Y. 1991. The relationship between beam transmission and concentration of suspended particulate material in the Neuse River Estuary, North Carolina. *Estuaries*, 14, 395-403.

Wolfe, D.A. (ed) 1986. *Estuarine Variability*. Academic Press, Inc., 509 pp.

Wolfe, D.A. and Kjerfve, B. 1986. Estuarine variability: an overview. In: Wolfe, D.A. (ed), *Estuarine Variability*, Academic Press, Inc., 3-17.

Chapter 8

MORPHODYNAMICS AND SEDIMENT FLUX IN THE BLYTH ESTUARY, SUFFOLK, UK

Conceptual modelling and high resolution monitoring

J.R. French, T. Benson and H. Burningham

Coastal and Estuarine Research Unit, University College London, Chandler House, 2 Wakefield St, London WC1N 1PF, UK

1. INTRODUCTION

Research into the dynamics of estuary morphology has been stimulated by increasing commercial, environmental and legislative pressures and by the accumulated impacts of human intervention (Roman and Nordstrom, 1996; Soulsby, 1997). Of particular concern in the UK is the impact on estuaries of sea-level rise and large-scale interventions associated with dredging and port development, flood defence and habitat restoration. Prediction of estuary response to such changes requires an understanding of present-day processes and their interaction with morphologies that are often shaped by past human activities. As Pye and Allen (2000) note, estuarine research has hitherto been characterised by narrow disciplinary foci, such that research fronts in engineering, geomorphology and Quaternary science have rarely converged. The UK Estuaries Research Programme (EMPHASYS, 2000; French *et al.*, 2002) has advocated a more holistic and interdisciplinary perspective, combining 'bottom up' studies of short-term hydrodynamics and sediment movement with 'top down' models of larger-scale morphodynamic behaviour, such that the predictive power of physically-based simulation may be realised within a conceptual framework provided by geomorphological analysis of longer-term sedimentary function. Few estuaries have been monitored in the spatial and temporal detail needed for integrated modelling of this kind and there is a need for intensive studies encompassing a greater variety of natural and anthropogenic contexts.

D.M. FitzGerald and J. Knight (eds.),
High Resolution Morphodynamics and Sedimentary Evolution of Estuaries, 143-171.

This paper examines the morphodynamics and sedimentary functioning of the Blyth estuary, Suffolk, southeast England. The natural regime of the Blyth has been transformed by dredging, training, and reclamation, and the estuary provides an excellent case study within which to investigate the interactions between a largely engineered morphology and contemporary physical processes. Geological and historical evidence, bathymetric surveys, and hydrographic data are synthesised into a conceptual geomorphological model based upon the identification of a series of morphodynamic units. Inferences concerning estuary sedimentary function are then tested with reference to high temporal resolution monitoring of tidal and meteorological processes. Fine sediment dynamics are shown to be mediated by the interaction between tidal processes and wave action over tidal flats formed after the abandonment of reclamations in the middle estuary. These findings have wider implications for estuarine flood defence management, especially in relation to the viability of large-scale managed realignment (Townend and Pethick, 2002) as an adaptive response to sea-level rise.

2. SCIENTIFIC APPROACH

Explicit prediction of estuarine morphological change over the 10 to 50 year timescales most relevant for management is not yet feasible. Coastal morphodynamic models which adjust the morphology at discrete time steps based on predicted rates of erosion and sedimentation have achieved some success, but their predictive horizon extends to periods of, at most, a few years (Ribberink *et al.*, 1995; Nicholson *et al.*, 1997; Cayocca, 2001). There are also practical and conceptual obstacles to the application of such models to systems like the Blyth. The high spatial resolution needed to define the interaction between hydrodynamic processes and a complex, largely intertidal, morphology makes longer-term simulation computationally prohibitive. More fundamentally, change in engineered morphologies is likely to be dominated by threshold effects, with abrupt changes resulting from, for example, modification of the estuary mouth, or inundation (possibly planned) of discrete flood compartments. Accordingly, the approach adopted here involves the synthesis of historical data, modern surveys and monitoring of contemporary geomorphic processes into a conceptual model of estuary function, aspects of which are then tested with reference to additional field observations. This provides a framework within which hydrodynamic simulations of specific management and environmental change scenarios can be planned, undertaken and interpreted (French, 2001). This paper focuses on the development and testing of the conceptual model: hydrodynamic modelling results will be presented elsewhere.

3. ENVIRONMENTAL CONTEXT

3.1 Physical setting

The Blyth is located in northern Suffolk (Fig. 1) in a rural catchment of 214 km^2. Catchment geology is dominated by till, crag and alluvium (Beardall *et al.*, 1991). The estuary has a tidal length of 10.7 km and the tidal range at Southwold averages 2.0 m at springs and 1.2 m at neaps (Hydrographer of the Navy, 2000). Mean sea-level (MSL), Mean High Water Springs (MHWS) and Highest Astronomical Tide (HAT) are approximately 0.2 m, 1.1 m and 1.6 m above Ordnance Datum (OD) respectively. Semi-diurnal tides are subject to a strong surge influence and the well-documented surge of 1953 reached 3.65 m OD at Southwold (French, 2001). Mean monthly river flow at Halesworth is 0.38 m^3s^{-1} (National Water Archive data; 1990-1996), and 90% of mean monthly flows are < 0.81 m^3s^{-1}. Given that tidal discharge can exceed 200 m^3s^{-1} at springs, the estuary is generally well-mixed.

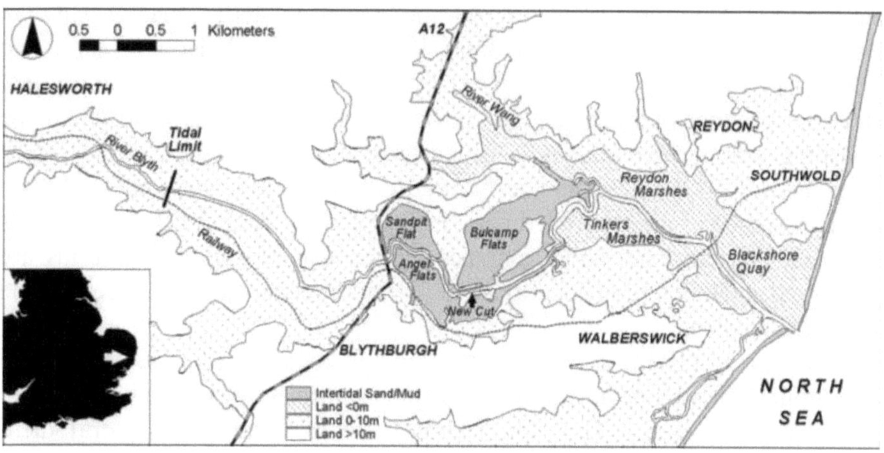

Figure 1. The Blyth estuary.

The Blyth is a barrier-enclosed estuary. A dune-backed shingle and sand beach extends southwards from an elevated headland on which Southwold is situated. The narrow estuary mouth is maintained by breakwaters, to the south of which extends a narrower shingle and sand beach, which is partially dune-backed. Wave-driven littoral sediment transport is predominantly from north to south (Motyka and Brampton, 1993). Estuary sediments are muddy, with some sand and gravel near the mouth.

3.2 Holocene stratigraphy

The Holocene evolution of the Blyth valley has been investigated in detail by Brew (1990) and Brew *et al.* (1992). Upstream of Blythburgh, the Holocene stratigraphy is thin (maximum 7.3 m) and dominated by biogenic sediments. The sequence in the outer estuary is thicker (up to 13.6 m) and dominated by marine clastic sediments, intercalated with thin biogenic horizons. Boreholes sunk into the reclaimed marshes of the outer estuary (Fig. 2) show that the underlying mid-Pleistocene Westleton Beds (pebbly and sandy gravels) are overlain by four main Holocene sedimentary units: a basal freshwater Lower Peat; an estuarine Lower Clay; a Middle Peat; and an Upper Clay. Radiocarbon dates from two boreholes place the Lower Peat between 6450 and 6750 ^{14}C years BP. Freshwater conditions appear to have been followed by a rapid marine incursion and estuarine silt/clay deposition until around 4500 ^{14}C years BP, when a transition to peat formation under freshwater, and later brackish, conditions occurred. In places, this Middle Peat is overlain by a thin horizon of silty clay with peat. Along the central axis of the valley, adjacent to the present subtidal channel, the top of the Middle Peat occurs at between –6.50 m and –4.25 m OD (e.g. cores 4 and 5 in Fig. 2). A second phase of marine incursion and estuarine silt/clay deposition commenced around 4300 ^{14}C years BP. A lack of sand and absence of fully marine micro-fossils within this Upper Clay unit imply that it is the product of inner estuary intertidal deposition at a time when the coastline was located seawards of its present position (see Steers, 1964). Brew *et al.* (1992) infer a southerly migration of the low water channel from the occurrence of uninterrupted clay, interpreted as channel fill, in two cores from the Reydon and Town marshes. Given that peat-clay transitions are preserved at intervening locations, migration appears to have occurred via discrete avulsions.

Brew *et al.* (1992) note that other Suffolk estuaries (the Deben, Orwell and Stour) exhibit a near-continuous sequence of marine silts and clays, with no intercalated peats. This demonstrates the importance of littoral drift, barrier growth and storm breaching as a source of estuarine environmental variability in the Blyth. Littoral sediment supply is more abundant in northern Suffolk and, historically, several smaller estuaries have been sealed by shingle and sand. There is historical evidence for the formation of a bar across the Blyth between 1500 and 700 BP (Parker, 1978). At this time, the mouth of the estuary was co-located with that of the Dunwich River several km to the south (Steers, 1964). This was blocked by shingle by a major storm in 1328, after which an artificial entrance was cut close to Southwold.

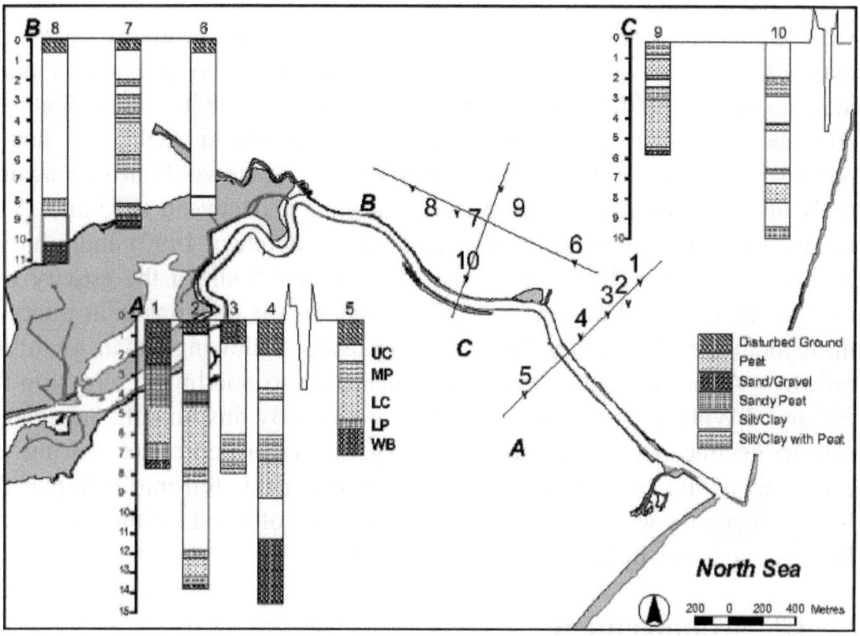

Figure 2. Simplified Holocene stratigraphy of the outer estuary. Vertical scale in metres; UC = Upper Clay; MP = Middle Peat; LC = Lower Clay; LP = Lower Peat and WB = Westleton Beds. Re-drawn and modified from multiple figures in Brew et al. (1992).

4. ANTHROPOGENIC CONTEXT

4.1 Dredging and reclamation

The present harbour at Southwold was excavated around 1630. After this date, it is difficult to disaggregate the natural geomorphological evolution of the estuary from the effects of navigational improvement and reclamation. Under the Southwold Harbour Act of 1741, the harbour entrance was fixed by training piers with the aim of maintaining a navigable depth through tidal scour. In 1757-58, a 7.5 km stretch of river upstream of Blythburgh was straightened and deepened and a quay constructed at Halesworth. Improvements in the middle estuary included excavation of the New Cut (Fig. 1) to remove a tortuous natural meander. The navigation handled a considerable trade in coal, grain and agricultural produce (Lawrence, 1990). Increasing agricultural prices encouraged the reclamation of the tidal floodplain and the valley of the River Wang tributary was enclosed in 1747.

Extensive marshes to the north of the tidal river either side of Blythburgh were embanked between 1770 and 1805. Along the southern shore, the marshes upstream of Blythburgh and the extensive Angel and Tinkers marshes in the middle estuary were all reclaimed by 1840. This effectively confined the estuary to a narrow channel. Reduction in tidal prism due to reclamation led to problems with the upkeep of Southwold harbour against the southward littoral transport of sand and shingle. Between 1805 and 1818, the mouth had to be dredged thirteen times (Lawrence, 1990) and siltation occurred along the Blackshore Quay. The outer 1.5 km of the estuary was dredged in 1829 and maintenance of the mouth was undertaken at intervals throughout the 19th century. Sea trade diminished following expansion of the railways. A narrow gauge line from Halesworth to Southwold, crossing the estuary at Walberswick, was completed in 1879. By this time the navigation was in decline, and agriculture also experienced a depression during the early part of the 20th century. After the embankments fell into disrepair the Sandpit marshes were abandoned in the 1920s, followed by the Angel and Bulcamp marshes in the 1940s (Simper, 1994).

4.2 Management

A variety of estuary uses and environmental functions and values are sustained within this inherited morphology and constrained process regime. Uses include small scale commercial fishing, recreational boating and water sport, and tourism. The estuary lies within the MAFF Suffolk River Valleys Environmentally Sensitive Area (established 1988) and the Suffolk Coasts and Heaths Area of Outstanding Natural Beauty. Extensive areas of intertidal and freshwater grazing wetland are managed for conservation by English Nature, including saltmarshes designated as internationally important under the 1973 RAMSAR convention.

Engineering issues include the stability of the estuary mouth and maintenance of 17 km of earth embankments which protect 670 ha of land and a few houses from tidal flooding. The mouth is defined by two piled breakwaters set 40 m apart. The northern breakwater was upgraded in the 1990s but the southern structure is in a poor state of repair. Although no major planform changes have occurred since the Bulcamp and Angel marshes were abandoned, the saltmarsh fronting the embankments is subject to erosional pressure. In places, loss of the saltmarsh has necessitated the installation of sheet piling or minor realignment of the defences.

The combination of small mouth cross-sectional area and large intertidal prism make the Blyth especially sensitive to sea-level rise. Post-1960 sea-level rise is 2.4 mm a^{-1} (data from Lowestoft, 20 km north of Southwold), and a rise of about 0.6 m is forecast for 2100 (MAFF, 1993). Upgrading of the flood defences to accommodate such a rise is likely to be prohibitively expensive. Whilst managed realignment has been advocated as an adaptive approach to estuarine flood defence management (Dixon *et al.*, 1998; Townend and Pethick, 2002), inundation of the large flood compartments of the Blyth would greatly increase the tidal prism and current velocities in the outer estuary. Furthermore, it might not achieve the desired environmental restoration outcomes on account of the low land levels (French and Reed, 2001). These factors highlight the importance of understanding the contemporary process regime and the morphodynamic changes likely to occur under joint environmental change and management scenarios.

5. DATA SOURCES

5.1 Morphology

Modern bathymetric charts cover only the outer estuary and its approaches. A land survey of 1840 provides elevations for each of the reclamations. These published sources have been supplemented with:

1. Twelve valley cross-sections surveyed in 1995 by the Environment Agency (EA). These include profiles across the mid-estuary tidal flats.
2. High resolution LIDAR altimetry (French, 2003) flown in 1998 by the EA National Centre for Environmental Data and Surveillance (2 m horizontal sampling interval; vertical accuracy ± 0.10 to 0.15 m).
3. Bathymetry acquired using a Raytheon single-beam depth sounder and Garmin differential global positioning system (estimated accuracy ± 0.1 m for depth and ± 1.5 m for horizontal position).
4. Vertical aerial photography (1:5,000 or 1:10,000) for 1974, 1981 (Cambridge Aerial Photography Unit) and 1997 (Natural Environment Research Council).

5.2 Hydrography

An EA-funded monitoring campaign in 1995-96 acquired hourly tidal level and three-hourly wave data at Southwold, Reydon, the western Bulcamp Flats and Blythburgh. Since 1997, the Coastal and Estuarine Research Unit (CERU) at UCL has acquired additional hydrographic data at various fixed and temporary stations (Fig. 3). Down-estuary profiles of salinity (using Valeport 6000Mk II CTD) were obtained between summer 1998 and spring 1999. Tidal levels were monitored using pressure recorders at Southwold and Blythburgh. Temporary gauges supported additional measurements between these stations. Current velocities were measured (using Valeport 104/105 self-recording flow meters) over sequences of spring and neap tides at up to five concurrent stations. Boat traffic restricted observations within the mouth to a series of timed float measurements over a spring tide. Most data were acquired at synchronised 15 minute intervals.

5.3 Sedimentation

Sedimentary processes within the abandoned reclamations were investigated by C.E. French *et al.* (2000). This work included measurement of tidal currents, waves and suspended sediment concentrations over the eastern part of the Bulcamp flats and within a small flood defence realignment adjacent to the Walberswick-Southwold bridge. Short-term instrument deployments were supported by Sedimentation Erosion Table (SET) measurements. The SET (Cahoon *et al.*, 2000) uses precisely-machined pins to record elevation changes relative to fixed benchmarks installed at points of interest (precision ± 1.5 mm). SET measurements have been made annually since 1997 at 18 locations within the saltmarsh fringing the low water channel and along the margins of the Angel and eastern Bulcamp flats. The difficulty of obtaining measurements without disturbance precluded more extensive sampling of the intertidal flats.

Intertidal sediments were sampled at numerous locations for grain size analysis. Vertical shear strength profiles were obtained in the Bulcamp flats using a Pilcon vane. Eight short cores were recovered from the bed of the low water channel between Reydon and the New Cut using a Glew gravity corer (locations given in Fig. 3).

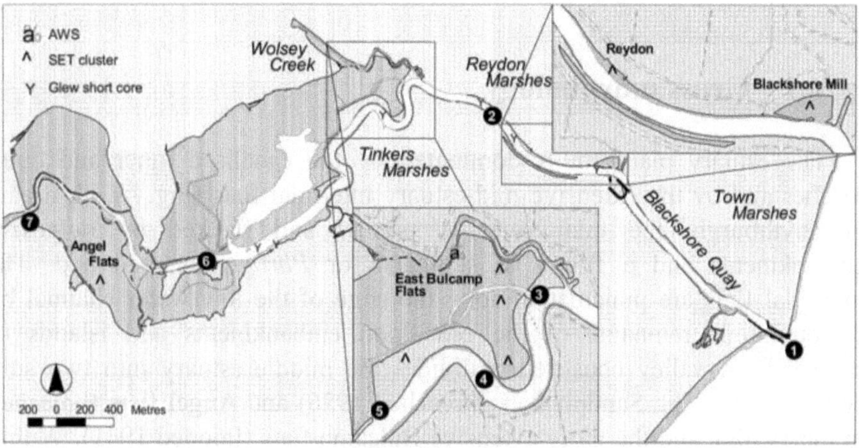

Figure 3. Location of sampling and monitoring stations. Key to numbered stations: 1 – Southwold (tidal levels; limited flow measurements); 2 – Reydon (tidal levels, currents, ADCP deployment); 3 – Wolsey Breach (currents); 4 – East Bulcamp (tidal levels, currents); 5 – Tinkers Breach (currents); 6 – New Cut (currents); 7 – Blythburgh (tidal levels, currents).

5.4 Monitoring of water and sediment flux

Exchanges of water and materials between the middle and outer estuary were monitored using a 1.2 MHz Acoustic Doppler Current Profiler (ADCP; RDI Workhorse) deployed near Reydon Quay (Fig. 3). The channel here is 45 m in width and has a mean depth of about 6 m at MHWS. The ADCP was deployed on a bottom frame placed at the channel centreline from June 2001 to June 2002. Vertical velocity profiles were recorded at 15 minute intervals (0.2 m bin width, 60 ping ensembles; standard error 18 mm s^{-1}). Tidal levels were recorded by an inbuilt pressure sensor. Cross-sectional transects with the ADCP in downward-looking mode were undertaken at 30 minute intervals over selected spring and neap tides. These yielded 'best estimates' of discharge from which linear transfer functions were obtained for the calibration of discharge estimated from the channel centreline data. In addition, ADCP backscatter intensities were calibrated against filtered bottle samples, to derive suspended sediment concentration (SSC). An automatic weather station (AWS) was maintained during this deployment.

6. MORPHOLOGY AND PHYSICAL PROCESSES

6.1 Gross morphology

The estuary planform is dominated by the canalised inner and outer reaches and by the extensive mid-estuary intertidal flats (Fig. 4). Upstream of Blythburgh, the tidal river is narrow and flanked by defensive embankments and a fringe of saltmarsh or *Phragmites* reed bed. The intertidal flats are punctuated along the edge of the low water channel by discontinuous remnants of the old flood embankments and islands of saltmarsh. A valley constriction divides the middle estuary into two sub-compartments: the Sandpit flats (breached 1926) and Angel flats (breached in the 1940s); and the more extensive Bulcamp flats (flooded 1941). Within the outer estuary, the low water channel is again constrained between extensive reclamations, which mostly lie below present MSL. A narrow and discontinuous strip of saltmarsh fronts the flood embankments. Minor realignments of the defences have been undertaken since the 1950s (e.g. along the Tinkers marsh frontage and adjacent to the old rail crossing).

Abrupt bathymetric transitions (Fig. 5) occur between the subtidal channel (the deeper parts of which reach –4.0 and –5.5 m OD, with localised scour to below –6.0 m OD) and the intertidal flats (<0.5 m OD) and saltmarsh (1.0 to 1.2 m OD). High resolution airborne LIDAR altimetry proved invaluable as a tool for mapping the reclaimed areas: these are extremely poorly covered by ground surveys. After filtering to remove unwanted high frequency detail and sub-sampling to reduce the data to manageable proportions (French, 2003), over 120×10^3 points were used to calculate flood compartment volumes. Individual compartments vary considerably in mean elevation, from 0.4 to 1.0 m OD upstream of Blythburgh to –0.5 m OD in the Reydon marshes. Estuary volume below MHWS (the sedimentary 'accommodation space') is 3.6×10^6 m^3 and mean tidal prism is 1.8×10^6 m^3 and 2.9×10^6 m^3 at neaps and springs respectively (Table I). A large potential tidal volume is contained within the reclaimed areas, especially in the outer estuary.

B

UNIT	1	2	3
Holocene sediments	thin (<8m)	moderate (> 8m)	thick (up to 13 m)
Tidal processes	weak	moderate	strong
Wave processes	negligible	moderate	weak
Morphodynamics	highly constrained	unconstrained, but retains influence of former engineering works	highly constrained

Figure 4. (a) Rectified and annotated air photo mosaic of the middle and outer Blyth estuary (NERC 1:5000 photography acquired July 1997). (b) Division of the estuary into Morphodynamic Units (for explanation see text).

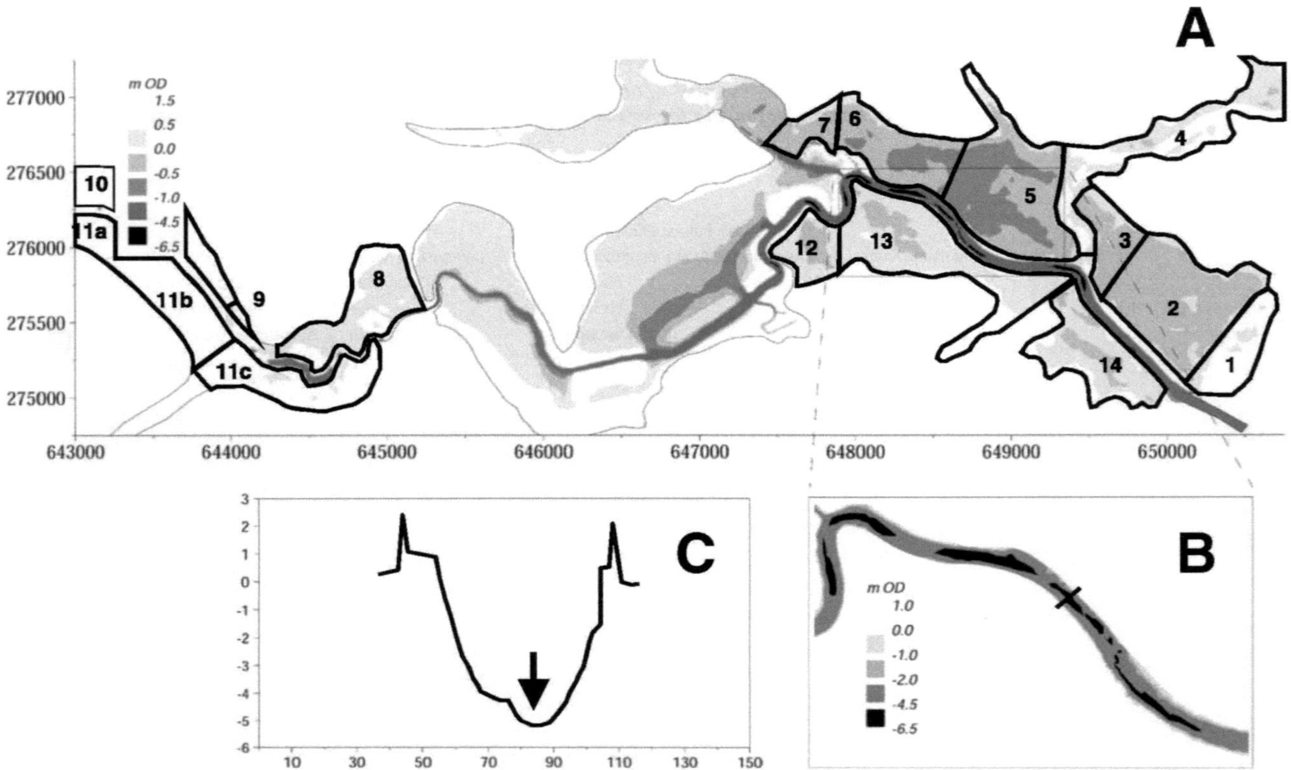

Figure 5. (a) Bathymetry and topography below 5 m OD contour. Numbered flood compartments are detailed in Table 3. (b) Inset showing detail of Reydon reach with ADCP section (solid line). (c) Cross-section at location of ADCP deployment (arrowed).

Table I. Spring tidal volume, tidal prism and potential tidal volume contained within the reclaimed areas, for present and forecast 2100 sea-level.

	Volume x 10^6 m^3			
2000 actual	Total	Unit 1	Unit 2	Unit 3
Volume at MHWS	3.58	0.16	2.58	0.84
Spring tidal prism	2.89	0.12	2.37	0.40
Neap tidal prism	1.80	0.07	1.49	0.24
Potential volume of reclamations at MHWS	3.88	0.59	0.00	3.29
2100 sea-level				
Volume at MHWS	5.32	0.21	4.13	0.98
Spring tidal prism	4.23	0.13	3.68	0.42
Potential volume of reclamations at MHWS	6.11	1.39	0.00	4.72

6.2 Hydrodynamics

Longitudinal salinity profiles generally showed essentially marine conditions (depth-mean salinities 30.7 to 33.9‰). Two profiles undertaken in March 1999 following heavy rainfall were the exception, showing mid-flood salinity dropping sharply to 28.7‰ at Blythburgh, and mid-ebb salinities as low as 26.3‰ at Blythburgh and <30‰ as far downstream as Reydon. Vertical stratification was evident at Blythburgh on this tide.

Tidal currents are strongly influenced by the morphological characteristics of a deep low water channel, extensive intertidal flats, and constricted mouth. Peak flood and ebb spring tidal currents between the mouth jetties are around 1.5 and 1.9 m s^{-1} respectively. Self-recording flow meter deployments (Table II) show that flows diminish in intensity up-estuary, but that strong flows are sustained within the two narrow saltmarsh channels (here termed the Wolsey breach and the Tinkers breach; Fig. 3) through which much of the tidal exchange with the Bulcamp Flats occurs. Most of the estuary is ebb-dominated (in terms of peak velocities), although the mid-estuary low water channel exhibits weak flood dominance.

Swell penetration occurs at Southwold, and a maximum wave height (H_{max}) of up to 1.27 m was recorded by the EA monitoring campaign of 1995-96. Wave heights are very low within the channelised estuary (H_{max} <0.17 m and <0.13 m at Blythburgh and Reydon Quay respectively). More significant waves were recorded over the western Bulcamp Flats (H_s<0.28 m and H <0.38 m). C.E. French *et al.* (2000) recorded significant wave heights (H_s) and H_{max} of up to 0.39 m and 0.46 m respectively in the eastern Bulcamp flats, where fetch exceeds 2 km under southwesterly winds.

Table II. Flood and ebb mean and peak spring tidal velocity magnitudes at monitoring stations indicated in Figure 3. (a) Short-term deployments, with flowmeters deployed 1.5 m above the bed. (b) Long-term ADCP deployent at Reydon Quay, with depth-mean velocities obtained from integration of vertical profiles.

A	Flood (mean)	(peak)	Ebb (mean)	(peak)	Flood-ebb Dominance
Southwold (mouth)	0.86 ms^{-1}	1.49 ms^{-1}	1.01 ms^{-1}	1.94 ms^{-1}	Ebb
Reydon Quay	0.41	0.69	0.41	0.79	Ebb
East Bulcamp	0.31	0.53	0.27	0.50	Flood
Wolsey Breach	0.36	0.66	0.44	0.73	Ebb
Tinkers Breach	0.18	0.34	0.35	0.61	Ebb
New Cut	0.29	0.51	0.27	0.44	Flood
Blythburgh	0.15	0.28	0.22	0.46	Ebb

B	Flood (mean)	(peak)	Ebb (mean)	(peak)	Flood-ebb Dominance
neaps	0.34	0.57	0.47	0.72	Ebb
springs	0.47	0.73	0.59	0.86	Ebb

6.3 Sediments

Intertidal sediments are muddy (d_{50} typically 8.5 to 16 μm; C.E. French *et al.*, 2000). There is a small accumulation of gravel at the mouth, and coarse rubble of anthropogenic origin mantles the bed within the Wolsey and Tinkers breaches and in parts of the low water channel near Blythburgh. Saltmarsh surface sediments exhibit bulk shear strengths in the range 15 to 25 kPa. Flat sediments are less well consolidated and surface shear strengths within the Bulcamp flats lie within the range 0.3 to 0.7 kPa. Shear strength increases with depth, typically up to around 5 to 10 kPa at 0.25 m, and more gradually to around 6 to 12 kPa thereafter (maximum sample depth 0.4 m). In some areas, the deposits are underlain at only 0.1 m by a more compacted clay with a high organic content (shear strength 12 to 16 kPa). This is probably a former agricultural surface from the time of reclamation. Short cores recovered from the middle estuary show a slight surficial concentration of shell (mainly bivalve mollusc fragments). Cores from the channel adjacent to the eastern Reydon marshes and from within the artificial New Cut were comprised of a crumbly humose peat with a substantial inclusion of clay and some shell material. Both this description and the bed elevations (between –3.5 m and –4.5 m OD) suggest that this represents the upper part of the Middle Peat unit documented by Brew *et al.* (1992).

6.4 Intertidal sediment dynamics

C.E. French *et al.* (2000) noted wave re-suspension of sediments within the eastern Bulcamp flats under a combination of high spring tides and strong, especially southwesterly, winds. Elevated suspended sediment concentrations (400 to 600 mg l^{-1}) over the flats correlated most strongly with increased wave heights and local fetch. In the absence of waves, tidal currents (<0.25 m s^{-1}) were insufficient to cause appreciable re-suspension.

The saltmarsh and tidal flat sites all show an increase in elevation (Fig. 6), mostly in excess of recent sea-level rise. Saltmarsh sedimentation exhibits a spatial pattern similar to that recorded in other UK saltmarshes (French and Spencer, 1993; Allen and Duffy, 1998): higher accumulation adjacent to the major channel and on lower, more recently formed, marsh. The only reed bed site shows very rapid accumulation (>10 mm a^{-1}), as do some of the intertidal flat sites, especially those within the small (<1 ha) Blackshore Mill sea defence realignment (Figs. 3 and 4). The latter is protected from wave action by a new sea wall and the remains of the old earth bank (C.E. French *et al.*, 2000). However, one site in the Angel Flats is accumulating at a rate insufficient to track sea-level rise (Fig. 6).

The fact that vertical saltmarsh growth is outpacing sea-level rise implies that sediment supply is sufficient to compensate for their infrequent inundation (most sites lie ≥ MHWS and are flooded by < 50 tides a^{-1}). Similarly high sedimentation rates have been reported elsewhere in southeast England (e.g. Cahoon *et al.*, 2000). Although it is difficult to draw similarly firm conclusions concerning the tidal flats owing to the sparseness of the data and the fact that only marginal locations were sampled, it is clear that sedimentation over the last 60 to 75 years has not been sufficient to allow a widespread transition to saltmarsh.

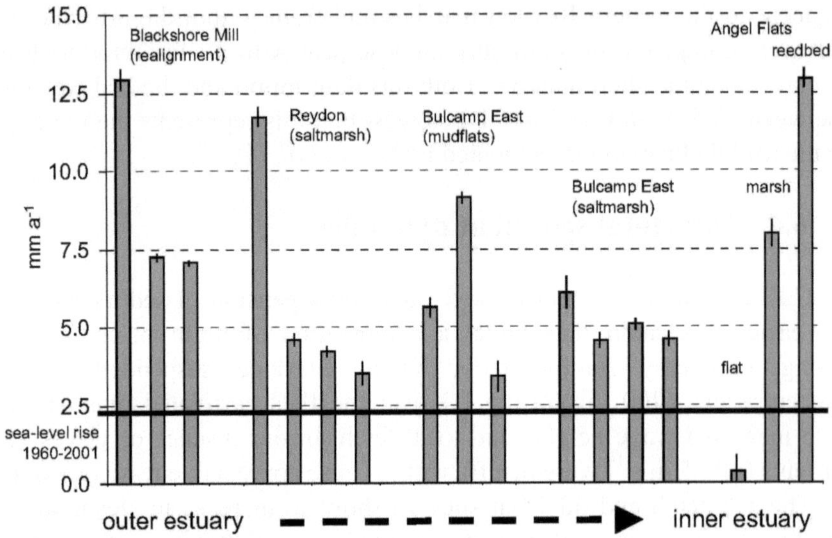

Figure 6. Annual elevation change recorded at SET sites, 1997-2001. Vertical bars represent 1 standard error.

Comparison of the 1840 land survey with modern high resolution bathymetry and LIDAR altimetry provides an insight into longer-term sedimentation over the intertidal zone as a whole. Table III shows mean elevations within the various reclamations in 1840 and in 1995-98. Interestingly, the reclamations varied considerably in elevation even in 1840, within a few decades of enclosure. Reclamation apparently included areas of intertidal flat or reedbed (in case of the Sandpit, Bulcamp New and Reydon 'marshes') as well as saltmarsh. Areas still reclaimed mostly show a small elevation loss over the last 160 years. The Reydon and Tinkers marshes are about 0.35 m lower now, on average, than in 1840. Such 'shrinkage' is well-documented within reclaimed areas. Allen (1991), for example, reports elevational 'deficits' of 0.88 m within 650 year old reclamations along the Severn estuary (southwest UK). Blyth estuary reclamations abandoned in the 1920s and 1940s show an elevational gain due to renewed sedimentation. This areally averaged accretion is lower than that indicated by the sparse SET data, although sedimentation within the Angel, Sandpit and old (upper) Bulcamp flats has either marginally exceeded or matched the estimated post-breaching sea-level rise of 0.1 m. The new (lower) Bulcamp flats show little net sedimentation.

Table III. Comparison of average flood compartment elevations in 1840 and 1995/1998. Shaded elements indicate abandoned reclamations. For locations see Fig. 5a.

Flood compartment	1840 level (m OD)	1998 level (m OD)	Change (m)
1. Havensbeach Marsh	0.76	0.61	-0.15
2. Town Marsh	-0.07	-0.15	-0.08
3. Woodsend	-0.07	-0.15	-0.08
4. Botany Marsh	0.16	0.09	-0.07
5. Reydon East	not disaggregated in 1840 survey		
6. Reydon Central	-0.15	-0.48	-0.33
7. Reydon West	not disaggregated in 1840 survey		
8. Union Farm 1	0.39	0.39	0.01
9. Union Farm 2	0.76	0.62	-0.15
10. Blyford	0.92	0.84	-0.08
11a. Blowers Marsh A	0.61	0.56	-0.05
11b. Blowers Marsh B	0.69	0.70	0.01
11c. Blowers Marsh C	1.07	1.03	-0.04
12. Tinkers Marsh West	0.39	0.01	-0.37
13. Tinkers Marsh	0.39	0.03	-0.36
14. Robinson's Marsh	0.16	0.02	-0.14
Angel	0.31	0.50	**0.19**
Sandpit	0.16	0.30	**0.14**
Bulcamp (old)	0.23	0.25	**0.02**
Bulcamp (new)	0.01	0.10	**0.09**

7. CONCEPTUAL MODEL

Based upon consideration of gross morphology, spatial contrasts in the intensity and relative dominance of fluvial, tidal and wave-related processes, and the degree of engineering intervention, three principal 'morphodynamic units' are identified (Fig. 4b). Unit 1 comprises the reclaimed inner estuary from the tidal limit at Blyford Bridge to the A12 highway causeway which crosses the valley at Blythburgh (a distance of 3.2 km along the channel centreline). This is the only part of the estuary where fluvial processes exert a significant influence. Both tidal and fluvial flows are weak, however, and apart from a few short stretches of steel sheet piling, there is little evidence for erosional pressure on the defensive embankments. At the downstream boundary of this unit, flows are routed through a concrete-lined channel under the A12 highway, which is occasionally overtopped by storm surge tides (Simper, 1994).

Unit 2 comprises the unconfined middle estuary, including the Angel, Sandpit and Bulcamp flats formed after breaching of the river walls between 1926 and 1941. Two 'sub-units' are separated by the Bulcamp headland. The Sandpit and Angel flats lie to the west, and the more extensive Bulcamp flats to the east. The downstream boundary of this unit lies near the confluence with Wolsey Creek, giving a tidal length of 4.2 km. This part of the estuary is characterised by strong tidal currents within the low water channel and in the narrow 'breaches' connecting it to the eastern Bulcamp flats. At high tide and under strong westerly/southwesterly winds, significant wave action occurs over the tidal flats. The erosional efficacy of these fetch and depth-limited waves is evident from erosion of the tidal flat face of the old earth embankments. Although sedimentation is occurring within the saltmarshes and in sheltered parts of the tidal flat, there is no evidence for net infilling of the intertidal accommodation space. Rather, there appears to be a delicate balance between accumulation of tidally-advected mud under calm conditions and wave-induced re-suspension under extreme wind events.

Unit 3 comprises the reclaimed middle and outer estuary. A spur of high ground and the embankment of the abandoned railway divides the unit into two sub-units. Upstream of the old rail crossing, tidal flows are constrained for 1.7 km between the Tinkers Marshes (managed as freshwater grazing marsh) to the south, and the low-lying Reydon Marshes (mixed arable and pasture) to the north. Downstream of the bridge, the outer 1.6 km of the estuary is relatively straight. Boat moorings and jetties extend over most of the northern bank (the Blackshore Quay) and along part of the southern shore. Unit 3 is characterised by strong tidal currents throughout, especially along the Blackshore Quay and at the mouth. Wave action is less significant than in Unit 2, although the saltmarsh is everywhere characterised by an erosional cliff. The channel is fixed in position between flood defences. These appear to be under erosional pressure at several locations, notably near the abandoned Reydon Quay where the saltmarsh fringe is absent.

The strong ebb dominance of the outer estuary is a consequence of the large mid-estuary intertidal prism (Table I) (see also, Speer and Aubrey, 1985). A large potential tidal volume is contained within the reclamations of Unit 3. Breaching of any or all of these flood compartments will have major hydrodynamic consequences in terms of the intensity of downstream flows. Furthermore, the hydrodynamic regime within Unit 3 is highly sensitive to sea-level rise, given that much of the upstream tidal prism in Unit 2 is intertidal. Neglecting the effect of any compensatory sedimentation, a rise in sea-level of 0.6 m (forecast for 2100) equates to a 48% increase in the present spring tidal prism (Table I).

Large-scale morphodynamic adjustment of the estuary is constrained by the combination of resistant Holocene sediments (compacted clays, silty peats and localised gravels) and fixed flood defence alignments. The morphology retains a strong anthropogenic influence, and exerts a persistent control on hydrodynamic processes. Unit 1 is essentially a stable conduit for the extremely low river inflow. Unit 2 is adjusting to the removal of infrastructural constraints in the 1920s-1940s. The nature of this adjustment is of interest since the abandoned reclamations provide a natural analogue for large-scale *managed* flood defence realignments, which are widely advocated as a sustainable response to sea-level rise (Townend and Pethick, 2002). Managed realignment is usually predicated on the assumption that flood defence and conservation benefits will accrue from landward relocation of existing defences, followed by a sequence of renewed tidal sedimentation, and re-establishment of saltmarsh (Dixon *et al.*, 1998). However, whilst saltmarshes in southeast England generally lie at or above MHWS (French and Reed, 2001) most estuarine reclamations are much lower in elevation. Experience from the Blyth suggests that, following inundation of such areas, episodic re-suspension during high windspeed events greatly reduces the rate of sedimentary infilling. The resistance, at depth, of sediments compacted under agricultural use, also impedes the evolution of tidal channel networks which are known to be important in facilitating both effective drainage and sediment supply (French and Reed, 2001). Post-1960 aerial photography for the Sandpit, Angel and Bulcamp flats reveals the persistence of early 20th century field drainage lines and minimal post-breaching development of natural channel networks.

The hydrodynamic impact of these *unmanaged* realignments is evident throughout the low water channel of Units 2 and 3. Preliminary hydrodynamic modelling (French, 2001) indicates that tidal current intensities have increased substantially since 1926-41. This has led to channel enlargment in the outer estuary. Active subtidal scour within Unit 3 is evidenced by the occurrence of peat particles in water samples obtained on the ebb of spring tides. Such particles are largely absent from flood and neap tide samples, when tidal bottom stresses are lower. Comparison of the modern bathymetry with the 1840 chart suggests slight deepening (<0.5 m) of the Reydon reach (Unit 3a), with deepening of the Blackshore reach (Unit 3b) by up to 1 m. The Reydon reach is more incised than elsewhere (Fig. 5b) and its scour zones penetrate the Middle Peat identified by Brew *et al.* (1992).

8. SEDIMENTARY FUNCTION

In a review of East Anglian coastal sediment sources and sinks, McCave (1987) argued that the Suffolk estuaries import sediment and that intertidal sedimentation keeps pace with sea-level rise. In the case of the Blyth, this implies an annual import of roughly 5,000 t. The preceding analysis suggests a more complex situation involving several source and sink terms and the possibility that the estuary may actually export sediment. Fluvial sources are clearly of minimal importance (we estimate <500 t a^{-1}) compared with a *potential* marine supply of the order 100 x 10^3 t a^{-1} (based upon mean tidal prism and a median flood tide SSC of 60 mg l^{-1}). Old seawalls constitute an internal sediment source. Around 5.5 km of embankment have been removed by erosion since the 1940s. These structures have a cross-sectional area of about 15 m^2 including the saltmarsh foundation. Assuming a dry bulk density of 1.7 t m^{-3} results in an estimated input of 140 x 10^3 t over about 65 years, equivalent to nearly 50% of McCave's budget for the same period.

Although the marine sediment supply is sufficient to drive infilling of the accommodation space, only the saltmarshes and the high and/or sheltered margins of the intertidal flats show sedimentation in excess of sea-level rise. The overall sediment budget appears to be mediated by the interplay of tidal advection and wave re-suspension of sediment within the intertidal flats of Unit 2, and erosional adjustment of the subtidal channel of Units 2 and 3. Insights into the direction, magnitude and timing of these fluxes can be obtained from the high resolution ADCP monitoring carried out at the interface between these morphodyamic units.

8.1 ADCP results

8.1.1 Tidal flows

Velocity data acquired by the ADCP deployed at Reydon (Fig. 3) are summarised in Table 2b. Plots of velocity and discharge versus stage (Fig. 7a,b) show distinct velocity and discharge peaks associated with the inundation and draining of the mid-estuary flats. Although this part of the estuary is usually ebb-dominated, there is a shift to weak flood dominance on the highest tides (Fig. 7c,d). This is attributed to the large proportional increase in mean depth (across the whole estuary) from 1.0 to 1.5 m between MHWS and HAT. Peak flood and ebb discharges at Reydon averaged 96 and 115 m^3s^{-1} respectively at neaps and 135 and 144 m^3s^{-1} at springs.

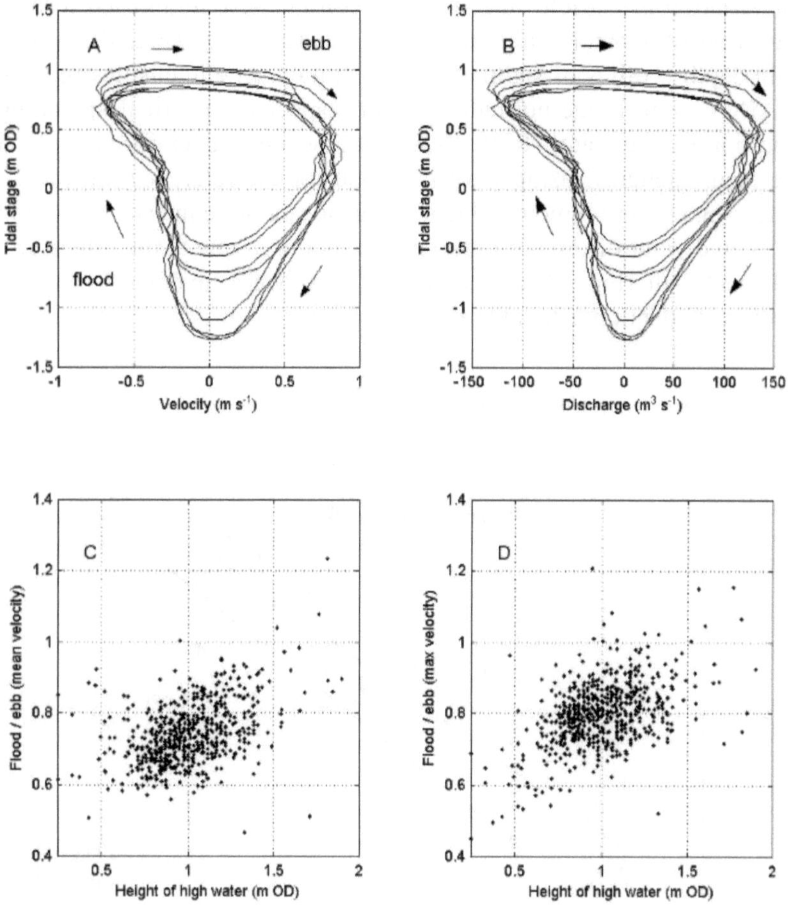

Figure 7. ADCP–derived (a) depth-mean velocity and (b) discharge as a function of stage for sample spring and neap tides. Lower plots show ratio of (c) mean and (d) peak flood and ebb velocities against the height of HW. Values of flood/ebb > 1.0 denote flood dominance.

8.1.2 Suspended sediment dynamics

The RDI ADCP also records the intensity of the backscattered signal for each of its four beams. This information can be calibrated to provide at least a semi-quantitative estimate of suspended sediment concentration (e.g. Reichel and Nachtnebel, 1994; Land and Bray, 1998; Holdaway *et al.*, 1999). In this study, calibration was undertaken using intensities for a single vertical bin and filtered bottle samples obtained from a co-located pump sampler intake. As reported by Alvarez and Jones (2002), a good correlation was obtained between Log(SSC) and backscatter intensity. The effects of tidal variation in particle size and composition were minimised through the application of separate calibrations for flood and ebb (Fig. 8). In order to obtain depth profiles of sediment concentration, raw intensities recorded by the ADCP were range-normalised with respect to spherical spreading and absorption by the water (RDI, 1999). Backscatter values for each depth bin were then calculated using the calibration for the reference bin level. Near surface values were removed due to acoustic side-lobe interference and wave effects, and values from the first bin were not used due to the effects of near-field backscattering. The remaining data were averaged to give an approximate depth mean value for SSC.

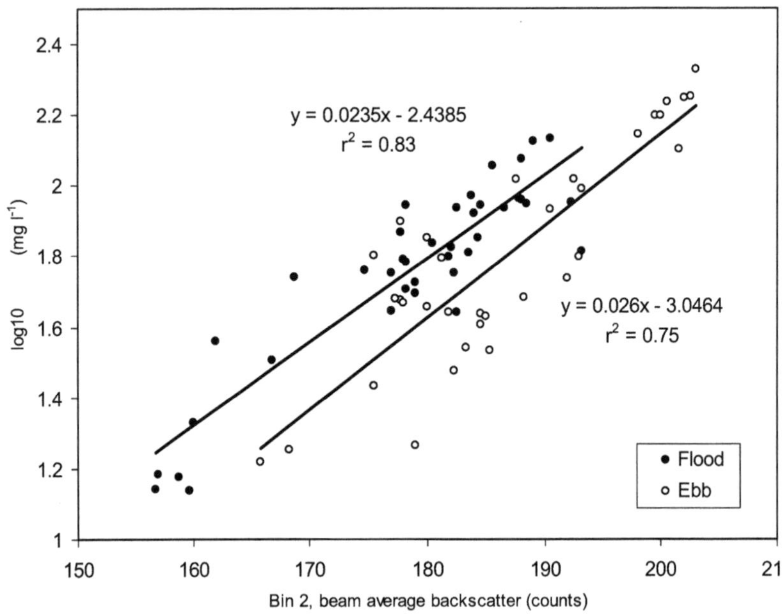

Figure 8. Calibration of ADCP backscatter against SSC.

The ADCP deployment yielded data for 650 complete tidal cycles. The median tide-averaged SSC was 60.2 mg l^{-1} for flood tides and 61.5 mg l^{-1} for ebb tides. The highest concentrations recorded were about 340 mg l^{-1} on both flood and ebb. Median ebb concentrations exceeded those of the preceding flood on approximately 55% of the tides monitored. Fig. 9 shows variation in depth-averaged SSC, tidal current velocity and water level over two spring tides. On the flood, minor variations in SSC are unrelated to current velocity. Settling occurs around High Water, followed on the ebb by a marked increase in velocity and, after a slight lag, SSC. SSC declines very sharply around Low Water.

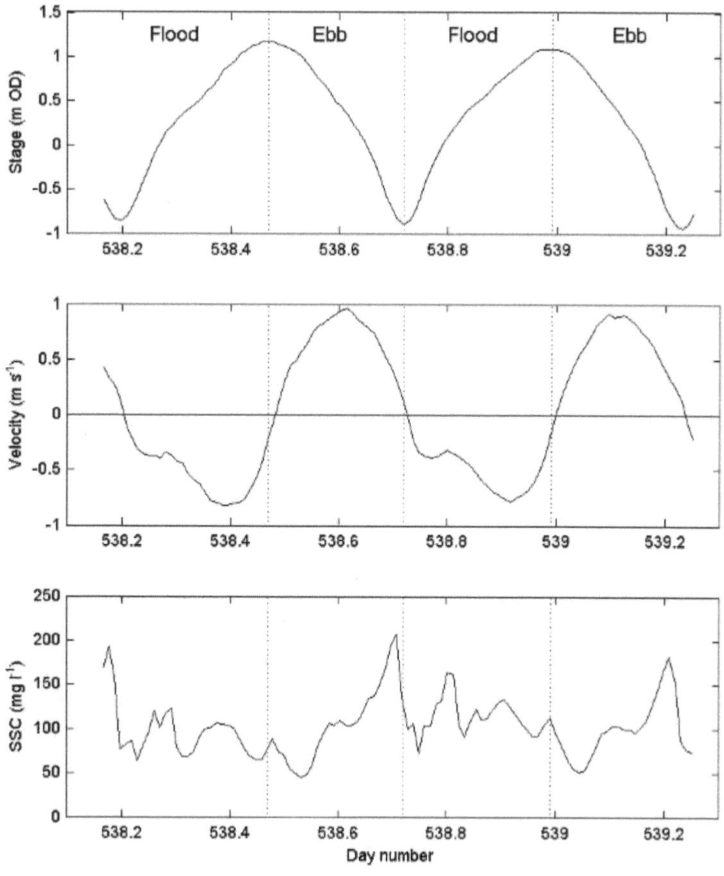

Figure 9. Spring-tidal variation in depth-averaged SSC, velocity and water level.

Fig. 10 presents suspended sediment and meteorological time-series for the whole deployment, averaged over individual tidal cycles. Equipment failure caused the loss of meteorological data for a two week period at the end of monitoring campaign. Apart from a clear spring-neap modulation, flood SSC is fairly consistent and only rarely exceeds 100 mg l^{-1}. The ebb series is characterised by larger and more numerous high concentration events and there is a strong visual correspondence between these events and periods of high windspeed. This provides direct evidence for resuspension of intertidal flat sediments and their export via the outer estuary. This interpretation is supported by the correlations between SSC, wind speed and current velocity (Table IV). Flood concentrations (both median and maximum values) are moderately correlated with peak tidal current speed but more weakly correlated with wind speed parameters. In contrast, ebb SSC values are strongly correlated with windspeed and only weakly correlated with tidal current speed. As expected, SSC exhibits significant strong autocorrelation.

Table IV. Correlations between flood and ebb SSC (median) and key physical parameters. SSC_{ebb} and SSC_{flood} = suspended sediment concentrations for preceding ebb and flood tide respectively; W_{mean} and W_{max} = mean and maximum windspeed over duration of flood or ebb; V_{mean} and V_{max} = mean and maximum current velocity. All values significant at $p = 0.01$.

	Tidal range	SSC_{ebb}	W_{mean}	W	V_{mean}	V_{max}
Flood SSC	0.32	0.70	0.24	0.26	0.47	0.47
		SSC_{flood}	W_{mean}	W_{max}	V_{mean}	V_{max}
Ebb SSC	0.18	0.57	0.60	0.52	0.12	0.15

8.2 Sediment flux

Quantification of the net sediment flux is difficult, given that even a small systematic bias in any of the component terms may lead to a large accumulated error in the total flux (see, for example, Lane *et al.*, 1997). Some check on the tidal discharge estimation is possible since the Blyth is gauged at Holton, just downstream of Halesworth. Although no data are available for the duration of the ADCP deployment, data for 1970-99 show freshwater inflow to be 9.5 to 12.6 x 10^6 m^3 a^{-1}. Integration of the tidal discharge data for Reydon results in a seawards directed annual residual flow of approximately 10 x 10^6 m^3 a^{-1}, within the observed range.

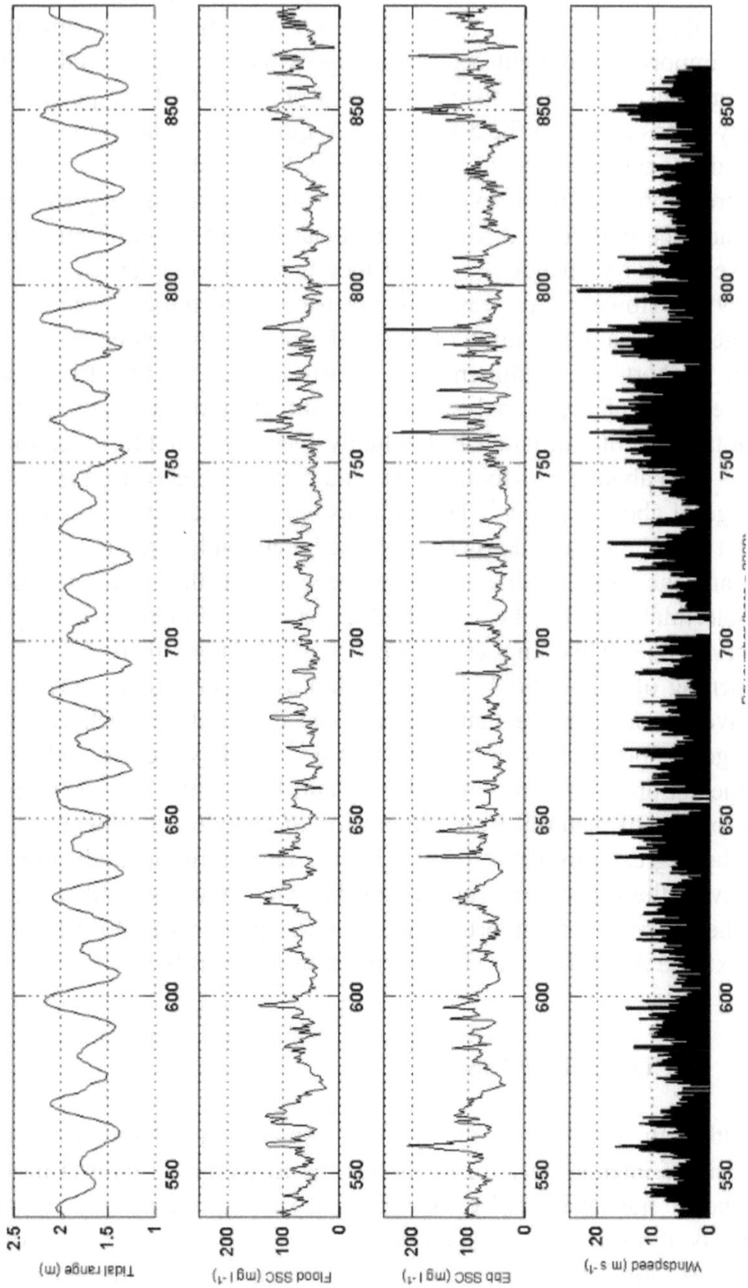

Figure 10. Time-series of tidal range, flood- and ebb-averaged SSC and windspeed. Note truncation of the windspeed record due to equipment failure. For reference, day 730 corresponds to 1 January 2002.

Integration of successive flood and ebb sediment fluxes results in an import of approximately 2.1 x 10^3 t a^{-1} from the estuary upstream of Reydon. A net import of sediment occurs on approximately 55% of individual tides. This annual flux represents a very small proportion of the potential marine supply and is less than the 5 x 10^3 t a^{-1} required to maintain the whole of the intertidal against present sea-level rise. The estimated flux should be interpreted with caution given the errors involved in gauging studies of this kind and the particular difficulties associated with the calibration of ADCP backscatter for suspended sediment measurement (see, for example, Lane *et al.*, 1997). However, the result is consistent with the analysis of historical changes and characterisation of physical processes presented earlier.

An import of sediment is clearly required to drive saltmarsh sedimentation. Recent analysis (French and Burningham, 2003) shows that the total area of saltmarsh has actually increased slightly over the last 30 years (from about 53 to about 61 ha) and that marsh elevation increases at an average of about 4 mm a^{-1}. This equates to a net accumulation of around 1.4 x 10^3 t a^{-1}. The main unknowns in the sediment budget relate to the intertidal flats and the subtidal channel. It has been shown that historical infilling of the intertidal flats has been negligible, barely sufficient to keep pace with sea-level rise. This implies an import of no more than 3.5 x 10^3 t a^{-1}. Deepening of the subtidal channel within the outer estuary does not appear to have averaged more than 0.5 m since the 1940s, which equates to an average export of less than 2 x 10^3 t a^{-1}. Of these terms, only saltmarsh sedimentation can presently be quantified with any certainty. The contribution of material from the erosion of abandoned seawalls is likely to have been greater in the past and is likely to be negligible at the present time. However, the other terms are reasonably well bounded and it seems clear that the net sediment import cannot be much greater than 3 x 10^3 t a^{-1}. In this context, the ADCP-derived flux seems entirely plausible.

9. CONCLUSIONS

Although it is one of the smallest estuaries in southeast England, the Blyth is of interest because of the degree to which its natural morphology has been transformed by navigational improvement and reclamation. Analysis of historical trends, observations of hydrodynamic and sedimentary processes and additional intensive monitoring have provided valuable insights into the physical functioning of the modern estuary and its sedimentary environments. Under the conceptual model presented here, abandonment of the mid-estuary reclamations has resulted in a distinctive planform shape and a highly compartmentalised process regime, under which morphodynamic adjustment is constrained by a combination of resistant sediments and fixed flood defence alignments. Long term, near-continuous monitoring of tidal exchanges between the middle and outer estuary shows that the system is characterised by a delicate sedimentary balance. Sediment import occurs under the influence of tidal processes, but this is punctuated by episodic export associated with wind-generated wave resuspension on the intertidal flats. The estuary does show a small net import of sediment, but this amounts to just 2% of the estimated potential supply and is inadaquate to allow intertidal sedimentation to exceed sea-level rise. Rapid sedimentation within the small area of established saltmarsh accounts for much of the sediment input, whilst the more extensive intertidal flats are accumulating sediment at a rate barely sufficient to match sea-level rise.

The Blyth estuary provides an important case study against which to evaluate the appropriateness of large-scale flood defence realignment as an adaptive response to sea-level rise. This approach is predicated on the assumption that landward relocation of existing defences will lead to renewed sedimentation and infilling of the intertidal accommodation space such that that significant re-establishment of saltmarsh will occur. This has not been the case in the Blyth, where unplanned realignment of a large part of the middle estuary created an 'event-driven' process regime, in which discrete episodes of wave resuspension largely counteract tidally-driven sedimentation. This calls into question the viability and appropriateness of large-scale realignments that do not incorporate measures to control post-breaching wave energy and, ideally, increase land levels prior to breaching.

10. ACKNOWLEDGEMENTS

Elements of this work were undertaken with financial support from the Jackson Environment Institute. Provision of bathymetric and LIDAR data by the Environment Agency and aerial photography by NERC (ARSF Award 97/07) is gratefully acknowledged. We thank C.J. Watson, C.E. French, N.J. Clifford, D. Gasca, D. Robinson and the Southwold harbourmaster, Mr K. Howells, for their assistance in the field.

REFERENCES

Allen J.R.L. 1991 Salt-marsh accretion and sea-level movement in the inner Severn Estuary: the archaeological and historical contribution. *Journal of the Geological Society, London,* 148, 485-94.

Allen J.R.L. and Duffy M.J. 1998 Medium-term sedimentation on high intertidal mudflats and salt marshes in the Severn Estuary, SW Britain: the role of wind and tide. *Marine Geology,* 150, 1-27.

Alvarez L.G. and Jones S.E. 2002 Factors influencing suspended sediment flux in the upper Gulf of California. *Estuarine, Coastal and Shelf Science,* 54, 747-59.

Beardall C.H., Dryden R.C., and Holzer T.J. 1991 *The Suffolk estuaries: report by the Suffolk Wildlife Trust on the wildlife and conservation of the Suffolk estuaries.* Ipswich, Segment Publications.

Brew D.S. 1990 *Sedimentary environments and Holocene evolution of the Suffolk estuaries.* Unpublished PhD thesis, University of East Anglia.

Brew D.S., Funnell B.M. and Kreiser A. 1992 Sedimentary environments and Holocene evolution of the lower Blyth estuary, Suffolk England, and a comparison with other East Anglian coastal sequences. *Proceedings of the Geologists Association,* 103, 57-74.

Cahoon D.R., French J.R., Spencer T., Reed D.J. and Moller I. 2000 Vertical accretion versus elevational adjustment in UK saltmarshes: An evaluation of alternative methodologies. In: Pye, K. and Allen, J.R.L. (eds) *Coastal and estuarine environments: sedimentology, geomorphology and geoarchaeology.* Geological Society of London Special Publication 175, 223-38.

Cayocca F. 2001 Long-term morphological modeling of a tidal inlet: the Arcachon Basin, France. *Coastal Engineering,* 42, 115-42.

Dixon M., Leggett D.J. and Weight R.C. 1998 Habitat creation opportunities for landward coastal realignment: Essex case studies. *Journal Institution of Water and Environmental Management,* 12, 107-12.

EMPHASYS Consortium 2000 *Modelling estuary morphology and process.* Final Report for MAFF Project FD1401. HR Wallingford, Report TR111.

French C.E., French J.R., Clifford N.J. and Watson C.J. 2000 Sedimentation-erosion dynamics of abandoned reclamations: the role of waves and tides. *Continental Shelf Research,* 20, 1711-33.

French J.R. 2001 *Hydrodynamic modelling of the Blyth estuary: impacts of sea level rise.* Report for Environment Agency, Anglian Region, 113pp.

French J.R. 2003 Airborne LIDAR in support of geomorphological and hydraulic modelling. *Earth Surface Processes and Landforms,* 28, 221-35.

French J.R. and Burningham, H. 2003 Tidal marsh sedimentation versus sea-level rise: a southeast England estuarine perspective. *Proceedings, Coastal Sediments '03,* Clearwater, Florida, 1-14.

French J.R. and Reed D.J. 2001 Physical contexts for saltmarsh conservation. In: A Warren and J.R. French (eds.) *Habitat conservation: managing the physical environment.* Chichester, Wiley, 179-228.

French J.R., Reeve D.E., and Owen M. 2002 *Estuaries Research Programme Phase 2 Research Plan.* Report for DEFRA, Project FD2115. London, DEFRA, 47pp.

French J.R. and Spencer T. 1993 Dynamics of sedimentation in a backbarrier saltmarsh. *Marine Geology,* 110, 315-31.

Holdaway G.P., Thorne P.D., Flatt D., Jones S.E. and Prandle D. 1999 Comparison between ADCP and transmissometer measurements of suspended sediment concentration. *Continental Shelf Research,* 19, 421-41.

Land J.M. and Bray R.N. 1998 Acoustic Measurement of Suspended Solids for Monitoring of Dredging and Dredged Material Disposal. In: *Proceedings of the 15th World Dredging Congress 1998*, Las Vegas, Western Dredging Association.

Hydrographer of the Navy 2000 *Admiralty Tide Tables: Volume 1. United Kingdom and Ireland including European Channel Ports*. Taunton, Hydrographer of the Navy.

Lane A., Prandle D., Harrison A.J., Jones P.D. and Jarvis C.J. 1997 Measuring fluxes in tidal estuaries: sensitivity to instrumentation and associated data analyses. *Estuarine Coastal and Shelf Science,* 45, 433-51.

Lawrence R. 1990 *Southwold River: Georgian life in the Blyth valley*. Southwold, Moxon, 150pp.

McCave I.N. 1987 Fine sediment sources and sinks around the East Anglian coast UK. *Journal of the Geological Society,* 144, 149-52.

MAFF 1993 *Strategy for flood and coastal defence in England and Wales*. MAFF/Welsh Office PB 1471, 39pp.

Motyka J.M. and Brampton A.H. 1993 *Coastal management: mapping of littoral cells*. HR Wallingford Report SR328, 102pp.

Nicholson J., Broker I., Roelvink J.A., Price D., Tanguy J.M. and Moreno L. 1997 Intercomparison of coastal area morphodynamic models. *Coastal Engineering*, 31, 97-123.

Parker R. 1978 *Men of Dunwich*. London, Collins, 272pp.

Pye K. and Allen J.R.L. 2000 Past, present and future interactions, management challenges and research needs in coastal and estuarine environments. In: K. Pye and J.R.L. Allen (eds) *Coastal and estuarine environments: sedimentology, geomorphology and geoarchaeology*. Geological Society, London, Special Publication 175, 435pp.

RDI 1999 River TRANSECT User's Manual. San Diego, RD Instruments.

Reichel G. and Nachtnebel H.P. 1994 Suspended sediment monitoring in a fluvial environment - Advantages and limitations of applying an acoustic-doppler-current-profiler. *Water Research,* 28, 751-61.

Ribberink J.S., Negan E.H. and Hartsuiker G. 1995 Mathematical modelling of coastal dynamics near a tidal inlet system. In: W. Dally and R.B. Zeidler (eds.) *Proceedings Coastal Dynamics '95*. Reston, American Society of Civil Engineers, 916-26.

Roman C.T. and Nordstrom K.F. 1996 Environments, processes and interactions of estuarine shores. In: K.F. Nordstrom and C.T. Roman eds *Estuarine shores: evolution, environments and human alterations*. Chichester, Wiley, 1-12.

Simper R. 1994 *Rivers Alde, Ore and Blyth*. Lavenham, Creekside Publishing, 84pp.

Soulsby R.L. 1997 *Estuaries: the case for research into morphology and process*. HR Wallingford, Report SR478.

Speer P.E. and Aubrey, D.G. 1985 A study of non-linear tidal propagation in shallow inlet/estuarine systems – II. Theory. *Estuarine Coastal and Shelf Science*, 21, 207-24.

Steers J.A. 1964 *The coastline of England and Wales*. Cambridge, Cambridge University Press, 750pp.

Townend I. and Pethick J. 2002 Estuarine flooding and managed retreat. *Philosophical Transactions Royal Society of London, A* 360, 1477-1495.

Chapter 9

CONTROLS ON ESTUARINE SEDIMENT DYNAMICS IN MERRYMEETING BAY, KENNEBEC RIVER ESTUARY, MAINE, U.S.A.

Michael S. Fenster[1], Duncan M. FitzGerald[2], Daniel F. Belknap[3], Brad A. Knisley[1], Allen Gontz[3] and Ilya V. Buynevich[4]

[1]*Randolph-Macon College,Environmental Studies Program, Ashland, VA 23005 USA*

[2]*Boston University, Department of Earth Sciences, Boston, MA 02215 USA*

[3]*University of Maine, Department of Geological Sciences, Orono, ME 04469-5790 USA*

[4]*Woods Hole Oceanographic Institute, Geology and Geophysics Department, Woods Hole, MA 02543-1598 USA*

1. INTRODUCTION

Over the past several decades, estuaries have earned a reputation as sediment sinks through the theoretical and empirical works of many scientists (e.g. Postma, 1967; Pritchard, 1967; Meade, 1969, 1972, 1982; Biggs, 1970; Biggs and Howell, 1984; Schubel, 1984; Knebel, 1989; Dalrymple *et al.*, 1990). These studies have documented the combined roles of sediment influx rates, sea-level rise, climate, and estuarine circulation as the dominant controls on estuarine infilling (Schubel, 1984). However, aspects of most of these models (e.g. distance-velocity asymmetry and settling lag and scour lag) only consider the movement of fine-grained sediments (<100 μm) capable of suspension or transport-limited systems in which estuarine sediment supply is greater than the transport capacity (Milliman and Meade, 1983). Much less is known about the dynamics (i.e. estuarine processes and time scales responsible for sediment fluxes) within fluvial-estuarine transition zones with respect to bedload sediment transport (Milliman and Meade, 1983).

D.M. FitzGerald and J. Knight (eds.),
High Resolution Morphodynamics and Sedimentary Evolution of Estuaries, 173-194.

Moreover, studies of bedload sediment transport near the mouths and entrances to the large estuaries have revealed a net landward bottom transport direction of coarse-grained sediments (Meade, 1969; Sherwood *et al.*, 1984; Knebel, 1989; Nichols *et al.*, 1991; Fenster, 1995), sand circulation cells (Ludwick, 1970, 1972), and/or mutually exclusive ebb-flood transport paths (Harris, 1988). Some studies conducted within estuary channels have indicated that even freshwater floods cannot provide enough energy to deliver sand to coastal or inner shelf environments (Bryce *et al.*, 1998).

Fewer studies have provided evidence that estuaries can serve as sand sources for mainland and barrier beaches (Horne and Patton, 1989; Cooper, 1993, 2002; Fenster *et al.*, 2001; FitzGerald *et al.*, 2004). Fenster *et al.* (2001) pointed out that variations in estuarine geometries and dynamics produce differences in bedload sediment transport regimes. In particular, ebb-dominance occurs primarily through spring snowmelt floods (freshets) in mesotidal, high-latitude, narrow, rock-bound estuaries (e.g. Fenster *et al.*, 2001). Similar results have been observed during extreme floods in microtidal, tide-dominated estuaries (Cooper, 2002). On the other hand, flood-dominance occurs within long, coastal plain estuaries where sea-level rise and bi-directional transport dominate (e.g. Knebel, 1989).

Empirical data from the lower 27 km of the Kennebec River estuary, Maine, USA, have shown that the estuary provides coarse-grained sediment to the nearshore and coastal region of south-central Maine (Fenster and FitzGerald, 1996; FitzGerald *et al.*, 2000; Fenster *et al.*, 2001). These studies showed that the relationship between freshwater discharge and tidal range could predict annual bedload transport fluxes through the Kennebec River estuary. In particular, for the Kennebec River estuary, seaward transport occurs when discharge values exceed the threshold range of 225-325 $m^3 s^{-1}$, independent of tidal range. These conditions occur during spring freshets which produce an ebb-dominated velocity asymmetry that flushes sediment from the Kennebec River estuary to the nearshore, and contributes to the formation and maintenance of barrier complexes (FitzGerald *et al.*, 2000). Other New England estuaries, such as the Saco, Merrimack, and Connecticut River estuaries, function similarly (Horne and Patton, 1989; FitzGerald *et al.*, 2004). Consequently, these findings provide empirical evidence that could augment existing conceptual and morphological models that are based on, but do not fully account for, net seaward sand transport in the relative contributions of marine and fluvial processes (Dalrymple *et al.*, 1992; Fenster and FitzGerald, 1996; Fenster *et al.*, 2001).

Less is known about sediment transport processes in the relatively low-energy central zone of net ebb-dominated estuaries. If net landward transport occurs in the marine-dominated, outer estuary and net seaward transport takes place in the river-dominated, inner estuary, then the mixed-energy, central zone should serve as an area of net convergence for sediment in the estuary (Dalrymple *et al.*, 1992) and therefore, the facies boundary between marine (tidally-influenced) and fluvial deposits (Pritchard, 1967). Thus, the central estuary acts as a sink, or a site of sediment deposition and long-term storage. This conceptual model begs the question, does the Kennebec River system have a central "storage" bay given a complex, bedrock-controlled channel geometry and a net downstream coarse-grained sediment transport regime? More broadly, can estuaries with net seaward sand fluxes temporarily store sediment within their confines and, if so, how and where? In addition, when and how is sediment released from storage?

In this paper we test the hypothesis that, during low river flow independent of tidal energy conditions, the Kennebec River estuary maintains a large sediment storage reservoir and its dynamics corroborate existing sediment transport models in which the central estuary serves as a sediment sink (i.e. Pritchard, 1967; Dalrymple *et al.*, 1992). To test this hypothesis, we selected a summer month characterized by low-freshwater flow, but high energy tidal phase. We expected this situation to maximize the potential for landward transport in this ebb-dominated transport system and sediment accumulation near the upstream limit of landward bottom flow (Meade, 1969).

2. STUDY AREA

The Kennebec River estuary is a relatively narrow, elongate, rock-bound estuary in west-central Maine (Fig. 1). The bedrock valley was produced initially by fluvial erosion of weathered metasedimentary rocks during the Tertiary Period and deepened later by Pleistocene glaciations. The estuary formed by drowning of the valley during Holocene sea-level rise (Belknap *et al.*, 1987; Kelley, 1987; Fenster and Fitzgerald, 1996; Barnhardt *et al.*, 1997).

The estuary receives freshwater input from Maine's largest rivers (Kennebec River and Androscoggin River) by combined discharge and drainage area ($\approx$ 24,460 km^2). Approximately 27 km from the mouth of the estuary, the Kennebec and Androscoggin join at the (appropriately named) Merrymeeting Bay (Fig. 1). These two river systems contribute abundant coarse-grained sediments to Merrymeeting Bay and lower Kennebec River estuary as they cut through unconsolidated ice-contact and periglacial deposits (Borns and Hagar, 1965; Thompson and Borns, 1985). The presence of coarse-grained deposits within this system differentiates it from neighboring estuaries which are floored mostly with mud (e.g. Penobscot River - Knebel, 1986; Sheepscot River - Kelley *et al.*, 1987; Damariscotta River - Belknap *et al.*, 1994). During low flow and low tidal elevations, some of the sand deposits emerge in the form of large bars in Merrymeeting Bay.

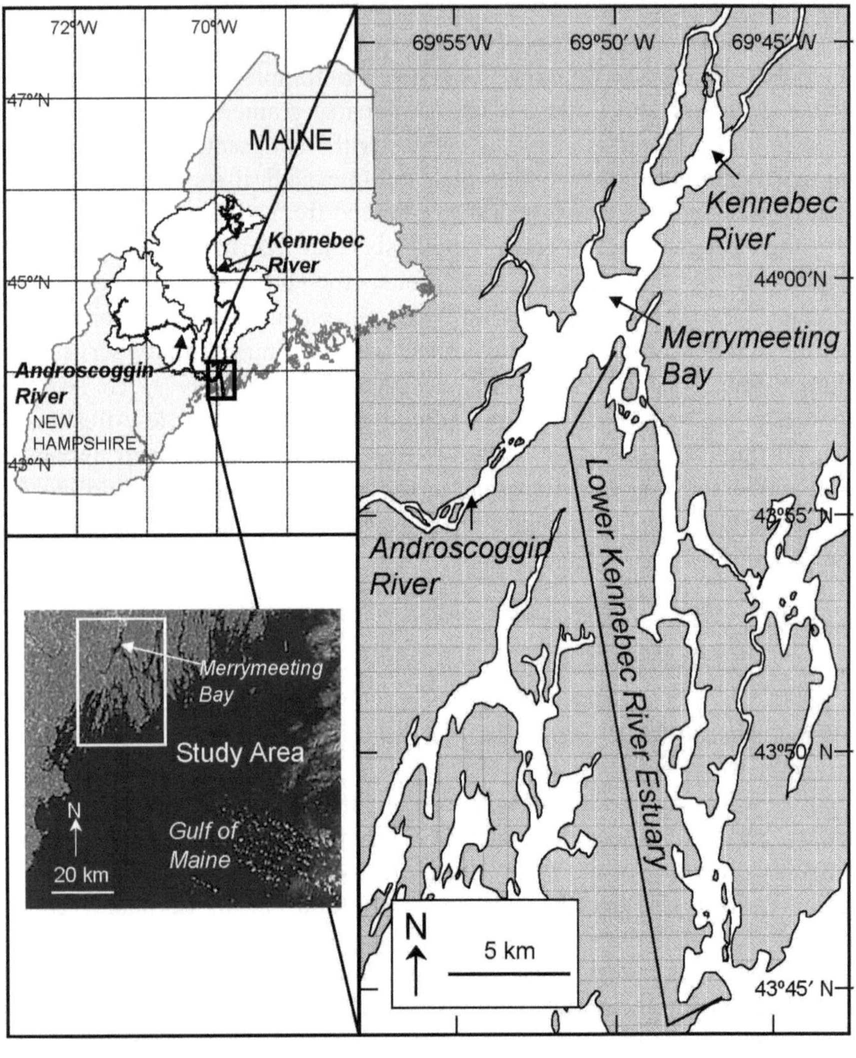

Figure 1. Map showing drainage basins areas of the Kennebec River and Androscoggin River (top left); satellite image of east-central coastal Maine (bottom left), and expanded view of the Merrymeeting Bay at the confluence of the Kennebec River and Androscoggin River (right).

Merrymeeting Bay is a shallow (depth ≤ 4 m), 2 km wide by 8 km long bay connected to the lower estuary by a narrow bedrock constriction (Figs. 1, 2). The lower estuary, south of the constriction known as the "Chops" (Fig. 2), is a partially mixed to stratified mesotidal estuary (Fig. 1; Fenster and FitzGerald, 1996). Fenster and FitzGerald (1996), Hannum (1996), and

Fenster *et al.* (2001) described the physical setting (including the hydrology) of the estuary in detail.

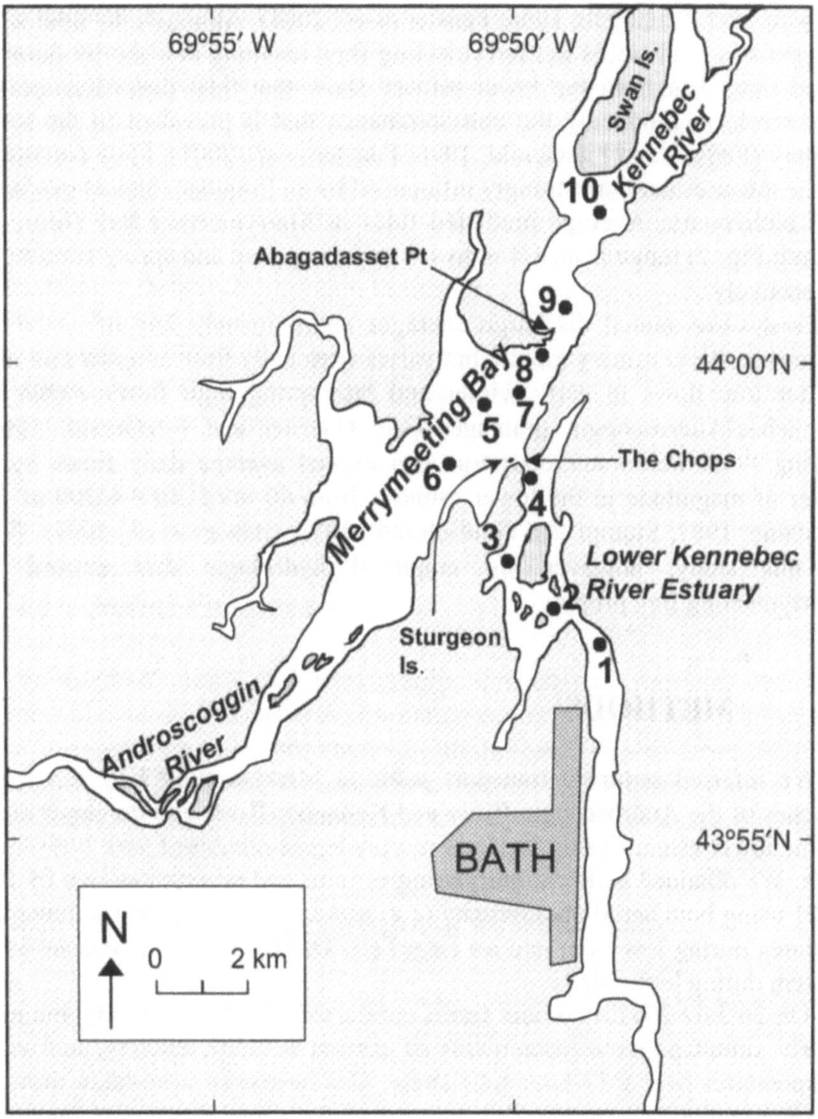

Figure 2. Map showing confluence of Androscoggin River and Kennebec River at Merrymeeting Bay and location of hydrography stations (numbered) used in this study.

The semidiurnal tidal range at the estuary mouth averages 2.6 m and increases to 3.5 m during spring tides. The tidal wave advances to the upper reaches of the lower estuary, 27 km north of the mouth, in approximately one hour (Fenster *et al.*, 2001). Although the limit of tidal influence extends to Augusta, Maine, nearly 65 km upstream of the mouth, previous studies have shown that salinity values diminish substantially at Merrymeeting Bay (Fenster and FitzGerald, 1996; Fenster *et al.*, 2001). Although the tidal wave dampens as it advances to Merrymeeting Bay, resulting in a shorter duration flood tide, data from the lower estuary show that tidal distortions play a minor role in producing the ebb-dominance that is prevalent in the lower estuary (Fenster and FitzGerald, 1996; Fenster *et al.*, 2001). Flow conditions in the lower estuary are strongly influenced by an irregular channel geometry and bathymetry. Average predicted tides in Merrymeeting Bay (Sturgeon Island; Fig. 2) range from 1.4 m to 1.9 m during neap and spring conditions, respectively.

Freshwater annual discharge averages approximately 261 m^3 s^{-1} at the Kennebec River estuary mouth, but varies seasonally from summer and mid-winter low flows to early winter and late spring high flows within the Kennebec-Androscoggin drainage basin (Fenster and FitzGerald, 1996). Spring flood freshwater discharge can exceed average daily flows by an order of magnitude in the lower estuary (from 40 m^3 s^{-1} to > 6,200 m^3 s^{-1}; Fontaine, 1987; Stumpf and Goldschmidt, 1992; Stewart *et al.*, 2001). Prior to this study, however, no empirical hydrologic data existed for Merrymeeting Bay proper.

3. METHODS

We inferred sediment transport paths in Merrymeeting Bay, the lower reaches of the Androscoggin River and Kennebec River and the upper reach of the lower estuary using bedform morphologies combined with hydrologic data. We obtained bedform morphologies (size and orientations) on 25 July 2001 using both aerial photography (e.g. an analysis of exposed sedimentary features during low tide) and an EdgeTech DF1000 digital side-scan sonar system during high tide.

On 26 July 2001, two boat teams conducted a hydrography by obtaining nearly simultaneous measurements of current velocity, salinity, and water temperatures over a 13-hour tidal cycle. Measurements were taken from 10 moored stations located in the thalweg of the upper portion of the lower estuary, Merrymeeting Bay proper and into the lower reaches of the Androscoggin River and Kennebec River (Fig. 2). Four stations were located in the lower Kennebec, four in the upper Kennebec River, one in the central Bay, and one in the Androscoggin side of the Bay (Fig. 2). The shallow depths in the Androscoggin River made boat passage impossible any farther upstream. At each station, current velocity (using a Marsh-McBirney Model 2000 Flo-Mate) and temperature and salinity measurements (using a YSI 30

Model 30 S-C-T meter) were taken near the bottom ($z \approx 1$ m), at mid-depth, and at the surface. Measurements were obtained at each station approximately once per hour. Because of the shallow depths, vertical profile measurements at Station 5 (Merrymeeting Bay) and Station 6 (Androscoggin River) contained only one or two measurements. Therefore, the depth-averaged measurements consist of an average of two data points or simply represent a single measurement when appropriate. Additionally, each boat team measured tide heights during every cycle using tide staffs that were installed on the shores of the estuary.

Freshwater discharge on the day of the hydrography was extrapolated using several gaging stations located within the Kennebec-Androscoggin watershed (no stations exist within the estuary). A strong log-log relationship between discharge and drainage area ($r = 0.99$) allowed accurate measurement of discharge at ungaged stations knowing the drainage area at the site of interest (Fenster and FitzGerald, 1996).

4. RESULTS

4.1 Aerial Photography

Aerial observations and photographs showed that much of the bay consists of sand bodies and fringing marsh exposed at low tide. The shoals were fairly uniform in length, linguoid (out of phase) in plan-form, and ebb-oriented (Fig. 3). A notable exception to the major ebb-oriented forms is a large flood-tide delta immediately north to northwest of the Chops (Fig. 4).

4.2 Side-scan Sonar: Bedform Orientations

A side-scan mosaic of the study area north of the Chops shows that asymmetrical, flood-oriented bedforms dominate the channel bottom of Merrymeeting Bay (Fig. 5). The bedforms range in height from 0.5 m to 2.0 m and wavelength from 10 m to 30 m. The arrows, drawn perpendicular to the bedform crests on Figure 5, show that sediment is dispersed in a variety of directions into the Bay.

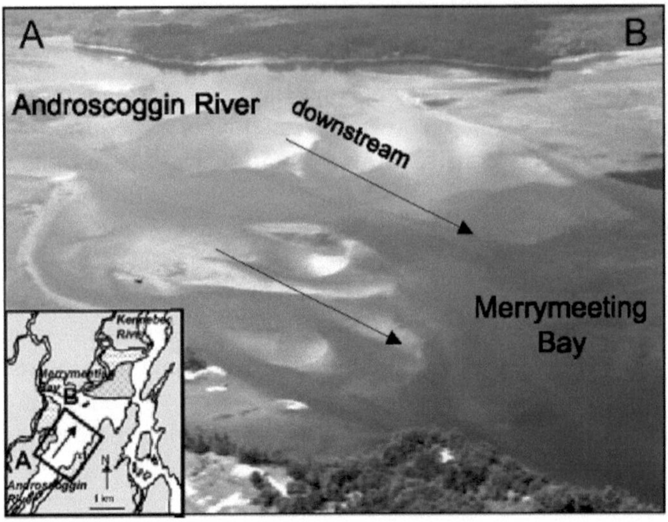

Figure 3. Oblique aerial photograph looking northwest of the lower Androscoggin River at Merrymeeting Bay. Note the exposed large-scale linguoid transverse bedforms with a downstream orientation.

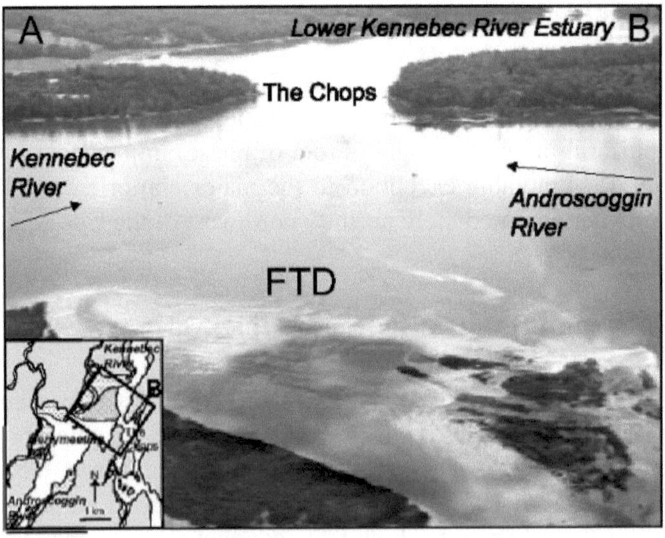

Figure 4. Oblique aerial photograph looking south of the exposed portion of the flood-tide delta (FTD) seaward of the lower Kennebec River estuary and the Chops and the confluence of the Androscoggin River and Kennebec River. The distance between the Chops is approximately 0.2 km.

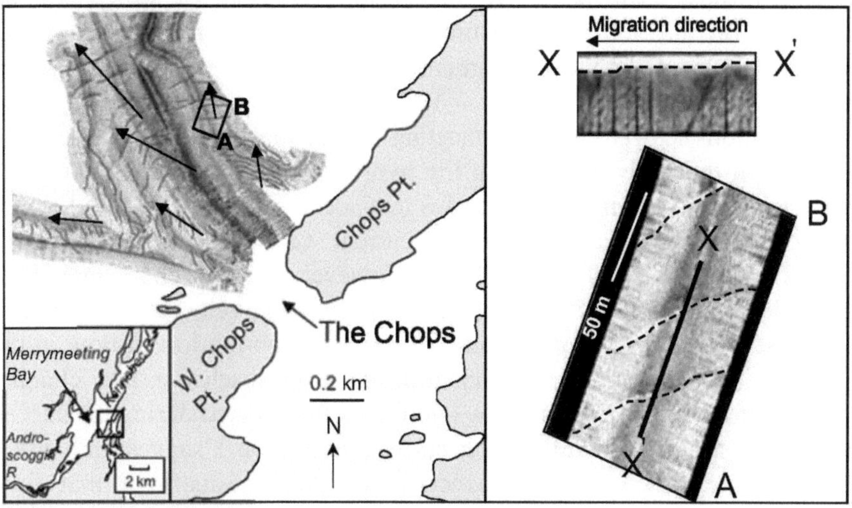

Figure 5. Side-scan sonar images showing flood-oriented bedforms directly landward of the Chops. Bedform crests are highlighted on the side-scan sonar record with arrows indicating migration direction as suggested by the bedform slip-faces. Insets show detailed images of a segment of the bay floor.

4.3 Tide Height

The observed tidal ranges landward and seaward of Merrymeeting Bay of 1.75 m and 2.10 m, respectively, compared well with the predicted tidal ranges of 1.93 and 2.33 at Sturgeon Island and Bath (NOAA, 2001). Likewise, the observed and predicted times of high and low slack waters corresponded well. Above Merrymeeting Bay (upper Kennebec), the first observed high and low slack waters occurred at 05:43 and about 12:25, respectively compared with predicted times of 05:08 and 11:40. Below Merrymeeting Bay (lower Kennebec), the first observed high and low slack waters occurred at 06:00 and 12:14 respectively, compared to the predicted times of 6:07 and 12:26. It should be noted that the hydrography did not capture the earliest peak high slack water tide height and therefore, the observed tidal range data slightly understate the predicted tidal range and timing of high tide.

4.4 Current Velocity

The current velocity results showed minor differences in depth-averaged and near-bottom values. However, these results do not necessarily indicate uniform vertical flow and may reveal the influence of water column turbulence on the velocity measurements. Consequently, because of our primary focus on bedload sediment transport, we report near-bottom measurements only.

The stations seaward of Merrymeeting Bay displayed the greatest current velocities in the study area during the hydrography. The strongest currents with a peak near-bottom velocity of 1.8 m s^{-1} occurred at Station 3 in the ebb direction (Fig. 6). Near-bottom ebb-current velocities exceeded flood-currents at Stations 2, 3, and 4 (Fig. 6). However, the peak ebb velocities at Station 1 were lower (0.8 m s^{-1}) than the peak flood velocities (1.3 m s^{-1}). These results demonstrate the strong influence of ebb-reinforced flow in the upper reach of the lower estuary (due to river discharge and channel geometry) for the low riverine flow conditions that existed during the survey period. The stronger flood than ebb currents at Station 1 suggest that either flood tides exert a greater influence in the lower estuary or that flow segregation precluded the capturing of maximum ebb flows. The slightly larger ebb than flood tidal range (2.1 m versus 2.0 m) in this reach demonstrates that the currents could, in part, result from greater ebb forcing. Overall, the maximum near-bottom current velocities at Stations 2, 3, and 4 exceeded the maximum flood currents by 38% to 78% with an average of 62%. Conversely, the maximum near bottom currents at Station 1 exceeded those of the ebb by 74%.

Station 5 in Merrymeeting Bay exhibited slightly stronger flood-directed currents than ebb-directed currents (Fig. 7). Near-bottom peak currents were nearly equal for the flood and ebb directions (0.80 m s^{-1} compared to 0.75 m s^{-1}, respectively), but near-bottom flood currents greater than 0.5 m s^{-1}, for example, persisted for more than twice the length of time as those for the ebb currents (4 hr 20 min versus 2 hr 0 min; Fig. 7). These data (i.e. longer flood-durations and greater flood-directed current velocities), combined with bedform orientation data and a large flood-tidal delta (Fig. 4), suggest that flood-directed transport patterns dominate central Merrymeeting Bay.

Maximum near-bottom flood current velocities for Station 6 on the Androscoggin River reached 0.90 m s^{-1} while those for the ebb reached 0.96 m s^{-1} (only one vertical reading was available from 12:27-16:58 due to the shallow water depths at the station; Fig. 7). The data show that the flow nearest the Androscoggin River (Station 6) maintains greater velocities than the flow in the Bay proper (Station 5) over almost the entire ebb cycle, but the converse is true for the flood cycle (Fig. 7). Moreover, although Station 6 displayed greater peak near-bottom flood current velocities than Station 5 (0.90 m s^{-1} versus 0.80 m s^{-1}, respectively), the maximum flood-directed velocities at Station 5 exceeded those at Station 6 for more than 4 hours during the flood cycle. In essence, the Androscoggin River displayed greater near bottom ebb-directed currents (0.96 m s^{-1} at Station 6 versus 0.75 m s^{-1} at Station 5) while the Bay proper displayed greater flood-directed currents (despite equal magnitude peak flood-directed velocities at each station). Thus, based on current data and bedform orientations, we infer that the Androscoggin River contributes sediment to the Bay even during low flow conditions.

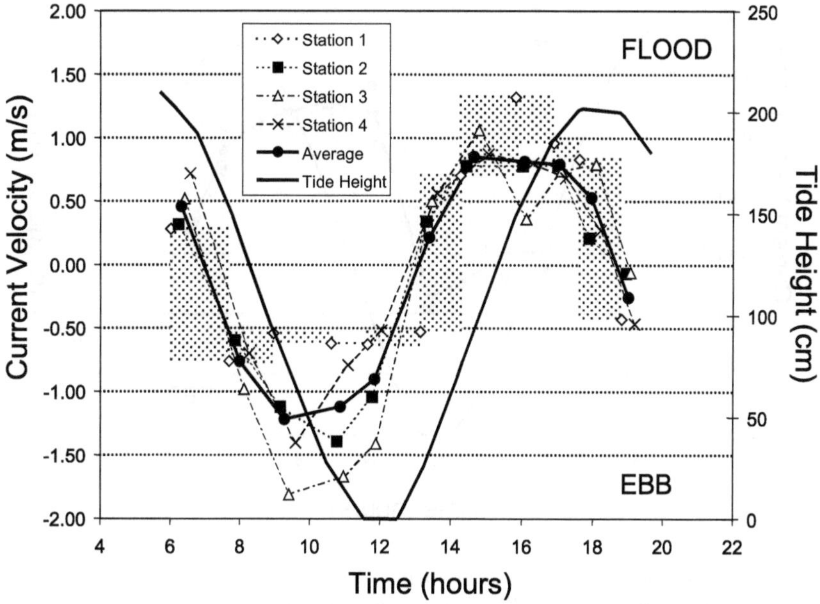

Figure 6. Near-bottom current velocities for the lower Kennebec River estuary stations labeled 1-4 (see Figure 2 for locations). Also shown is the tide height during the survey period. Note the ebb-dominance at all stations except Station 1.

Two of the four stations (Stations 7 and 8) on the Kennebec River reach exhibited ebb-dominated velocities and two stations displayed less pronounced flood-dominance (Stations 9 and 10). However, comparison of the mean peak near-bottom ebb velocity to the mean peak flood velocity at all four stations points toward general ebb-dominance over the entire reach (Fig. 8). In particular, ebb-directed velocities exceeded flood-directed velocities at Stations 7 and 8 by 49% and 36% near the bottom, respectively (Figs. 8A, 8B). Station 8 displayed the greatest ebb-directed velocities for near-bottom velocities (0.76 m s^{-1}). These higher velocities are expected because of the station's position downstream of a narrow bedrock constriction at Abagadesset Point (Fig. 2). Likewise, the (slightly and anomalously) higher flood-directed velocities farther upstream at Station 9

(0.86 s^{-1} versus 0.55 m s^{-1}) and Station 10 (0.61 s^{-1} versus 0.38 m s^{-1}) are expected. These results are also a product of their location upstream of the same bedrock constriction and the forcing of a larger volume of water from Merrymeeting Bay through a narrow constriction into the Kennebec River during the flood tide.

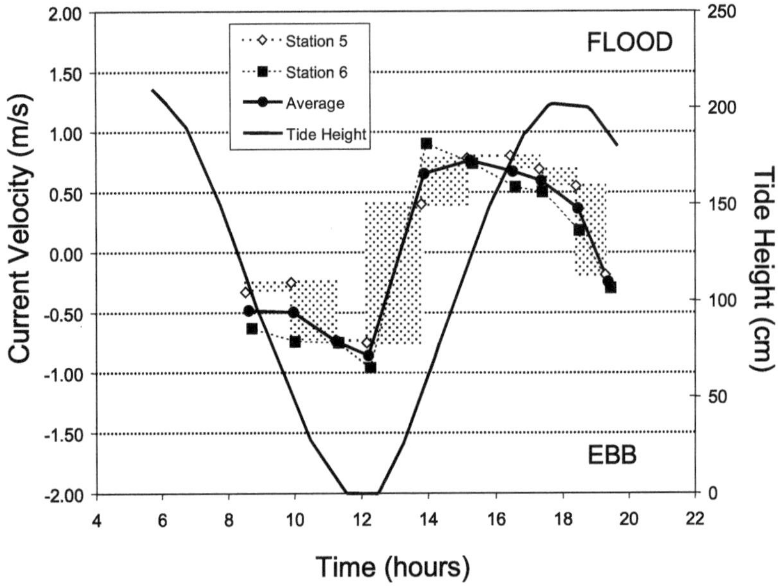

Figure 7. Near-bottom current velocities within Merrymeeting Bay (station 5) and at the lowest extent of the Androscoggin River (station 6) (see Figure 2 for locations). Also shown is the tide height during the survey period. Note the flood-dominance at Station 5 and slightly larger ebb-directed currents than flood at Station 6.

As expected, ebb-directed velocities generally decrease upstream from, for example, 0.67 m s^{-1} (Station 7) to 0.27 m s^{-1} (Station 10) at 10:45 (Fig. 8). However, flood-directed velocities do not increase downstream, primarily because of the stronger influence of channel geometry on flood flows than on ebb flows (e.g. dispersion of water from the Kennebec River into Merrymeeting Bay). The results from the Kennebec River portion of the study area provide evidence that, although the Kennebec River displays a general ebb-dominance, channel geometry strongly influences flow strength.

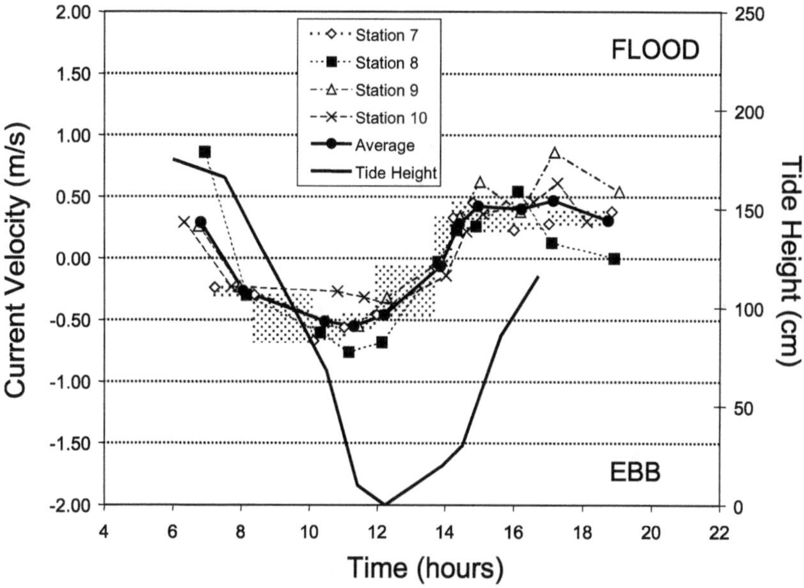

Figure 8. Near-bottom current velocities for the upper Kennebec River stations labeled 7-10 (see Figure 2 for locations). Also shown is the tide height during the survey period.

4.5 Salinity and Temperature

As expected, the salinity values for all stations were greater at flood tide than at ebb tide and were greater in the estuary seaward of Merrymeeting Bay than landward of the Bay (Fig. 9). At high tide, the values ranged from 1.93 ppt (Station 10) to 8.47 ppt (Station 4), whereas at low tide the values ranged from 0.10 ppt (Station 10) to 6.47 ppt (Station 1). At both high and low tide, the salinity values are an order of magnitude lower at the stations landward of Merrymeeting Bay than at the stations seaward of the Bay. These results suggest that a significant change in salinity occurs in Merrymeeting Bay and that the bay is near the upstream limit of saltwater intrusion. Finally, although the study area showed relatively low salinity values, the low vertical variability in salinity values from the bottom to the surface indicates that partially-mixed to mixed conditions existed during the hydrography.

The temperature readings at any given station and at any given time never varied by more than 1°C from bottom to top. The high temperatures ranged from 21°C in the lower estuary to 26°C in the upper estuary. The low temperatures ranged from 15°C in the lower estuary to 22°C in the upper estuary.

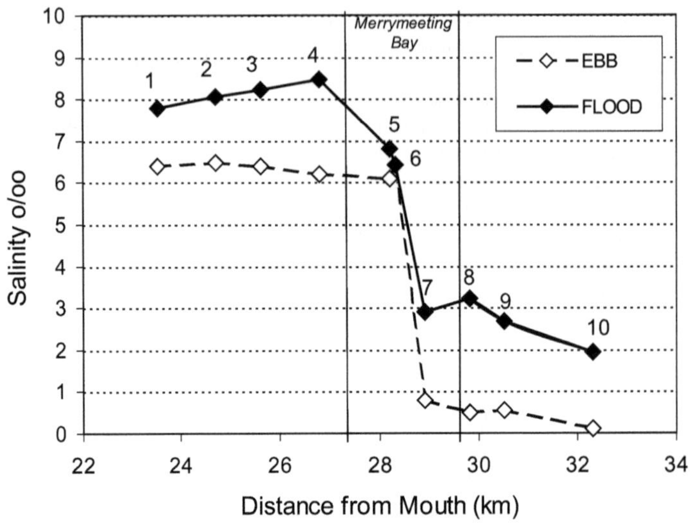

Figure 9. Longitudinal salinity values for ebb- and flood-tides at survey stations (numbered). Note that Merrymeeting Bay appears to be the upstream limit of saltwater incursion.

4.6 Freshwater Discharge

The extrapolated freshwater discharge value of 212 $m^3 s^{-1}$ versus the mean monthly value of 252 $m^3 s^{-1}$ in Merrymeeting Bay on the day of the hydrography confirmed that the survey took place during a low freshwater discharge period. In addition, the discharge on 26 July 2001 at the stream gage located nearest Merrymeeting Bay on the Androscoggin River (gage 590 = 117 $m^3 s^{-1}$) fell below the mean monthly value of any month (Fig. 10; Stewart *et al.*, 2001). Moreover, July possessed the second lowest average monthly discharge value compared to all the months (e.g. 143 $m^3 s^{-1}$ compared to 137 $m^3 s^{-1}$ for August; Fig. 10).

5. DISCUSSION

As demonstrated by Fenster and FitzGerald (1996) and Fenster *et al.* (2001) and supported by the results of this study, the Kennebec River estuary does not fit neatly into existing estuarine morphologic classification schemes (e.g. Pritchard, 1967; Dalrymple *et al.*, 1992; Cooper, 1993) because of its (1) net ebb-dominated bedload (coarse-grained) transport system; (2) highly variable, non-idealized morphology and geometry; (3) bedrock-cut valley that confines fluvial-tidal flow and provides pinning points for barrier development and growth; and (4) bay-head flood tidal delta. Moreover, the Kennebec River estuary does not display the characteristic "straight-meandering-straight" tidal-fluvial channel system and attendant bank-attached bars in its central, mixed-energy zone (Dalrymple *et al.*, 1992). Instead, large seaward-oriented linguoid transverse sand bars in the lower Androscoggin River; ebb-dominant currents within the Kennebec River, Androscoggin River and lower Kennebec River estuary; a flood-tide delta in Merrymeeting Bay; and the flood-dominated flow on the flood ramp of the flood-tide delta (Station 5) suggest that a different and possibly more complex sediment transport regime exists within the Kennebec River estuary than accounted for by existing models.

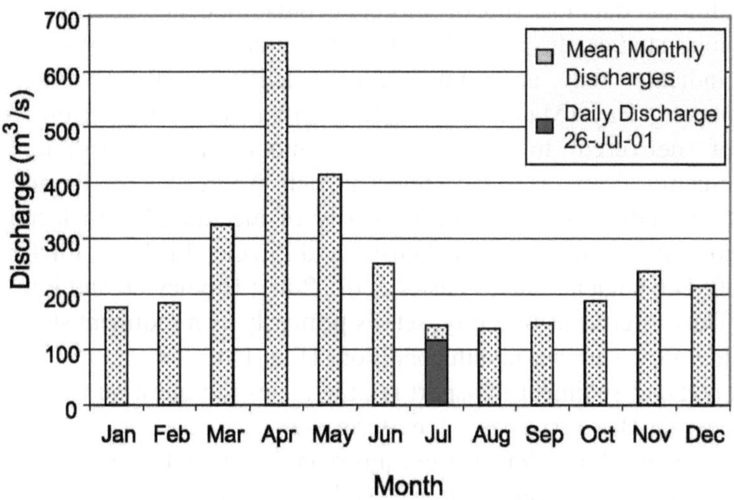

Figure 10. Mean discharge on the day of the hydrography (26 July 2001) compared to monthly averages at gage 590 on the Androscoggin River. Note the low riverine flow conditions present during the survey day.

Although a sediment choked basin of the Kennebec River estuary at Merrymeeting Bay does conform with the low, mixed energy region of net sediment accumulation in the estuarine end members outlined by Dalrymple *et al.* (1992), the spatial and temporal mechanisms that lead to infilling within Merrymeeting Bay differ considerably. Moreover, the variable energy distribution within Merrymeeting Bay produces substantially different sediment delivery and transport regimes, morphological elements, and potentially different facies.

In particular, bedload sediment from upstream sources moves into the basin via riverine flow reinforced by ebb-tidal currents (Fig. 11). During the flood-tidal stage, this sediment is intercepted and recirculated by flood-tidal currents and deposited onto and along the periphery of a large flood-tidal delta (surface area ≈ 1 km^2). Consequently, this delta serves as a reservoir for sediment during low riverine flow conditions. During the ebb-tidal cycle (and low discharge periods), a minor amount of the bedload material coming from the rivers moves seaward around the flood-tide delta and through the Chops into the lower estuary. However, some of this river-borne sand remains in the Bay's river entrances. The Androscoggin River bedload sediment enters from the southwest into a shallow, low-gradient basin choked with large ebb-oriented transverse bars that are exposed at low tide (Figs. 3, 11). From the opposite direction, Kennebec River bedload material is transported to the Bay from the northeast and moves through the Chops via a 275 m wide ×9 m deep channel connection to the lower estuary (Fig. 11). Additional data that support this conceptual model come from a number of previous studies. In particular, the texture and mineralogy of estuarine sands indicate that upstream sources deliver bedload sediment to Merrymeeting Bay (Malone, 1997; Kniskern *et al.*, 1998). However, the sediment delivered to the Kennebec mouth and nearshore comes predominantly from the quartz-feldspar rich Androscoggin River. Lesser amounts of sediment, consisting primarily of metamorphic rock-fragments come from the Kennebec River (Fenster and FitzGerald, 1996; FitzGerald *et al.*, 2000; Buynevich, 2001; Fenster *et al.*, 2001; Buynevich and FitzGerald, 2003). Thus, Merrymeeting Bay, serves primarily as a sediment storage area during low flow summer month conditions (Fig. 11).

This bedload sediment transport regime raises the question as to whether Merrymeeting Bay serves as a long-term *net*
estuary or a sink for sediment from upstream? Although Merrymeeting Bay does serve as a temporary sediment sink during low freshwater discharge summer months, previous studies conducted along the entire reach of the estuary seaward of the Bay have shown that spring freshets can flush coarse-grained sediment from the lower estuary ("outer" estuary of Dalrymple *et al.*, 1992) to the nearshore. Thus, a major role of Merrymeeting Bay may be to serve as a sediment source to the lower estuary during high discharge conditions. The net seaward movement of sediment derived from the estuary contrasts with the fundamental requirement of Dalrymple *et al.* (1992) to differentiate between a delta and an estuary; namely, the net transport direction of bed material. In particular, Dalrymple *et al.* (1992) use the "net landward movement of sediment derived from outside the estuary mouth (averaged over a period of several years) [as] one of the primary features that distinguishes estuaries from delta distributaries where the net sediment transport is seaward." Moreover, because Merrymeeting Bay is located well seaward of the limit of tidal influence, the facies boundary between marine

(tidally-influenced) and fluvial sediments within such paraglacial, rock-bound estuaries may occur well seaward of the tidal limit that is characteristic of wide-mouth, funnel-shaped embayments as proposed by Pritchard (1967).

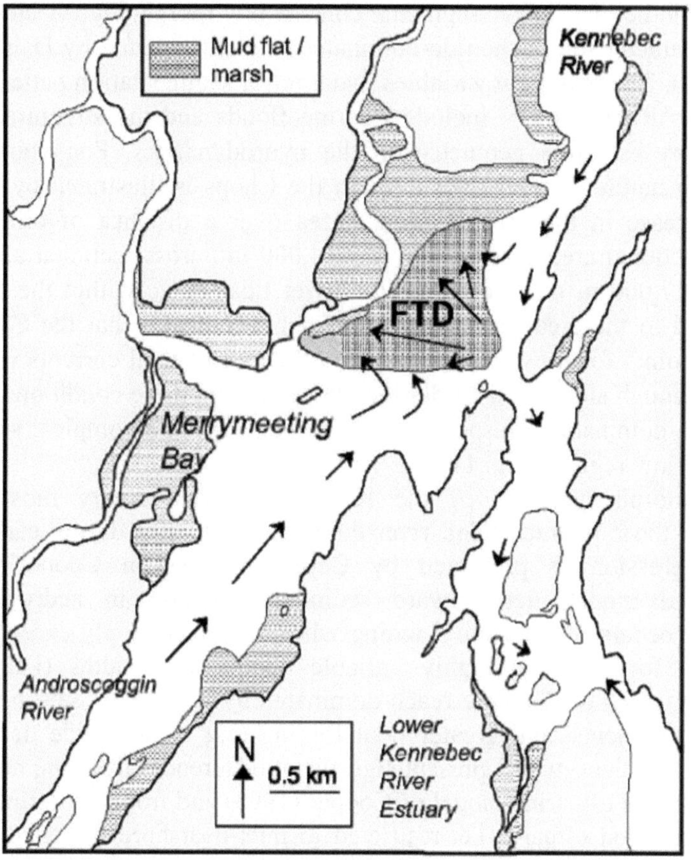

Figure 11. Map of Merrymeeting Bay and vicinity showing extent of flood-tide delta (stippled region labeled FTD) and inferred sediment transport paths based on hydrographic and bedform orientation data (arrows).

This study illustrates how an estuary can both resemble and differ from the wave-dominated and tide-dominated end-member estuaries according to the classification proposed by Dalrymple *et al.* (1992). First, the large flood-tide delta in Merrymeeting Bay resembles most closely that of the bay-head delta found in the river-dominated zone of the *wave*-dominated estuary.

However, the bay-head delta in the Kennebec River is *flood*-oriented and not ebb-oriented. Second, the "central basin" of the Kennebec River is dominated by medium- to coarse-grained sand as opposed to the fine-grained organic muds accumulating in the prodelta region of the bay-head delta. Third, although the Kennebec River estuary is a tide-dominated estuary, its upper, middle, and lower segments contain few morphological elements or facies characteristic of the tide-dominated model presented by Dalrymple *et al*. (1992). The two main variables that control sedimentation patterns in the Kennebec River estuary include riverine floods and the structural control exerted by estuarine geometry on the hydrodynamics. For example, the confining nature of tidal flow through the Chops is illustrated by the two-fold decrease in the cross-sectional area over a distance of only 900 m (cross-sectional area near Station 4 $\approx$ 14,000 m^2; cross-sectional area at the Chops $\approx$ 7,000 m^2). In addition, the lower tidal range within the basin (as compared to the area seaward of the Chops) indicates that the Chops is a "choke point" for flow and lessens the influence of tidal currents within the Bay. Although situated in a tide-dominated setting, these conditions result in a net ebb-dominated transport system situated within a complex, seasonally variable flow regime (Fig. 11).

The morphodynamics of the Kennebec River estuary most closely resemble those of microtidal, river-dominated estuaries during early stages of transgression as presented by Cooper (1993). In Cooper's (1993) conceptual model, net seaward sediment transport in bedrock valley estuaries occurs by episodic flooding when sediment supply exceeds valley volume. However, the highly variable depths and widths (i.e. channel geometry), lack of a middle reach dominated by suspension settling of fine-grained sediments, and greater tidal forcing (e.g. a flood-tide delta at the tidal limit of the estuary) present significant differences from the microtidal, funnel-shaped estuarine model of Cooper (1993) and from the definition of a river-dominated estuary (i.e. restricted to inlet-nearshore morphodynamics) given by Cooper (2002). Instead of low water volumes in river-dominated estuaries minimizing tidal influences, large floods in the Kennebec River estuary that are reinforced by ebb-tidal currents episodically supplant the tidal prism to produce a net ebb-dominated (river-dominated) estuary. In fact, Cooper (1991) points out that narrow bedrock valleys with an abundant sediment supply can remain fluvially dominated throughout its evolution.

The results from this study also point out the necessity to capture data over a variety of temporal scales. The weak flood dominance in the upper part of the Kennebec River (within the study area) and weak ebb dominance in the Androscoggin River suggest that downstream sand transport may not occur during low-flow summer months. Instead, downstream transport must occur over longer time scales, during high-discharge flood events, or both.

Finally, although new estuarine models have aided our understanding of these systems in general, refinement of the models should be made to consider controls other than those based on the relative importance of marine and fluvial processes. In particular, regulated flows through bedrock-constricted basins may influence facies distributions, flow patterns, and net sediment transport trends (e.g. Cooper, 1993). Further hydrographic and detailed geophysical survey work during spring freshets will test the hypothesis that, although Merrymeeting Bay acts as a depocenter during low flow summer months, coarse-grained sediment can be flushed from the Bay during spring freshets to contribute bedload sediment to the lower estuary. In this manner, the sediment convergence zone in Merrymeeting Bay acts only as a temporary sink.

6. CONCLUSIONS

This study on sediment and flow dynamics in Merrymeeting Bay on the Kennebec River estuary shows that, in contrast to current tide-dominated estuarine models (e.g. "straight-meandering-straight"), the lower Kennebec River estuary ultimately receives bedload sediment and freshwater discharge from two relatively large rivers that enter a wide basin from opposite directions. Estuarine geometry and structural control via bedrock-confined channels can offset predicted sedimentation patterns by amplifying flow through bedrock constrictions and diminishing flow in wide embayments. Even during low river discharge conditions, ebb-directed flows dominate the lower reaches of the Kennebec River and Androscoggin River as a function of channel and estuarine geometry and riverine flow reinforced by tidal currents. As would be expected, the flood ramp of the flood-tidal delta in Merrymeeting Bay displays greater flood- than ebb-directed flows.

The ubiquitous and large ebb-oriented transverse bars at the confluence of the Androscoggin River with the Kennebec River at Merrymeeting Bay, the flood-oriented bedforms and flood-tidal delta within Merrymeeting Bay, the structural fabric of the region, and flow measurements indicate that during the flood-tidal stage, landward-directed currents intercept and transport downstream moving bedload sediment along the periphery of the flood-tide delta and/or deposit the sediment onto the flood-tide delta located within Merrymeeting Bay (Fig. 11). Additionally, during low discharge periods, most river-borne bedload sediment remains in the entrances to Merrymeeting Bay. However, during the ebb-tidal cycle a minor amount of the bedload material coming from the rivers moves seaward around the flood-tide delta and through the Chops into the lower estuary.

During low river flow, the Kennebec River estuary maintains a large sediment storage reservoir independently of the degree of tidal energy. The presence of this sediment sink approximately 27 km from the estuary's mouth corroborates existing sediment transport and estuarine models that predicts net bedload convergence, but is not found in the lowest-energy meandering zone (i.e. Pritchard, 1967; Dalrymple *et al.*, 1992). Future work will test the hypothesis that the depocenter that resides in Merrymeeting Bay during low flow conditions becomes a sediment source to the lower estuary during high flow conditions.

Estuarine models should incorporate additional parameters that control flow and net sediment transport patterns in rock-bound, high-latitude estuaries in addition to the relative importance of marine and fluvial processes. These parameters include variable riverine and tidal flow conditions, bedrock confining of flow, and gross and net sediment transport patterns.

7. ACKNOWLEDGMENTS

The Virginia Foundation of Independent Colleges provided funding for this project. We thank Joseph Nielson and Gregory Stewart of the Maine Geological Survey for providing provisional discharge data and Dr. Helene Burningham, Dr. Andrew Cooper, and Dr. Jasper Knight for their excellent reviews of the original manuscript.

REFERENCES

Barnhardt, W.A., Belknap, D.F. and Kelley, J.T. 1997. Stratigraphic evolution of the inner continental shelf in response to Late Quaternary relative sea-level rise, northwestern Gulf of Maine. *Geological Society of America Bulletin*, 109, 612-630.

Belknap, D.F., Kelley, J.T. and Shipp, R.C. 1987. Quaternary stratigraphy of representative Maine estuaries. In: FitzGerald, D.M. and Rosen, P.S. (eds) *Glaciated coasts.* Academic Press, San Diego, 143-176.

Belknap, D.F., Kraft, J.C. and Dunn, R. 1994. Transgressive valley fill lithosomes: Delaware and Maine. In: Boyd, R., Zaitlin, B.A. and Dalrymple, R. (eds) *Incised Valley Fill Systems: Origin and Sedimentary Sequences.* SEPM Special Publication 51, 341-354.

Biggs, R.B. 1970. Sources and distribution of suspended sediment in northern Chesapeake Bay. *Marine Geology*, 9, 187-201.

Biggs, R.B. and Howell, B.A. 1984. The estuary as a sediment trap; alternate approaches to estimating its filtering efficiency. In: Kennedy, V.S. (ed) *The estuary as a filter.* Academic Press, New York, 107-129.

Borns, H.W., Jr. and Hagar, D.J. 1965. Late glacial stratigraphy of a northern part of Kennebec River valley, western Maine. *Geological Society of America Bulletin*, 76, 1233-1250.

Bryce. S. Larcombe, P. and Ridd, P.V. 1998. The relative importance of landward-directed tidal sediment transport versus freshwater flood events in the Normanby River estuary, Cape York Peninsula, Australia. *Marine Geology*, 149, 55-78.

Buynevich, I.V. 2001. *Fluvial-marine interaction and Holocene evolution of sandy barriers along an indented paraglacial coastline.* Unpublished Ph.D. dissertation, Boston University, Boston, Massachusetts.

Buynevich, I.V. and FitzGerald, D.M. 2003. Textural and compositional characterization of recent sediments along a paraglacial estuarine coastline, Maine, USA. *Estuarine, Coastal and Shelf Science*, 56, 139-153.

Cooper, J.A.G. 1991. *Sedimentary models and geomorphological classification of river-mouths on a subtropical, wave-dominated coast, Natal, South Africa.* Unpublished PhD thesis, University of Natal, Durban.

Cooper, J.A.G. 1993. Sedimentation in a river dominated estuary. *Sedimentology*, 40, 979-1017.

Cooper, J.A.G. 2002. The role of extreme floods in estuary-coastal behaviour: contrasts between river- and tide-dominated microtidal estuaries. *Sedimentary Geology*, 150, 123-137.

Dalrymple, R.W., Knight, R.J., Zaitlin, B.A. and Middleton, G.V. 1990. Dynamics and facies model of a macrotidal sand bar complex, Cobequid Bay-Salmon River estuary (Bay of Fundy), Canada. In: Smith, D.G., Reinson, G.E., Zaitlin, B.A. and Rahmani, R.A. (eds) *Clastic Tidal Sedimentology.* Canadian Society for Petroleum Geology, Memoir 16, 137-160.

Dalrymple, R.W., Zaitlin, B.A. and Boyd, R. 1992. Estuarine facies models: conceptual basis and stratigraphic implication. *Journal of Sedimentary Petrology*, 62, 1130-1146.

Fenster, M.S. 1995. *The Origin and Evolution of the Sand Sheet Facies: Eastern Long Island Sound.* Unpublished Ph.D. dissertation, Boston University, Boston, Massachusetts.

Fenster, M.S. and FitzGerald, D.M. 1996. Morphodymanics, stratigraphy, and sediment transport patterns in the Kennebec River estuary, Maine, U.S.A. *Sedimentary Geology*, 107, 99-120.

Fenster, M.S., FitzGerald, D.M., Kelley, J.T., Belknap, D.F., Buynevich, I. and Dickson, S.M. 2001. Net ebb sediment transport in a rock-bound, mesotidal estuary during spring-freshet conditions: Kennebec River estuary, Maine, U.S.A. *GSA Bulletin*, 113, 1522-1531.

FitzGerald, D.M., Buynevich, I., Fenster, M.S. and McKinlay, P. A. 2000. Sand dynamics at the mouth of a rock-bound, tide-dominated estuary. *Sedimentary Geology*, 131, 25-49.

FitzGerald, D.M., Buynevich, I.V., Fenster, M.S., Kelley, J.T. and Belknap, D.F., 2004. Contribution of coarse-grained sediment to the nearshore and inner shelf by large estuaries: New England, USA. In: FitzGerald, D.M. and Knight, J. (eds) *High Resolution Morphodynamics and Sedimentary Evolution of Estuaries*, Kluwer Academic Publishers, this volume.

Fontaine, R.A. 1987. *Flood of April 1987 in Maine, Massachusetts, and New Hampshire*. U.S. Geological Survey Open-File Report 87-460, 33pp.

Hannum, M.B. 1996. *Late Quaternary evolution of the Kennebec and Damariscotta River estuaries, Maine*. Unpublished M.S. Thesis, University of Maine, Orono, ME.

Harris, P.T. 1988. Large-scale bedforms as indicators of mutually evasive sand transport and the sequential infilling of wide-mouthed estuaries. *Sedimentary Geology*, 57, 273-298.

Horne, G.S. and Patton, P.C. 1989. Bedload sediment transport through the Connecticut River estuary. *Geological Society of America Bulletin*, 101, 805-819.

Kelley, J.T. 1987. Sedimentary environments along Maine's estuarine coastline. In: FitzGerald, D.M. and Rosen, P.S. (eds) *Glaciated coasts.* Academic Press, San Diego, 151-176.

Kelley, J.T., Belknap, D.F. and Shipp, R.C. 1987. *Geomorphology and sedimentary framework of the inner continental shelf of south central Maine.* Maine Geological Survey, Open-File Report 87-19, 76pp.

Knebel, H.J. 1986. Holocene depositional history of a large glaciated estuary, Penobscot Bay, Maine. *Marine Geology*, 73, 215-236.

Knebel, H.J. 1989. Modern sedimentary environments in a large tidal estuary, Delaware Bay. *Marine Geology*, 86, 119-136.

Kniskern, T.A., Buynevich, I.V., FitzGerald, D.M. and Peters, J.L. 1998. Sedimentological characteristics of fluvial-estuarine deposits in the Merrymeeting Bay, Maine: Implications for sediment transport. *Geological Society of America Northeastern Section Abstracts with Programs*, 30, 30.

Ludwick, J.C. 1970. Sand waves and tidal channels in the entrance to Chesapeake Bay. *Virginia Journal of Science*, 21, 178-184.

Ludwick, J.C. 1972. Migration of tidal sand waves in Chesapeake Bay entrance. In: Swift, D.J.P., Duane, D.B. and Pilkey, O.H. (eds) *Shelf Sediment Transport: Process and Pattern.* Dowden, Hutchinson, and Ross, Inc., Stroudsburg, PA, 377-410.

Malone, J.T. 1997. *Determining provenance of river sediment: Kennebec and Androscoggin Rivers, Maine.* Unpublished M.S. thesis, University of Maine, Orono, Maine.

Meade, R.H. 1969. Landward transport of bottom sediments estuaries of the Atlantic coastal plain. *Journal of Sedimentary Petrology*, 39, 222-234.

Meade, R.H. 1972. Transport and deposition of sediments in estuaries. In: Nelson, B.W. (ed) *Environmental framework of coastal plain estuaries.* Geological Society of America Memoir 133, 91-120.

Meade, R.H. 1982. Source, sinks, and storage of river sediment in the Atlantic drainage of the United States. *Journal of Geology*, 90, 235-252.

Milliman, J.D. and Meade, R.H. 1983. World-wide delivery of river sediment to the oceans. *Journal of Geology*, 91, 1-21.

National Oceanic and Atmospheric Administration (NOAA), National Ocean Service, Center for Operational Oceanographic Products and Services (CO-OPS), n.d. Available from http://www.co-ops.nos.noaa.gov/tide_pred.html, retrieved 26 July 2001.

Nichols, M.M., Johnson, G.H. and Peebles, P.C. 1991. Modern sediments and facies model for a microtidal coastal plain estuary, the James River estuary, Virginia. *Journal of Sedimentary Petrology*, 61, 883-899.

Postma, H. 1967. Sediment transport and sedimentation in the marine environment. In: Lauff, G.H. (ed) *Estuaries.* AAAS Publication 83, Washington, DC, 158-179.

Pritchard, D.W. 1967. What is an estuary? Physical viewpoint. In: Lauff, G.H. (ed) *Estuaries.* AAAS Publication 83, Washington, DC, 3-5.

Schubel, J.R. 1984. Estuarine circulation and sedimentation: an overview. In: Hag, B.U. and Milliman, J.D. (eds) *Oceanography of Arabian Sea and Coastal Pakistan.* Van Nostrand Reinhold Co. Inc., New York, 114-136.

Sherwood, C.R., Creager, J.S., Roy, E.H., Gelfenbaum, G. and Dempsey, T. 1984. *Sedimentary Processes and Environments in the Columbia River Estuary.* Final Report on the Sedimentation and Shoaling Work Unit of the Columbia River Estuary Data Development Program, University of Washington, 183pp.

Stewart, G.J., Nielsen, J.P., Caldwell, J.M. and Cloutier, A.R. 2001. *Water Resources Data Maine, Water Year 2001.* Water-Data Report ME-01-1, 232pp.

Stumpf, R.P. and Goldschmidt, P.M. 1992. Remote sensing of suspended sediment discharge into the western Gulf of Maine during the April 1987 100-year flood. *Journal of Coastal Research*, 8, 218-225.

Thompson, W.B. and Borns, H.W., Jr. 1985. *Surficial geologic map of Maine.* Maine Geological Survey, scale 1:500,000.

Chapter 10

COARSE-GRAINED SEDIMENT TRANSPORT IN NORTHERN NEW ENGLAND ESTUARIES: A SYNTHESIS

Duncan M. FitzGerald[1], Ilya V. Buynevich[2], Michael S. Fenster[3], Joseph T. Kelley[4], Daniel F. Belknap[4]
[1]*Department of Earth Sciences, Boston University, Boston, MA 02215, USA*

[2]*Geology & Geophysics Dept., Woods Hole Oceanographic Institution, Woods Hole, MA 02543, USA*

[3]*Department of Environmental Studies, Randolph-Macon College, Ashland, VA 23005, USA*

[4]*Department of Geological Sciences, University of Maine, Orono, ME 04469, USA*

1. INTRODUCTION

Although it is widely stated in the literature that estuarine river mouths are sediment sinks, northern New England estuaries are an exception to this model because they export coarse-grained sediment to the nearshore. The traditional view is that estuaries fill with sediment ranging from mud to gravel derived from fluvial or upland sources as well as from the inner continental shelf and adjacent shorelines. Fluvial sediments are deposited primarily in the inner and central portions of an estuary, although fluvial mud can be deposited in the outer estuary in some tide-dominated systems (e.g. sections of the Gironde River (France), Allen, 1991; Fly River (Papua New Guinea), Harris *et al.*, 1993). The deposition of sediment in the inner and central portions of an estuary is due to the combined influences of a downstream decrease in the riverine current strength and clay flocculation

D.M. FitzGerald and J. Knight (eds.),
High Resolution Morphodynamics and Sedimentary Evolution of Estuaries, 195-213.

produced by fresh and saltwater mixing. In estuaries having high sediment loads, fluidized mud can be an important component of estuarine sedimentation (Wells, 1983, 1995). Marine sediments enter the outer estuary due to residual, flood-oriented bottom currents and stronger flood than ebb tidal currents. This former flow pattern is caused by the seaward-flowing freshwater advecting the underlying saltwater producing a mass balance deficit of saltwater (Dyer, 1973). With respect to coarser-grained, bedload sediment, net landward transport results from deformation of the tidal wave as it propagates upstream. The shoaling and steepening of the tidal wave causes a shorter flood than ebb duration resulting in stronger flood than ebb-tidal currents. Because bedload sediment transport is proportional to the cube or higher power of velocity, a slightly stronger flood current results in net landward bedload transport.

During the Holocene transgression increasing accommodation space and an abundant supply of sediment led to the deposition of thick estuarine sedimentary sequences in many coastal plain and other passive margin settings, as described in a compendium of incised-valley fill studies (Boyd *et al.*, 1994). The valley-fill deposits in northern New England estuaries are up to 40 m in thickness and overlie glacial sediments or bedrock (Rhodes, 1973; Belknap *et al.*, 1986, 1987, 1994; Boothroyd and FitzGerald, 1989). These estuarine sediments range in texture from mud and muddy sands (e.g. Damariscotta River estuary; Belknap *et al.*, 1994) to coarse-grained sediments including sand and gravel (e.g. Kennebec River estuary; Belknap *et al*,. 1989; Fenster and FitzGerald, 1996).

The landward movement of sand into estuaries has been documented in river systems throughout the world including in the Ord River estuary, Australia (Wright *et al.*, 1973), Chesapeake Bay (Colman *et al.*, 1992) the Cobequid Bay-Salmon River estuary, Canada (Dalrymple *et al.*, 1990), and the Gironde Estuary, France (Allen, 1991) among others. The influx of marine sediment into estuaries in the form of washovers and flood-tidal deltas at wave-dominated settings and by bedload transport in tidal channels at tide-dominated settings is a major tenet of the established estuarine models of Dalrymple *et al.* (1992) and Boyd *et al.* (1992). Despite the general acceptance of this paradigm for estuary sedimentation (Harris *et al.*, 2002), recent studies of estuaries in northern New England suggest a different model; one in which bedload transport is controlled by episodic high discharge events in a high-energy tidal regime (Manthorp, 1995; Fenster and FitzGerald, 1996; Hannum, 1996; Fenster *et al.*, 2001). Similar results have been observed during extreme floods in microtidal estuaries (Cooper, 2002). The major objectives of this paper are to summarize various datasets for northern New England estuaries and to demonstrate that these systems regularly export bedload to the nearshore and contribute sand to the maintenance of coastal barriers. This paper focuses on the Merrimack River in northern Massachusetts, Saco River in southern Maine, USA, and the Kennebec River in central Maine (Fig. 1).

2. PHYSICAL SETTING

One of the major attributes that distinguishes estuaries in northern New England from those in coastal plain settings, such as on the East and Gulf Coasts of the United States, is the control that bedrock exerts on the location and geometry of the river as well as the distribution of sand bodies at the estuary mouth. The course of the Merrimack, Saco, and Kennebec Rivers is largely a function of the general north-south structural grain imprinted on the region during Paleozoic tectonism (Osberg *et al.*, 1985). The lower Kennebec River, for example, is situated in a bedrock-cut valley that formed due to the preferential weathering of weak rocks within a north-southward trending, isoclinally-folded metasedimentary belt. Like the Saco and Merrimack Rivers to the south, major offsets in the course of the Kennebec River are related to local fault patterns. Fault control is particularly well revealed in the sharp-angle bends of the lower Saco River (Fig. 1). Bedrock also influences sedimentation patterns at the mouth of northern New England estuaries (Fenster *et al.*, 2001). It has been argued by FitzGerald *et al.* (2000) that the strength of the tidal currents at the tide-dominated entrance to the Kennebec River would be unlikely to permit the existence of sandy barriers were it not for the presence of large bedrock outcrops, which have served as anchoring sites for sand accumulation.

Although there are minor differences in the geologic setting of the river systems discussed in this paper, such as the size of their drainage areas, freshwater discharge, and extent of human alterations (Table I), estuarine dynamics and sediment transport in the rivers are quite similar. Because all the rivers experience approximately the same climate and amount of precipitation, their freshwater discharge is closely related to the size of their drainage areas. The Saco River has the smallest catchment basin, and thus its mean and peak discharges are only about one third as large as that of the Kennebec and Merrimack Rivers (Table I). The three estuaries have approximately the same spring tidal range (2.9-3.1 m) and mean shallow-water wave height (0.4 m), however their tidal prisms vary by an order of magnitude due to differences in the extent of tidal flooding. The narrow channel of the Saco River and truncation of the tidal wave by a dam located 8 km upstream of its mouth yield a relatively small tidal prism of 8×10^6 m^3. Conversely, the Kennebec River has a comparatively wide channel, expansive tidal flats, marshes, bays, and other open-water environments, which combine to produce the largest tidal prism of the three (101 x 10^6 m^3). The Merrimack River with a physiography transitional between the other two has a tidal prism of 30 x 10^6 m^3.

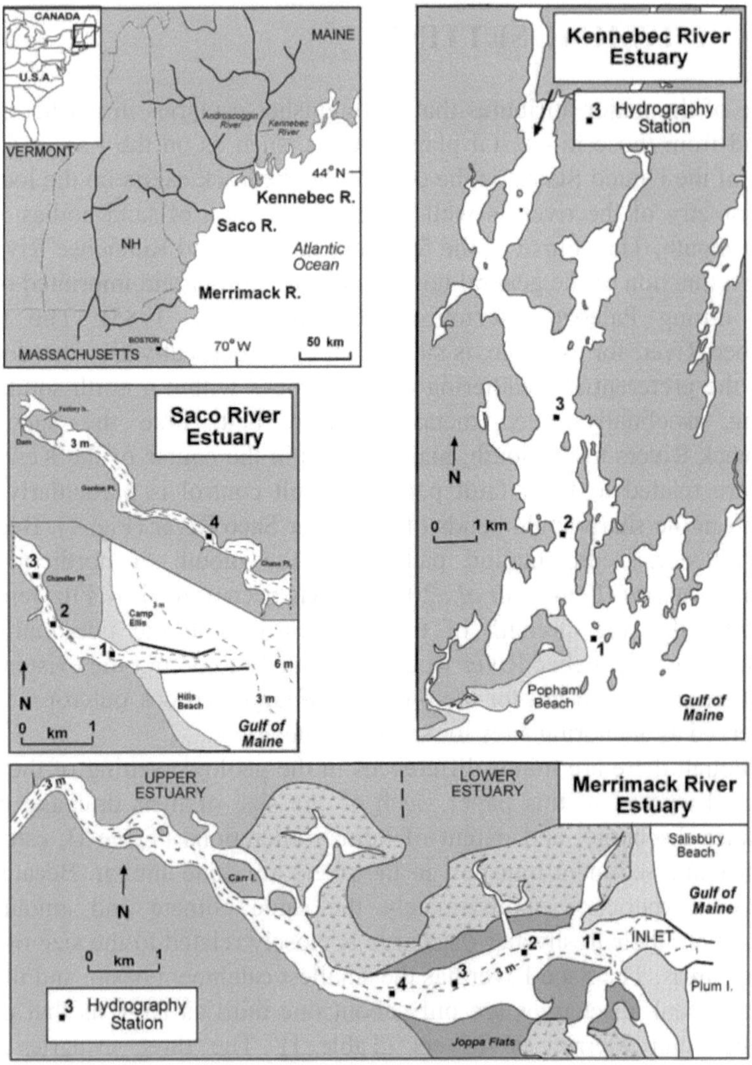

Figure. 1. Location of the Kennebec, Saco, and Merrimack Rivers and locations of current stations. Velocity stations in Table II are shown on the map.

During spring tides, average freshwater discharge of the rivers is small in comparison to their saltwater tidal prism (Kennebec 6%, Saco 14%, Merrimack 15%). The tidal dominance of the lower estuary during prevailing flow conditions is indicated by the fact that the channel cross-sectional area at the river mouths is governed by its tidal prism according to O'Brien's (1969) relationship. During moderate to high freshwater discharge conditions, the Merrimack and Saco River estuaries are vertically well-mixed (Farrell, 1970; Hartwell, 1970), becoming partially mixed at low discharge. Estuarine conditions are different in the lower Kennebec River due to its high saltwater tidal prism compared to its freshwater discharge. Strong tidal currents in the lower river, combined with large-scale turbulence imparted to the water column by bedrock constrictions of the channel, result in a partially to vertically mixed estuary. During high flood stage in the rivers, the freshwater may completely supplant the salt water in the estuary (Manthorp, 1995; Fenster *et al.*, 2001).

3. COASTAL SEDIMENTARY ACCUMULATION FORMS

Northern New England has experienced several episodes of Pleistocene glaciation that effectively stripped away most of the sediment cover in the coastal region producing an irregular rocky shoreline. As a result, sediment accumulation along the coast is compartmentalized. The barrier systems that do exist are proximate to major river systems and occur within broad embayments or are adjacent to peninsulas (Fig. 2, Table I). The barrier chain in the Wells Embayment may be an exception to this general trend, but these barriers are narrow (width <100 m), relatively thin (thickness <4-5 m), and contain an order of magnitude less sediment than the barriers associated with the major rivers (Montello, 1993; Mills, 1997, Kelley *et al.*, 2003). However, even the Wells barriers may have been partially supplied with sediment from inland deltaic deposits via small rivers (Montello, 1993; Tary *et al.*, 2002).

It has been suggested that the barriers in northern New England may have been directly supplied with sediment from the rivers during their evolution (FitzGerald and van Heteren, 1999; Buynevich and FitzGerald, this volume). The long-term bedload sediment contribution of many northern New England rivers to the coast during the Late Pleistocene and throughout Holocene is evidenced by the presence of lowstand deltas and sandy parasequences on the inner continental shelf off many of the region's major rivers (Oldale *et al.*, 1983; Belknap *et al.*, 1986; Kelley *et al.*, 1989; Belknap and Shipp, 1991; Barnhardt *et al.*, 1997; Belknap *et al.*, 2002, Kelley *et al.*, 2003). These offshore sand sources may have been partially reworked onshore during the Holocene transgression supplying additional sediment during the barrier progradational phase (Belknap *et al.*, 1989, Barnhardt *et al.*, 1997, Kelley *et al.*, 2003). The Piscataqua River estuary, located on the

New Hampshire-Maine border, is a major exception to this trend and has delivered little coarse sediment to the inner shelf since deglaciation. Most of the coarse-grained sediment load of the Piscataqua River has been deposited in the quiet waters of Great Bay Estuary before reaching the coast (Ward, 1992).

Figure. 2. The distribution of sandy glacial deposits in the drainage basin of

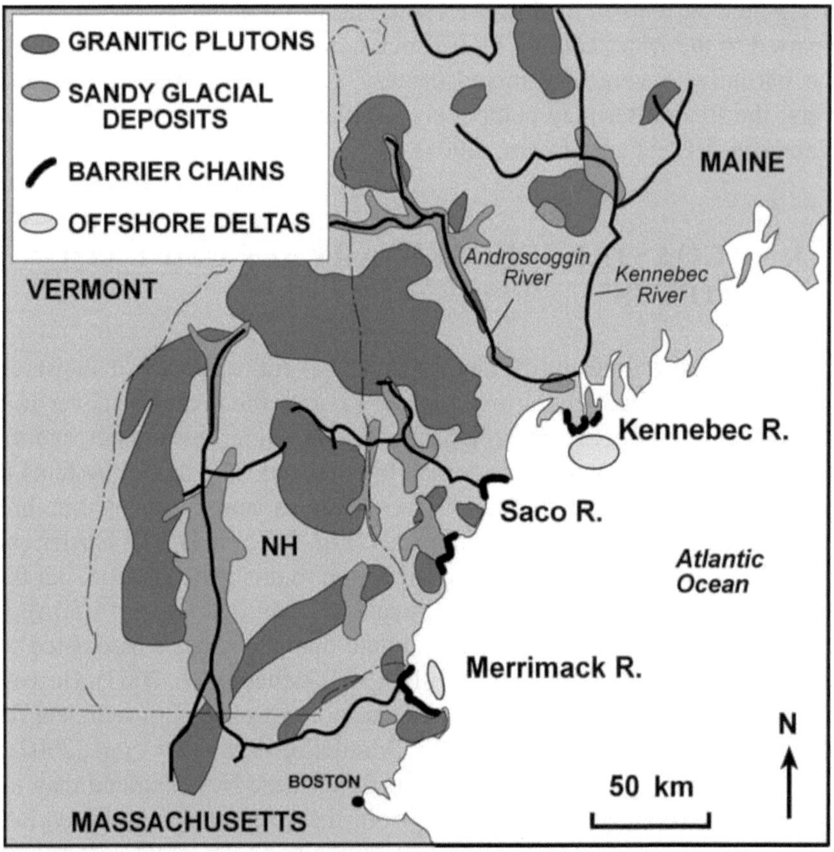

the Kennebec, Saco, and Merrimack Rivers as well as the presence of associated barriers and offshore deltas is related to the size and extent of exposed granitic plutons within the drainage system.

The close correspondence between the major rivers and the presence of nearby sandy barriers onshore and large sand deposits immediately offshore (Table I) is related to voluminous glacio-fluvial and glacio-marine deposits within the river's drainage basin (Fig. 2). Following deglaciation the rivers and their tributaries incised highstand and regressive deltas, outwash, eskers, and submarine fans delivering large quantities of sediment to the coast. We will make a case that this process continues albeit at a lower rate to the present time. The ultimate source of most of the sand is the granitic plutons. Hanson and Caldwell (1989) have presented a convincing argument that physical and chemical weathering of granitic bedrock and glacial excavation of saprolite were responsible for the formation of many sand-rich proglacial deposits. It should be noted that despite volumetrically large quantities of sand in certain areas of coastal New England (e.g. Pineo Ridge, eastern Maine; Ashley *et al.*, 1991) there is an absence of barriers in these regions due to the lack of rivers to deliver sand to the coast (Fig. 2).

Table I. Physiography, hydraulics and sedimentology of northern New England estuaries discussed in the text.

	Kennebec	Saco	Merrimack
1. Physiography			
Geologic setting	Peninsula, deep embayment	Bedrock valley	Drowned river valley, upper bedrock valley
A_d (km^2)	14,775	4403	12,976
$Q_{w\ mean}$ (m^3s^{-1})	278	77	203
Spring freshet peak Q_w (date)	6347 (4/87)	1320 (3/36)	4902 (3/36)
Spring tidal range (m)	3.1	3.0	3.0
Shallow water wave height (m)	0.4	0.4	0.4
Tidal prism (m^3)	101×10^6	8.1×10^6	30×10^6
2. Sedimentology			
Bedload	Medium sand to granules	Medium sand to pebbles	Medium to coarse sand
Bedforms	Megaripples, sand waves, transverse bars	Megaripples, sand waves	Megaripples, sand waves
Sediment source	Eskers, outwash plains	Eskers, outwash plains	Eskers, outwash plains
Offshore sediment accumulation forms	Paleo-delta lobes at -20-30 m, -30-40 m, -50-60 m (volume $2.1 \times 10^9\ m^3$)	Scattered	Paleo-delta at –50 m (volume $1.3 \times 10^9\ m^3$)
3. Associated barrier system			
Barrier chain	Reid State Park, Popham barriers	Saco Bay barriers	Merrimack embayment barriers
Thickness (m)	10.5	11.3	20.5
Length (km)	11	10	21
Volume (m^3)	24×10^6	22×10^6	115×10^6
4. Estuary type	Partially to vertically mixed	Partially to vertically mixed	Partially to vertically mixed
5. Alterations	Dams, dredging	Dams, jetties, dredging	Dams, jetties, dredging
6. References	Fenster and FitzGerald, 1996; Fenster *et al.*, 2001; Belknap *et al.*, 2002; Kelley *et al.*, 2003	Farrell, 1970; Manthorp, 1995; Kelley *et al.*, 2003	Hartwell, 1970; Oldale *et al.*, 1983

4. ESTUARINE SEDIMENT TRANSPORT

A comprehensive dataset was collected in the three fluvial-estuarine systems at different times during a period of approximately 10 years (Manthorp, 1995; Fenster and FitzGerald, 1996, Fenster *et al.*, 2001; FitzGerald *et al.*, 2002). The data include river stage recordings, current magnitudes and directions, bedform geometry and orientations, grain-size and mineralogical data of bottom sediments, and other estuarine measurements (Tables I, II). Additional information was obtained from several Masters and PhD theses focusing on the hydraulic and sedimentologic character of the estuaries, including Hartwell (1970), Anan (1971), Farrell (1970), Hubbard (1975), Barber (1995), Manthorp (1995), and Hannum (1996).

4.1 Sedimentological Trends

A summary of textural data for the channel bottom sediments in the three estuaries is presented in Figure 3. Each data point on the graph represents the mean grain size of one to three sampling sites across the channel thalweg. The sediments comprising the estuarine channels are compositionally submature and contain significant proportions of feldspar grains and/or rock fragments (>30%). Mica flakes are common throughout the estuaries, reflecting the granitic bedrock source. Most of the sediment grains are angular to sub-rounded, indicating the lack of abrasion and reworking.

Fine sand and mud are found within the estuaries in low energy areas such as those forming expansive tidal flats and marsh regions (e.g. Joppa Flats, Merrimack River; Atkins Bay, Kennebec River; Fig. 1). The lower portion of each of the estuaries is floored by coarse-grained sand. An exception to this trend is the medium sand found in a 2-km reach upstream from the mouth of the Kennebec River, where the bedrock-framed channel significantly widens causing a reduction in tidal current velocities (Fenster *et al.*, 2001). In contrast, bottom sediments coarsen (to very coarse sand and granules) at the entrance to the Merrimack River estuary where the channel is narrowed by jetties. Tidal current velocities in this region exceed 2.0 ms^{-1} (Hubbard, 1975). Very coarse sediment (medium sand to granules and small pebbles) is also found in the Merrimack River upstream from a series of dams that are located 35 km from the river mouth, suggesting that the coarse sediment flooring the estuary is sourced from the upstream drainage system. Similarly, coarse sediment has been reported behind dams in the Kennebec River (Dudley *et al.*, 1999).

Figure. 3. Grain size trends in the Kennebec, Saco, and Merrimack River

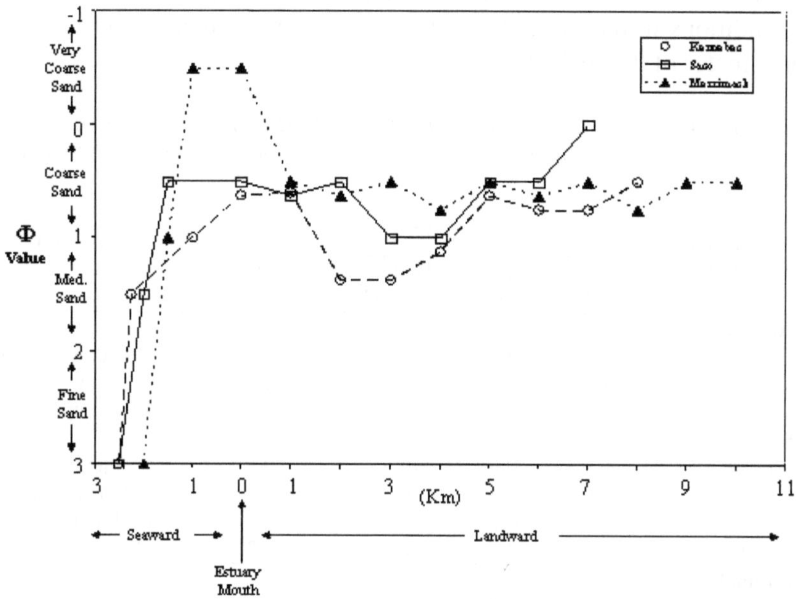

estuaries. Note the fining grain size trend seaward of the mouth of the estuaries.

The overall coarse-grained nature of the channel bottom changes drastically seaward of the estuary mouths. The coarse sand grades into fine sand in the distal portion of the ebb-tidal deltas, reflecting the expansion of the ebb jet and concomitant decrease in competency of the tidal currents. Within this overall seaward fining trend a pathway of coarse sediment can be traced from the inlet throat at both the Kennebec and Merrimack Rivers through the main channel to a subtidal bar that parallels the downdrift beach. In both estuaries the textural character of the bar sands are identical to those of the entrance channel (FitzGerald *et al.*, 2000, 2002; Buynevich and FitzGerald, 2003).

4.2 Summary of Estuarine Hydraulics

A summary of the hydrologic parameters for each of the estuaries is given in Tables I and II. The mean freshwater discharge among the estuaries varies by an order of magnitude and the saltwater tidal prism by two orders of magnitude. Discharge and tidal prism are greatest in the Kennebec River estuary and smallest in the Saco River estuary. Examples of current velocities in the lower 5 km of the estuaries are presented in Table II (Hartwell, 1970; Manthorp, 1995; Fenster *et al.*, 2001). Maximum current velocities represent depth-integrated values that were measured during mean tidal conditions in the Kennebec and Merrimack Rivers and during spring tidal conditions in the Saco River. Bottom current velocities were recorded within a meter of the seabed.

Table II. Hydraulic data (from Hartwell, 1970; Manthorp, 1995; Fenster et al., 2001) for selected New England estuaries at locations with respect to channel mouth.

River and station (tidal range m)	Velocity				Location
	Max ebb	Max flood	Bottom max ebb	Bottom max flood	
Kennebec (2.7)					
Station 1	1.7	0.7	1.6	0.8	1 km downstream
2	1.9	0.5	1.9	0.8	1.2 km upstream
3	1.3	1.0	1.3	1.4	4.5 km upstream
Saco (3.0)					
Station 1	0.9	0.6	0.5	0.6	Inlet throat
2	0.9	0.5	0.2	0.5	1 km upstream
3	0.6	0.7	0.5	0.7	1.8 km upstream
4	1.2	0.3	0.6	0.3	3 km upstream
Merrimack (2.8)					
Station 1	1.5	0.9	0.2	1.0	1.2 km upstream
2	0.9	0.8	0.9	0.8	1.4 km side of channel
3	1.6	0.6	NA	NA	2 km
4	1.3	0.6	1.3	0.8	3 km

Tidal currents in the lower portion of the rivers exhibit typical time-velocity asymmetry for estuarine settings (Dronkers, 1986). Maximum flood currents occur late in the tidal cycle when channels are at high stage and tidal flats are covered with water. Conversely, maximum ebb-tidal currents occur near low water when strong tidal flow is confined to the deep channels. This hydraulic regime leads to ebb-dominated estuarine channels in which the depth-averaged maximum ebb current velocities exceed flood velocities by 0.15 to 1.4 ms^{-1} (Table II). It should be noted that the tidal current asymmetry measured in the Kennebec River would have been even greater if the flood-tidal range had not been 0.3 m greater than the ebb range. An exception to the trend of ebb dominance occurs at Station #3 in the Saco River, possibly due to the sheltering of ebb currents by an upstream bedrock obstruction. Although the estuaries are strongly ebb-dominant suggesting net seaward sediment transport, bedload transport is actually controlled by the magnitude, duration, and direction of the bottom currents. Furthermore, the coarse sand flooring most of the channel thalweg in the estuaries indicates that most sediment in the channels moves as bedload. The bottom current data for the estuaries show highly variable velocity asymmetries due to the influence of the salinity stratification, which is a function of the freshwater discharge, tidal regime, magnitude of turbulence, and other factors (Table II). The limited short-term data in Table II do not allow a definitive determination of net bedload sediment transport in these estuaries. Moreover, the dataset does not include the hydraulic conditions that exist during major floods, which have been shown to have a significant impact on the sediment transport patterns (Fenster *et al.*, 2001; FitzGerald *et al.*, 2002).

4.3 Bedforms

The alignment and geometry of bedforms provide a direct means of determining the direction of net bedload sediment transport. In a bi-directional tidal current regime large bedforms provide the most reliable proxy for sediment transport because large bedforms are less likely to change orientation due to tidal forcing. Of the three estuaries, the Kennebec River has the most comprehensive bedform dataset, consisting of 12 fathometer and side-scan sonar surveys that were run along the length of the lower river between 1986 and 1989 as well as multiple fathometer, side-scan, bathymetric, and LIDAR (LIght Detection And Ranging) surveys taken at the estuary mouth during the past 20 years (Fenster and FitzGerald, 1996; FitzGerald *et al.*, 2000). In addition, there exists detailed bedform maps of the ebb-tidal delta region for the Saco and Merrimack River estuaries (Manthorp, 1995; FitzGerald *et al.*, 2002). Analysis of the Kennebec data indicates that large-scale bedforms[1], including sandwaves and megaripples, change their orientation and are strongly influenced by seasonal variations in freshwater discharge that drives the estuarine circulation (Fenster and FitzGerald, 1996). In the spring when river discharge reaches a maximum, sandwaves and megaripples are ebb-oriented, whereas during the summer when river discharge is at a minimum the bedforms are flood-oriented to symmetrical in form. It was also determined that these subordinate bedforms are superimposed on transverse bars having wavelengths 0.4 to 1.2 m and heights up to 10 m. The bars have a persistent ebb-orientation and are overprinted by the migration of subordinate bedforms. Thirteen bars have been identified along the lower 20-km of the Kennebec River.

At the mouth of the Kennebec River strong bottom currents and an abundant sand supply produce large bedforms in the entrance channel and in a large spillover lobe channel that exists between two islands (FitzGerald *et al.,* 2000). Similar conditions and bedforms occur in the jettied channel of

[1] According to the classification scheme of Ashley et al. (1990), transverse bars = two-dimensional, very large, compound subaqueous dunes; sandwaves = two-dimensional large to very large subaqueous dunes; and megaripples = two and three dimensional, small to medium subaqueous dunes.

the Saco River and at the mouth of the Merrimack River. During periods of spring tides and/or high river discharge, sandwaves and megaripples are ebb-oriented. On the ebb-tidal delta of the Merrimack River estuary, mutually evasive sand transport pathways exist in which flood-oriented sandwaves move into the estuary along the shallow southern portion of the delta (Fig. 4). Countering the landward transport of the sand into the inlet are the ebb-oriented sandwaves that migrate out of the northernmost two thirds of the channel. These sandwaves continue their movement across the ebb-tidal delta and are gradually deflected to the south due to the influence of northeast storms and the southerly longshore currents they generate (FitzGerald *et al.*, 2002).

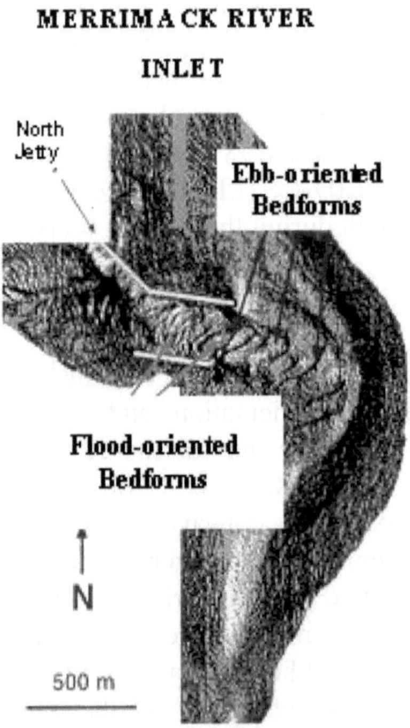

Figure. 4. LIDAR map for the Merrimack River ebb-tidal delta system. Note the sandwaves migrating out the inlet (from FitzGerald et al., 2002). Location shown on Figure 1.

4.4 Role of Spring Freshets

Major river floods in New England occur during spring freshets or intense rainfall events associated with hurricanes and extratropical storms. Spring freshets take place in late March and April when overland flow derived from a combination of melting snow and rainfall significantly increases freshwater discharge in a river, commonly by more than an order of magnitude. Hydrographs of the Saco River measured at the Cornish, Maine gaging station (Fig. 1) in 1987 and 1993 illustrate the effects of freshets (Fig. 5). During non-flood conditions average river flow at Cornish is about 36 m^3s^{-1}. During the freshets of 1987 and 1993 large snowmelts coupled with moderate rainfall produced peak discharges of 930 m^3s^{-1} and 566 m^3s^{-1} respectively (Fig. 5, Manthorp, 1995). Under average flow conditions the saltwater tidal prism for the Saco River estuary is about 11 times that of the freshwater discharge for a half tidal cycle. However, during the freshet of 1993, which lasted from late March through early May, the freshwater discharge in the Saco River completely supplanted the saltwater tidal prism (Manthorp, 1995). Tidal currents and salinity measured during the freshet at five stations downstream of the dam showed that the estuary was completely dominated by freshwater discharge. Over a complete tidal cycle ebb flow and freshwater conditions were recorded at all stations, except for the seawardmost station. At the inlet throat weak flood-tidal currents (<0.4 ms^{-1}) with salinity values ranging from 15 to 20‰ were measured at three-quarters flood (1.5 h period) when tidal waters were rising steeply. By using the river stage at the Cornish gaging station as a proxy for flow conditions in the estuary, it was determined that ebb currents persisted in the Saco River for a period of at least 40 consecutive days (FitzGerald *et al.,* 2002). During this time a layer of coarse sand and fine gravel was deposited in the thalweg of the estuary, whereas fine sand was deposited seaward of the jettied entrance (Manthorp, 1995).

Similar freshet conditions occurred in the Kennebec River during the spring floods of 1987 (Fenster and FitzGerald, 1996). During average flow conditions the freshwater discharge in the Kennebec is only 6% of the tidal prism. However, during the 1987 flood freshwater discharge replaced the saltwater tidal prism from 30 March through 3 April producing continuous ebb flow in the river. Fenster and FitzGerald (1996) determined that when freshwater discharge exceeds 225 to 325 m^3s^{-1} bedforms in the river remain ebb-oriented and bedload sediment moves downstream. A lesser magnitude freshet monitored in 1997 produced a strongly ebb-dominant current regime (Fenster *et al.*, 2001).

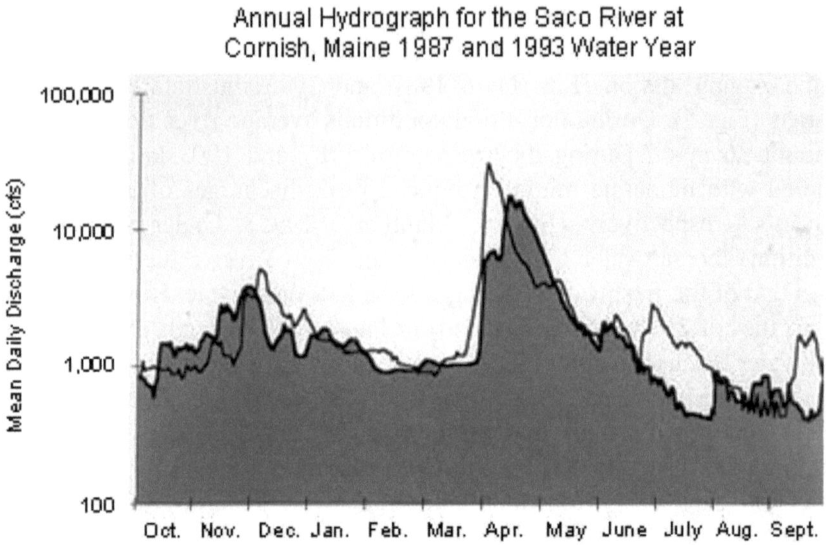

Figure. 5. Hydrographs for the Cornish, Maine gaging station of the Saco River for 1987 and 1993 (from Manthorp, 1995). Location of gaging station is shown on Figure 1.

5. DISCUSSION AND SUMMARY

The totality of information for the three estuaries strongly suggests that each exports sand to the nearshore and to the barrier systems at their mouths. The rivers were responsible for the formation of voluminous deltas and the deposition of other inner shelf sand bodies during lower stands of sea level during the early Holocene. Moreover, there is no evidence to suggest that discharge of sand by the rivers has ceased at the present day. All the estuaries are floored by coarse-grained sediment, which can be traced along the entire length of their thalwegs to their ebb-tidal deltas. The sediment is compositionally and texturally immature consisting of a large percentage of feldspar and rock fragments (>30%) as well as mica flakes. Individual sand grains are angular to subrounded (FitzGerald *et al.*, 2002; Buynevich and FitzGerald, 2003). The offshore and distal portion of the rivers' ebb-tidal deltas are composed of quartz-rich, medium to fine sand that contains little mica and low a percentage of feldspar and rock fragments (<20%). Thus, the sand within the estuary appears to have been sourced from upstream.

Bedform data in the Kennebec River indicate that the orientation of megaripples and sandwaves changes seasonally. However, these subordinate forms are superimposed on larger-scale transverse bars that have a persistent ebb-orientation. In the ebb-tidal delta region of the estuaries, the main channel of the Merrimack River and spillover channel of the Kennebec River contain a field of ebb-oriented sandwaves. The bedforms extend to subtidal bars located along the downdrift region of the terminal lobe. Their persistent ebb orientation is evidence that the sandwaves are migrating seaward and feeding sand to these downdrift bars, which in turn deliver sediment to the adjacent beaches (FitzGerald *et al.*, 2002).

The depth-averaged current data suggest that the estuaries are strongly ebb-dominant, however the bottom velocity current measurements are less conclusive. Despite the moderate tidal ranges and large tidal prisms, the overall hydraulic regime and bedload transport in the estuaries are controlled by large flood events that are caused by spring freshets and intense rainfall associated with hurricanes and extratropical storms. It is during these events that the bi-directional tidal currents are replaced by strong unidirectional freshwater flow that may last from a few days to several weeks. These floods have been responsible for blanketing the Saco River estuary with coarse sediment (Manthorp, 1995) and for the delivery of sand to the crests of the transverse bars (Fenster and FitzGerald, 1996).

A final indication of the downstream transport of sand to the estuary mouths comes from dredging records. The mouths of all three estuaries are dredged on a regular basis to provide a navigable entranceway to harbors and shipyards; these programs have been in practice for more than sixty years (FitzGerald *et al.*, 2002). The sediment removed from the channels consists of coarse sand and fine gravel and is much coarser-grained than the sediment found beyond the estuary mouth. Although the analysis is highly qualitative, the dredging records have been used to calculate potential sediment contribution by the Saco River, which is approximately 1.2×10^6 m^3/century (Barber, 1995; Kelley *et al.*, 2003). When this value is extrapolated over the past 5000 years, which is coincident with the late Holocene deceleration of the rising sea level and development of coastal barriers in Maine (van Heteren, 1995), we can account for 77% the volume of sand contained in the Saco Bay nearshore and barrier system (Fig. 6; Barber, 1995; Kelley *et al.*, 2003). The same type of computations using dredging records have been employed to explain the volume of sand deposited within the Merrimack Embayment (FitzGerald *et al.*, 2001).

In summary, major northern New England estuaries cannot be characterized as sediment sinks, rather they are active contributors of sand to the nearshore and associated barrier systems. During the Holocene transgression the estuarine channels may have accreted vertically due to increasing accommodation space, but the riverine supply of sediment has been sufficient to continually discharge sand. Infrequent large magnitude flood events are the primary factor controlling bedload sediment transport through the estuaries, yielding substantial volumes of coarse-grained material to the coastal ocean.

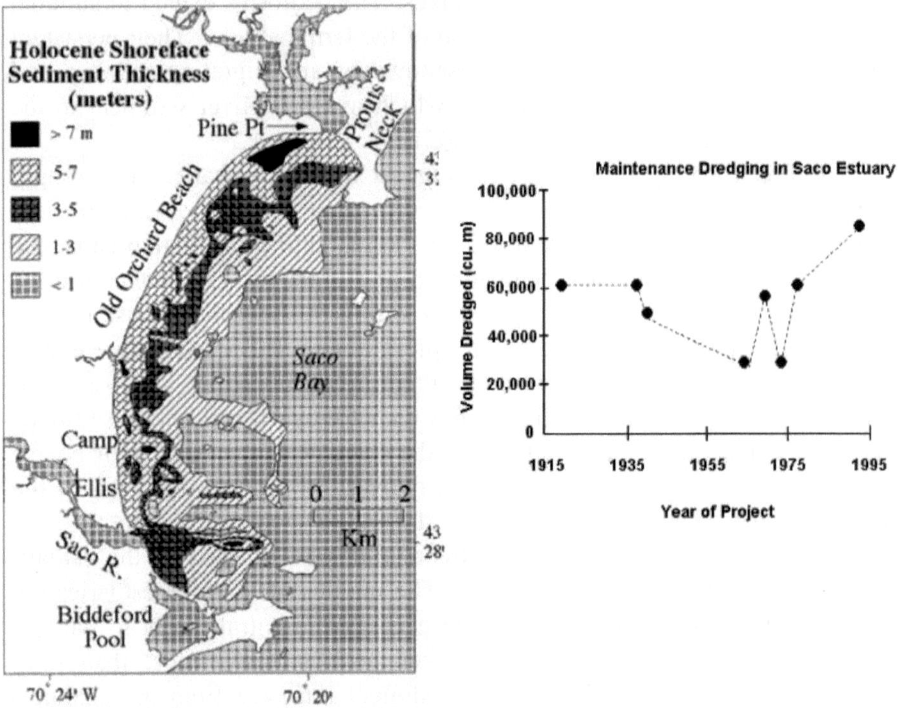

Figure 6. Sand budget for the Saco Embayment region. A. Isopach map of sediment comprising the nearshore zone indicates a total volume of 56 x 10^6 m^3 (Barber, 1995). Added to the shoreface and ebb-tidal delta sediment is the sand contained in the onshore barrier lithosome, which was computed to be 22 x 10^6 m^3 (van Heteren et al, 1996) producing a total volume of 78 x 10^6 m^3 of sediment in the embayment. B. Using dredging records at the mouth of the Saco River as a proxy, the sediment discharged by the Saco River since the mid-Holocene can account for 60 x 10^6 m^3 of sediment contained in the barrier and shoreface (Barber, 1995).

REFERENCES

Allen, G.P. 1991. Sedimentary processes and facies in the Gironde estuary: a recent model of macrotidal estuarine systems. In: Smith, G.D., Reinson, G.E., Zaitlin, B.A. and Rahmani, R.A. (eds) *Clastic Tidal Sedimentology*, Canadian Society of Petroleum Geologists Memoir, 16, 29-40.

Allen, J.R.L. 1990. The Severn Estuary in southwest Britian: its retreat under marine transgression, and fine sediment regime. *Sedimentary Geology*, 66, 13-28.

Anan, F.S. 1971. *Provenance and statistical parameters of sediments of the Merrimack Embayment, Gulf of Maine.* Unpublished PhD dissertation, University of Massachusetts, Amherst, 376pp.

Ashley, G.M., Boothroyd, J.C., and Borns, H.W., Jr. 1991. Sedimentology of late Pleistocene (Laurentide) deglacial-phase deposits, eastern Maine; An example of a temperate marine grounded ice-sheet margin. In: Anderson, J.B. and Ashley, G.M. (eds) *Glacial Marine Sedimentation; Paleoclimatic Significance.* Geological Society of America, Special Paper 261, 107-125.

Barber, D.C. 1995. *Holocene depositional history and modern sand budget of inner Saco Bay, ME*. Unpublished MS thesis, University of Maine, Orono, 178pp.

Barnhardt, W.A., Belknap, D.F. and Kelley, J.T. 1997. Stratigraphic evolution of the inner continental shelf in response to late Quaternary relative sea-level change, northwestern Gulf of Maine. *Geological Society of America Bulletin*, 109, 612-630.

Belknap, D.F. and Shipp, R.C. 1991. Seismic stratigraphy of glacial marine units, Maine inner shelf. In: Anderson, J.B. and Ashley, G.M. (eds) *Glacial Marine Sedimentation; Paleoclimatic Significance*. Geological Society of America Special Paper 261, 137-157.

Belknap, D.F., Kelley, J.T. and Shipp, R.C. 1987. Quaternary stratigraphy of representative Maine estuaries: initial examination by high-resolution seismic reflection profiling. In: FitzGerald, D.M. and Rosen, P.S. (eds) *Glaciated Coasts.* Academic Press, 177-207.

Belknap, D.F., Shipp, R.C., Kelley, J.T. and Schnitker, D. 1989. Depositional sequence modeling of the late Quaternary geologic history, west-central Maine coast. In: Tucker, R.D. and Marvinney, R.G. (eds) *Maine Geological Survey, Studies in Maine Geology,* 5, 29-46.

Belknap, D.F., Shipp, R.C. and Kelley, J.T. 1986. Depositional setting and Quaternary stratigraphy of the Sheepscot Estuary, Maine: a preliminary report. *Géographie physique et Quaternaire*, 40, 55-69.

Belknap, D.F., Kraft, J.C. and Dunn, R. 1994. Transgressive valley fill lithosomes: Delaware and Maine. In: Boyd, R., Zaitlin, B.A. and Dalrymple, R. (eds) *Incised Valley Fill Systems: Origin and Sedimentary Sequences.* SEPM Special Publication, 51, 341-354.

Belknap, D.F., Kelley, J.T. and Gontz, A.M. 2002. Evolution of the glaciated shelf and coastline of the northern Gulf of Maine, USA. *Journal of Coastal Research Special Issue*, 36, 37-55.

Boothroyd, J.C. and FitzGerald, D.M. 1989. *Coastal geology of the Merrimack Embayment: SE New Hampshire and NE Massachusetts.* SEPM Eastern Section Fieldtrip Guidebook, 45pp.

Boyd, R., Dalrymple, R. and Zaitlin, B.A. 1992. Classification of clastic coastal depositional environments. *Sedimentary Geology*, 80, 139-150.

Boyd, R., Zaitlin, B.A. and Dalrymple, R. (eds) 1994. *Incised Valley Fill Systems: Origin and Sedimentary Sequences.* SEPM Special Publication, 51, 401pp.

Buynevich, I.V. and FitzGerald, D.M. 2003. Textural and compositional characterization of recent sediments along a paraglacial estuarine coastline, Maine, U.S.A. *Estuarine Coastal, and Shelf Science*, 56, 139-153.

Colman, S.M., Halka, J.P. and Hobbs, C.H. 1992. Patterns and rates of sediment accumulation in the Chesapeake Bay during the Holocene rise in sea level. In: Fletcher, C.H. and Wehmiller, J.F. (eds) *U.S.A. Quaternary Coastal Systems.* SEPM/IGCP Special Publication, 101-112.

Cooper, J.A.G. 2002. The role of extreme floods in the estuary-coastal behavior: contrasts between river- and tide-dominated microtidal estuaries. *Sedimentary Geology*, 150, 123-137.

Dalrymple, R.W., Zaitlin, B.A. and Boyd, R. 1992. Estuarine facies models: Conceptual basis and stratigraphic implications. *Journal of Sedimentary Petrology*, 62, 1130-1146.

Dalrymple, R.W., Baker, E.K., Hughes, M. and Harris, P.T. 1996. Geomophology and sedimentology of the muddy, tide-dominated, Fly River delta, Papua New Guinea. *Geological Society of America, Northeastern Section, 31st annual meeting; abstracts*, 28 (3), 47.

Dalrymple, R.W., Knight, R.J., Zaitlin, B.A. and Middleton, G.V. 1990. Dynamics and facies model of a macrotidal sand-bar complex, Cobeuid Bay-Salmon River Estuary (Bay of Fundy). *Sedimentology*, 37, 577-612.

Dronkers, J. 1986. Tidal asymmetry and estuarine morphology. *Netherlands Journal of Sea Research,* 20, 117-131.

Dudley, R.W., Kelley, J.T. and Dickson, S.M. 1999. Mapping riverbed sediments in the Kennebec River, Maine using side scan sonar and ground-penetrating radar. *Geological Society of America Abstracts with Programs*, 31, A13.

Dyer, K.R. 1973. *Estuaries: A Physical Introduction.* Wiley, London, 140pp.

Farrell, S.C. 1970. *Sediment distribution and hydrodynamics: Saco and Scarboro Rivers Estuaries, Maine.* Unpublished MS thesis, University of Massachusetts, Amherst, 129pp.

Fenster, M.S. and FitzGerald, D.M. 1996. Morphodynamics, stratigraphy, and sediment transport patterns of the Kennebec River estuary, Maine, USA. *Sedimentary Geology*, 107, 99-120.

Fenster, M.S., FitzGerald, D.M., Kelley, J.T., Belknap, D.F., Buynevich, I.V. and Dickson, S.M. 2001. Net ebb sediment transport in a rock-bound, mesotidal estuary during spring freshet conditions: Kennebec River estuary, Maine, U.S.A. *GSA Bulletin*, 113, 1522-1532.

FitzGerald, D.M. and van Heteren, S. 1999. Classification of paraglacial barrier systems: Coastal New England, USA. *Sedimentology*, 46, 1083-1108.

FitzGerald, D.M., Buynevich, I.V., Fenster, M.S. and McKinlay, P.A. 2000. Sand circulation at the mouth of a rock-bound, tide-dominated estuary. *Sedimentary Geology*, 131, 25-49.

FitzGerald, D.M., Buynevich, I.V., Davis, R.A., Jr. and Fenster, M.S. 2002. New England tidal inlets with special reference to riverine associated inlet systems. *Geomorphology*, 48, 179-208.

FitzGerald, D.M., Buynevich, I.V., Fenster, M.S., Kelley, J.T. and Belknap, D.F. 2001. Contribution of coarse-grained sediment to the nearshore and inner shelf by large estuaries: New England, USA. *GSA Abstracts with Programs*, 33.

Hannum, M.B. 1996. *Late Quaternary evolution of the Kennebec and Damariscotta River estuaries, Maine*. Unpublished MS thesis, University of Maine, Orono, 126pp.

Hanson, L.S. and Caldwell, D.W. 1989. The lithologic and structural controls on the geomorphology of the mountainous areas in north-central Maine. In: Tucker, R.D. and Marvinney, R.G. (eds) *Maine Geological Survey, Studies in Maine Geology,* 5, 147-167.

Harris, P.T., Baker, E.K., Cole, A.R. and Short, S.A. 1993. A preliminary study of sedimentation in the tidally dominated Fly River, Gulf of Papua. *Continental Shelf Research*, 13, 441-472.

Harris, P. T., Heap, A., Bryce, S., Smith, R., Ryan, D. and Heggie, D. 2002. Classification of Australian coastal depositional environments based upon a quantitative analysis of wave, tidal and fluvial power. *Journal of Sedimentary Research*, 72, 858-870.

Hartwell, A.D. 1970. *Hydrography and Holocene Sedimentation of the Merrimack River estuary, MA*. Technical Report CRG-5, Geology Dept., University of Massachusetts, 166pp.

Hubbard, D.K. 1975. Morphology and hydrodynamics of the Merrimack River ebb-tidal delta. In: Cronin, L.E. (ed) *Estuarine Research*, v II. Academic Press, New York, 253-266.

Kelley, J.T., Shipp, R.C. and Belknap, D.F. 1989. Sedimentary framework of the southern Maine inner continental shelf: influence of glaciation and sea-level change. *Marine Geology*, 90, 139-147.

Kelley, J.T., Dickson, S.M., Belknap, D.F., Barnhardt, W.A. and Barber, D.C. 2003. Sand volume and distribution on the paraglacial inner continental shelf of the northwestern Gulf of Maine. *Journal of Coastal Research*, 19, 41-56.

Manthorp, P.A. 1995. *Estuarine circulation and sediment transport in the Saco River estuary, Maine*. Unpublished MS thesis, Boston University, MA, 230pp.

Mills, S. 1997. *The stratigraphy and evolution of three barrier systems in the northern Wells embayment, Maine*. Unpublished MS thesis, Boston University, MA, 231pp.

Montello, T.M. 1993. *Stratigraphy and evolution of the barrier system along the Wells - Ogunquit embayment in southern Maine*. Unpublished MS thesis, Boston University, MA, 211pp.

O'Brien, M.P. 1969. Equilibrium flow areas of inlets on sandy coasts. *Journal of Waterways, Harbors, and Coastal Engineers, ASCE*, 95, 43-55.

Oldale, R.N., Wommack, L.E. and Whitney, A.B. 1983. Evidence for a postglacial low relative sea-level stand in the drowned delta of the Merrimack River, western Gulf of Maine. *Quaternary Research*, 19, 325-336.

Osberg, P.H., Hussey, A.M. and Boone, G.M. 1985. *Bedrock geologic map of Maine*. Maine Geological Survey, Augusta, MA.

Rhodes, E.G. 1973. *Pleistocene-Holocene sediments interpreted by seismic refraction and wash-bore sampling, Plum Island- Castle Neck, Massachusetts*. US Army Corps of Engineers Technical Memorandum, #40, 75pp.

Tary, A.K., FitzGerald, D.M. and Buynevich, I.V. 2001. Late Quaternary evolution of a marine-limit delta plain, southwest Maine. In: Weddle, T.K. and Retelle, M.J. (eds) *Deglacial History and Relative Sea Level Changes, Northern New England and Adjacent Canada*. Geological Society of America Special Publication, 351, 125-149.

van Heteren, S., FitzGerald, D.M., Barber, D.C., Kelley, J.T. and Belknap, D.F. 1996. Volumetric analysis of a New England barrier system using ground-penetrating radar and coring techniques. *Journal of Geology*, 104, 471-483.

Ward, L.G. 1992. Estuarine geomorphology. In: Short, F.T. (ed) *The Ecology of the Great Bay Estuary, New Hampshire and Maine: An Estuarine Profile and Bibliography*. NOAA, Coastal Ocean Program Publication, 39-43.

Wells, J.T. 1983. Dynamics of coastal fluid muds in low-, moderate-, and high-tide-range environments. *Canadian Journal of Fisheries and Aquatic Science*, 40 (supp. 1), 130-142.

Wells, J.T. 1995. Tide-dominated estuaries and tidal rivers. In: Perillo, G.M.E. (ed) *Geomorphology and Sedimentology of Estuaries*. Developments in Sedimentology, 53, 179-205.

Wright, L.D., Coleman, J.M. and Thom, B.G. 1973. Process of channel development in a high-tide range environment: Cambridge Gulf-Ord River Delta, western Australia. *Journal of Geology*, 81, 15-41.

Chapter 11

MORPHODYNAMIC BEHAVIOUR OF A HIGH-ENERGY COASTAL INLET: LOUGHROS BEG, DONEGAL, IRELAND

Helene Burningham
Coastal and Estuarine Research Unit, University College London, Chandler House, 2 Wakefield St, London WC1N 1PF, UK

1. INTRODUCTION

Morphological monitoring is fundamental to the understanding of coastal morphodynamics, and should provide a comprehensive awareness of coastal behaviour in response to storms, climate, sea-level change and human activities on different scales, when supported by historical (meso-) scale examinations of coastal change.

The understanding and documentation of contemporary morphodynamics on high-energy coastlines is relatively limited, particularly in the context of the paraglacial coastlines of north-west Europe. Whilst a notable body of work exists concerning inlet and estuarine morphology and hydrodynamics within the North Sea (e.g. Buller *et al.*, 1975; Ciavola, 1997; Van der Spek, 1997) and Irish Sea basins (e.g. Haynes and Dobson, 1969; Jago, 1980; Pye, 1996), the coasts facing the north-east Atlantic remain somewhat unknown in comparison. Research on the west and south-west coasts of Ireland (Delaney and Devoy, 1995; Orford *et al.,* 1999) has alluded to the importance of episodic storm events on coastal morphodynamics during the

D.M. FitzGerald and J. Knight (eds.),
High Resolution Morphodynamics and Sedimentary Evolution of Estuaries, 215-243.

late-Holocene. However, there is still a lack of published examples of morphological change in Ireland, particularly in the north-west.

The aims of this paper are: 1) to present evidence of behaviour of a meso-tidal inlet on the north-west coast of Ireland over a series of timescales; 2) to determine the mechanisms associated with this behaviour; and 3) to demonstrate the difficulties associated with the integration of data on a variety of scales/resolution. Ultimately, this contribution acknowledges a need for increased higher resolution monitoring of dynamic coastal systems.

2. STUDY AREA

The coastline of north-west Ireland has an irregular configuration, characterised by alternating headlands/peninsulas and bays/inlets. The west-facing Loughros Beg system represents one of 25 estuaries on the County Donegal coastline (Fig. 1). Coastal physiography is dominated by regional geological structure, which dictates the size and orientation of the embayments. Sedimentary systems throughout the region comprise accumulations of glacigenic sands, transported onshore during the Holocene marine transgression (Shaw and Carter, 1994).

Donegal estuaries can be fundamentally described as comprising narrow inlet regions, bounded by either dune barriers or rock headlands, that open to wider estuarine basins, extensive intertidal sand flats and sinuous ebb channels. Upper estuary reaches tend to be narrow and fluvial input is relatively small.

The Loughros Beg estuary is accommodated within a ~ 9 km long, WNW–SSE trending, elongate, glacially-scoured bedrock valley. Steep hinterland to the south (Slievetooey and Glengesh), which rises over 300 m within 1 km of the estuary margins, is flanked at lower elevations by drumlins. The solid geology of the bedrock valley is dominated by quartzite, with minor beds of pelite and limestone present on the Loughros Point peninsula. The tidal inlet is located 2 km landward of the natural terminus of the bedrock valley. Here, the Maghera dune system projects from the southern estuary margin, constraining tidal flow toward the northern margin: a soil horizon within the dune strata has been dated to 3415 ± 135 years BP (Carter, 1982).

Fluvial influence within the system is limited: several small streams drain the ~ 48 km^2 catchment area, although fluvial input is focused at the estuary head (Bracky River), and the backbarrier environment of Maghera, close to the inlet (Owenwee River) (Fig. 1). Loughros Beg has a mean spring tidal prism of ~ 1×10^7 m^3.

The climate of north-west Ireland is windy, wet and stormy. The north-east Atlantic wind regime is dominated by south to south-westerlies. Winds recorded at Malin Head and Belmullet (north-west Ireland: Fig. 1) have an average annual speed (1956–1998) of 15.5 knots (~ 8 ms^{-1}) and 12.5 knots (~ 6.4 ms^{-1}) respectively. The wind climate exhibits a strong directional bias, whereby 37% of winds come from the south-west quadrant, and 60% have a westerly component (Fig. 2): average annual speed increases to 17.3 knots (~ 8.9 ms^{-1}) and 13.8 knots (~ 7.1 ms^{-1}) from a westerly direction at Malin Head and Belmullet respectively (Burningham, 1999).

Wind speeds of gale force speeds and above (> 34 knots (~ 17.5 ms^{-1})) represent, on average, ~ 2% of the wind speed record and are generally from the west (onshore within the study area). Storminess of the wind regime on the west coast increases northwards, where increased peak winds and periods of sustained strong winds are greatest at Malin Head (MacClenahan *et al.*, 2001). Furthermore, it has been shown that decaying east US coast and mid-Atlantic hurricanes travel as extra-tropical storms, eastward across the North Atlantic toward the west coast of Ireland (Cooper and Orford, 1998).

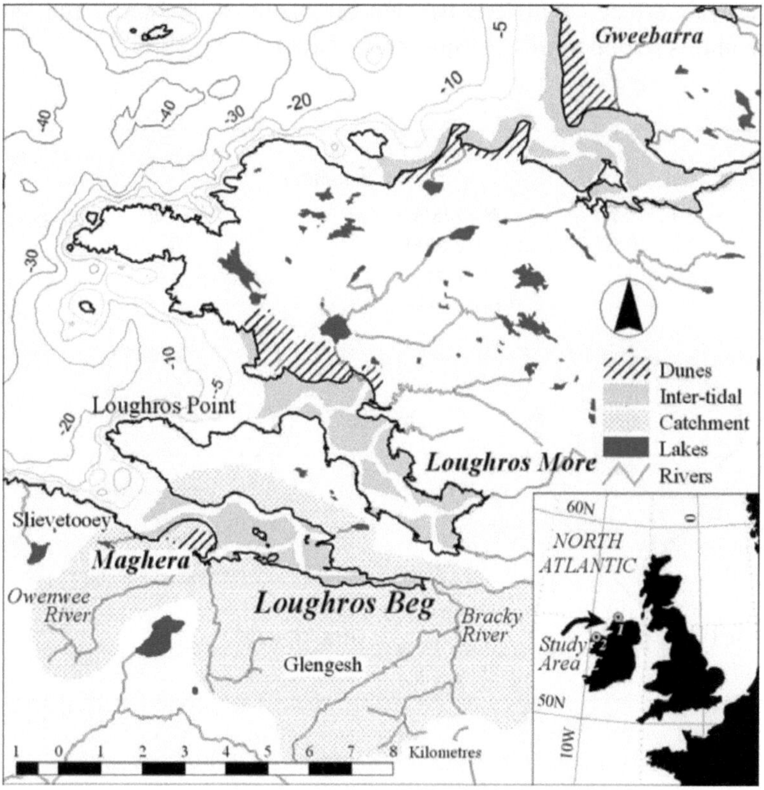

Figure 1. Regional setting of the Loughros Beg estuary and location of the estuary mouth, tidal inlet area at Maghera: bathymetric contours are in metres (OD Dublin). Inset are shown Irish Meteorological Stations (1) Malin Head, (2) Belmullet.

Wave climate is similarly high energy: Bacon and Carter (1991) reported annual average (1982–1988) offshore wave heights of 3.75 m from Ocean Weather Station Lima (650 km west of Donegal). Tides on the mid-west coast of Donegal are semi-diurnal and meso-tidal: mean tidal range varies from 1.6 m at neaps to 3.5 m at springs.

There is a significant paucity of sea level records in Ireland which pre-date the mid-20th century, and subsequent records do not provide conclusive evidence of significant change. Scott (1996) calculated an overall rise of 0.18 mm yr^{-1} in the 1960–1991 tidal records of Portmore, near Malin Head, whereas sea level records for Malin Head (PSMSL, 2002) suggest a fall of 0.12 mm yr^{-1} over the 1958-1997 period. Both datasets exhibit considerable temporal fluctuations in excess of the overall suggested trend, but equally they both suggest a negligible change in sea level over the latter half of the 20th century. Over a longer time scale, Pugh (1982) estimated a relative sea level rise in the north of Ireland of ~ 0.08 m between 1845 and 1980 (0.6 mm yr^{-1}), although this is not necessarily representative of the north-west as the effect of differential glacial histories, isostatic and eustatic factors in this region varies considerably (Taylor *et al.*, 1986).

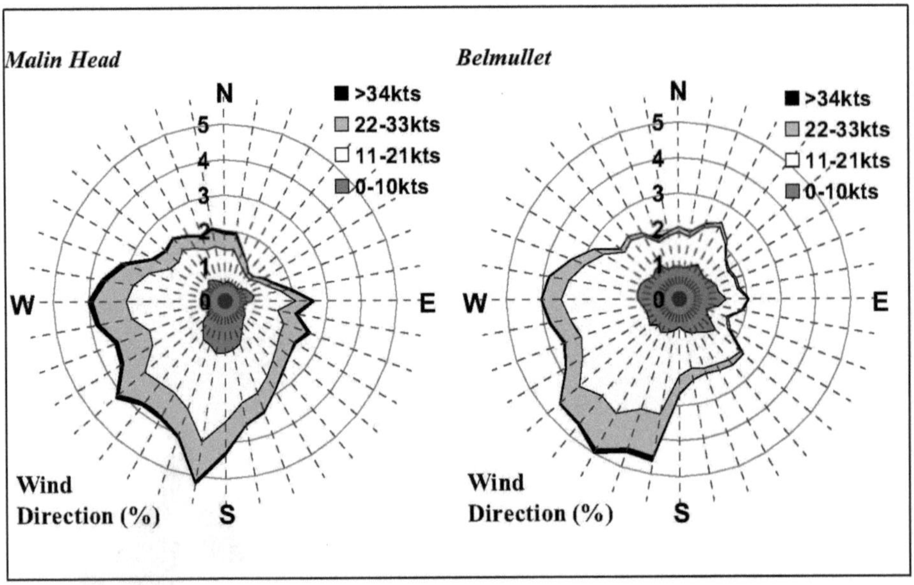

Figure. 2. Wind climate (1956-1998) at Malin Head and Belmullet (modified from Burningham, 1999).

3. METHODS

The morphology of the Loughros Beg inlet was examined over a series of timescales, varying from bimonthly to half-century, using published map surveys, aerial photograph data and topographical field surveys.

3.1 Map and Aerial Photograph Data

Morphological character and changes over the historical or meso-scale (10–100s years) were assessed through digital shoreline analysis of existing published maps and aerial photographs. Map data were digitised directly into ARC-INFO, whilst aerial images were scanned and geo-referenced within ERDAS Imagine. Further digitising and analysis were performed within ArcView, with a final spatial accuracy of the digitised coverages of ± 15 m. Specific features delineated from each data source included high and low water marks and dune and salt-marsh boundaries.

The west coast of Donegal was first mapped by the Ordnance Survey (OS) between 1833 and 1836, toward the production of the first series of *6-inch to one mile* (1:10,560 scale) maps. Although the map published from this survey (1835) contains minimal shoreline information (detail is given of towns, land-use boundaries and roads only), high- and low-water mark, rivers, delineation of dune and bog systems are included. This survey was revised during 1845–1852 (Andrews, 1993) and the 'revised First Edition' map (1853) displays a more complete survey, with the inclusion of any landscape changes. The Second Edition of the six-inch map series for the study area was published in 1907 and incorporated revisions surveyed in 1904–05. Unfortunately, there have been no subsequent surveys or revisions at this scale. The more recent 1:50,000 series is based on the 1907 six-inch map with the addition of road and town developments obtained from aerial photographic surveys. Minimal attention has been given to the coastline and coastal environments, and therefore map evidence of morphological change is limited to the 1835–1907 series of six-inch maps. The absence of a grid coordinate system on all of the published six-inch maps necessitated the establishment of a local grid coordinate framework. Reference markers used for rectification purposes were surveyed in the field with differential GPS, and an arbitrary metric grid was derived.

Aerial surveying of the west Donegal region did not commence until the 1950s, and consequently there are no overlaps between the data formats of map and aerial survey. The Irish Air Corps completed a survey covering the Loughros Beg region in 1951. The OS of Ireland commissioned surveys in 1977 and 1995 and the Office of Public Works obtained aerial images in 1994.

3.2 Ground Survey Data

As part of a wider study (Burningham, 1999), the Loughros Beg inlet was topographically surveyed bimonthly between March 1996 and June 1998. Field surveys, based on shore-normal profiles, were conducted at low-water spring tides and included the identification of tidal and debris lines (a proxy for the high-water mark), low-tide position (proxy for the low-water mark) and foreshore bars, dune front and edge of vegetation. A similar survey was also performed in September 2001. Surveys were performed using electronic distance measurement (EDM) and differential GPS techniques, and referenced to the same arbitrary framework grid established for the map/photo analysis.

Short-term (months–years) morphological changes in inlet character were extracted from the topographical data through the delineation of low- and high-water marks, and the identification of specific intertidal sediment bodies. Survey data were imported into ArcView to allow direct comparison with map/photo coverages. Although these features were often fully surveyed and delineated in the field, interpolation of the outlines was required on those surveys consisting of profile topography alone.

3.3 Supplementary Observations

Throughout the field surveys of 1996–1998 and 2001, ground photography was employed to supplement the topographical surveys. These photographs provide further information regarding the presence, absence, size and orientation of sediment bodies within the Loughros Beg inlet, and the determination of inlet configuration was supported by this evidence.

4. RESULTS

4.1 Meso-scale Behaviour

Inlet morphology in 1835 is characterised by a single ebb channel and extensive inter- and supratidal sand deposits (Fig. 3). From the east, the ebb channel enters the inlet region close to the northern margin of the estuarine basin. To the west, the ebb channel exits into Loughros Beg Bay close to the southern margin, diagonally dividing the inlet shoals and thereby defining an intertidal longshore barrier. The supratidal system of Maghera, which covers over 650,000 m^2, extends ~ 850 m from the southern margin: map annotation indicates that this region comprises vegetated dunes.

The morphology of Loughros Beg in 1853 is almost identical to that observed in 1835, and hence is not included in Figure 3. It is difficult to assess whether this lack of change is entirely natural or is an indication that, unlike other regions on this coastline, the Loughros system was not re-surveyed between the *First Edition* and *revised First Edition.*

The basic configuration of the intertidal deposits and ebb channel within the inlet presents little change between 1835/1853 and 1907. A small flood barb (*sensu* Robinson (1960)) present on the seaward side of the intertidal barrier in 1835/1853 and missing from the 1907 map, corresponds to the only significant change in the low-water mark, and the high-water mark to the north of Maghera is depicted as lying slightly closer to the ebb channel.

With respect to the supratidal region of Maghera, although elevations defined on the 1853 and 1907 maps appear unchanged, the general annotation of this area has been modified. A large portion of the supratidal system of Maghera has lost the vegetated dune status previously suggested by the 1835 and 1853 map annotation. Whilst this could indicate a change in mapping practices between the First and Second Editions, archaeological publications from the turn of the 19th–20th century corroborate the interpretation of a dune system experiencing considerable denudation. D'Evelyn (1933: 88) in an account of the archaeological finds of Maghera, stated:

"*At the time of my first visit* [~ 1896 (Knowles, 1901)] *there was a considerable area of sand hills, but these have now* [~ 1933] *been almost all blown away and much of the ground over which I used to collect is flat and covered by the sea at high tide.*"

In addition, archaeological visits in 1898 by Knowles (1901: 342) revealed that Maghera at that time was characterised by:

"*a considerable extent of sand hills which had suffered greatly from denudation, the old surface exposed in many places, and some parts entirely broken up and the contents scattered about*"

Both descriptions suggest that the map observations are correct, and that during the 1890s there was a significant change in the supratidal character of Maghera. Based on mapped evidence, ebb channel and intertidal shoal configuration, however, experienced negligible change during the 19th century.

The 1951 aerial photograph of the Loughros Beg inlet displays a smaller dune system fronted by a large supratidal flat, similar to the character indicated by the 1907 map. The photograph provides considerably more information concerning the nature of the inter- and supratidal deposits than can be obtained from published maps. Aeolian bedforms are clearly visible on the photograph and the delineation between supra- and intertidal deposits is similarly distinct.

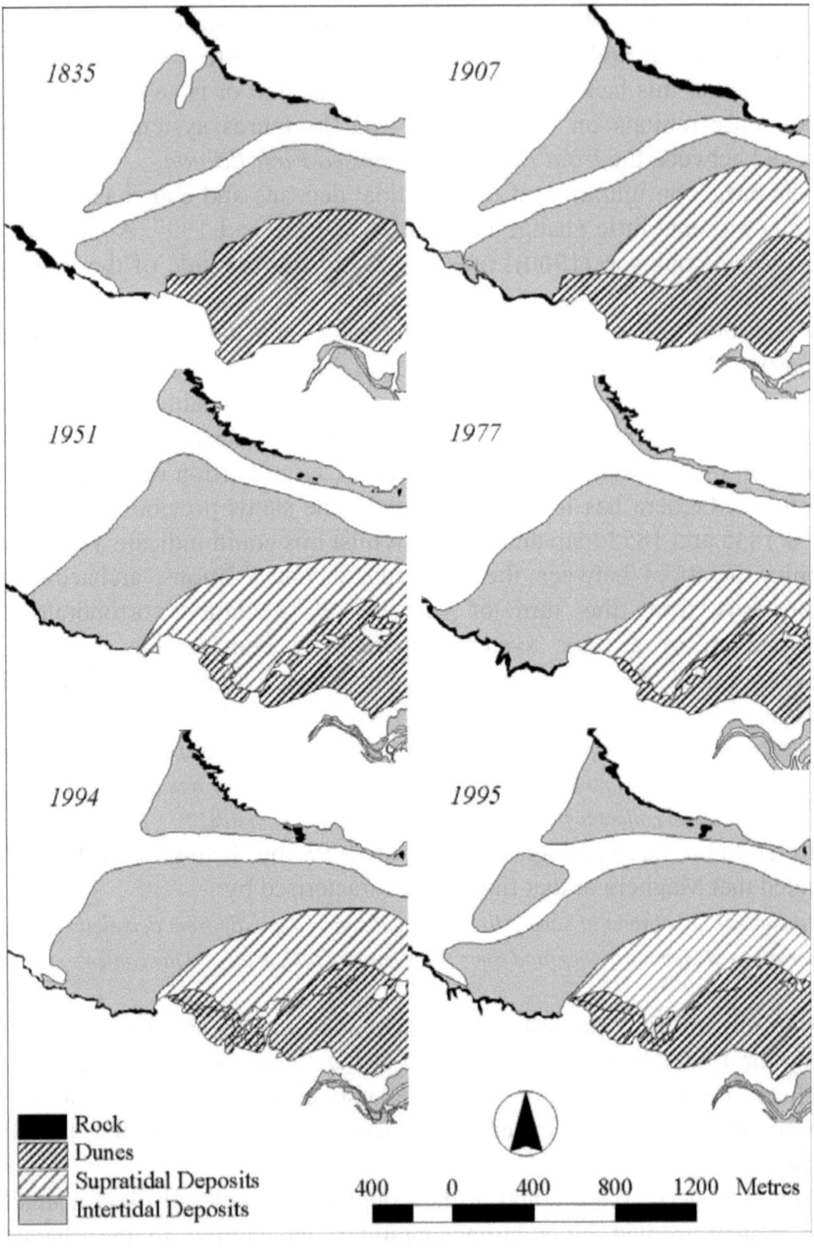

Figure. 3. Historical changes in the configuration of the Loughros Beg inlet. Delineations within the supratidal deposits represent boundaries between vegetated dunes and unvegetated supratidal strand.

The Maghera dune system has a denuded appearance, marked with several small blowouts, and dune vegetation is covered in a thin deposit of wind blown sand. On their seaward margin, the dunes grade into the supratidal strand flat via a narrow zone of embryo dunes.

The dominant change observed from the 1951 photograph is in the position of the ebb channel within the inlet region. The channel no longer crosses the intertidal shoals of the mouth but maintains a position close to the northern margin throughout the length of the inlet. This position allows a very wide intertidal zone to front the supratidal flat of Maghera.

In 1977, channel position continues to follow the northern margin throughout the inlet, whilst inter- and supratidal zones on the southern margin maintain similar areal coverage to that in 1951. Blowouts within the Maghera dune system are still visible, although a decrease in dune sand redistribution is evident with a reduction in the surficial coverage of wind blown deposits. The foredunes have prograded slightly, shifting the zone of embryo dunes fractionally seaward and extending the main dune area.

The ebb channel is presumed to have maintained a similar course between 1951 and 1977. An oblique aerial photo of the Loughros Beg inlet from 1965 (J. K. St. Joseph – Cambridge University Collection of Air Photographs: ALQ79) indicates a northerly channel. King (1965: 48) also noted that "*the river now drains out on the north side of the bay*". Furthermore, a SPOT satellite image from 1984, whilst inadequate for shoreline analysis due to low resolution, indicates that the channel was still close to the northern margin of the inlet.

Inlet morphology in 1994 implies that configuration has entered a new phase: the ebb channel deviates southward from the 1951/1977 positions as it exits the inlet. This mid-inlet position is notably different to that of the 1835/1907 configurations. There is evidence from the 1994 aerial photographs that the dune environment has continued to stabilize, with increased vegetation coverage (with the exception of two blowouts on the northern rim) and further establishment, growth and progradation of foredunes on the western margins of the system. The supratidal region overall has remained a similar size.

Further changes in inlet configuration are evident from the 1995 aerial photographs. The main ebb channel occupies a similar position to that shown in 1994, however just before exiting the mouth it bifurcates. The secondary channel flows further south, in a position similar to that of the 1907 ebb channel. The bifurcation defines a small intertidal, lower foreshore bar in a mid-inlet position.

Relative changes in the position and delineation of the ebb channel through the inlet clearly show significant degrees of both stability and movement over the 160-year period (Fig. 4). Channel position to the immediate north of the Maghera dune system shows considerable stability, whereas directly west, as the channel exits the inlet, channel migration exhibits an amplitude or swathe covering over 70% of the full bedrock valley width (Fig. 4). With the addition of field survey data from 1998 and 2001, the shift in channel position since the mid 20th century has tended towards a return to the 19th century configuration.

The 1998 and 2001 inlet morphologies are not, however, identical to those of 1835/1853 or 1907: the ebb channel has a distinct curvature that was absent in the 19th century morphology. The course of the 1998 ebb channel flows ~ 275 m further west, in comparison to the 19th century course, before taking the south-west trending route to cut across the mouth of the estuary. As the channel exits the inlet close to the southern margin of the valley, the 1835/1853, 1907 and 1998 courses concur, although the 1998 seaward intertidal barrier extends beyond that of the earlier years. The 2001 configuration presents a further deviation, whereby the mouth of the ebb channel and extremes of the intertidal barrier have shifted ~ 125 m landward.

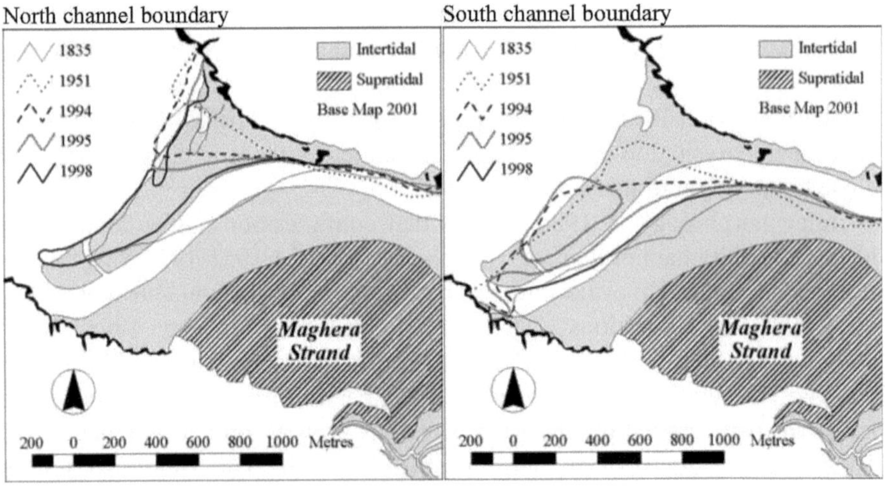

Figure. 4. Superimposed inlet configurations (delineation of ebb channel) derived from historical maps/aerial photographs and field observations/surveys.

4.2 Short-term Behaviour

Whilst complete translation of the ebb channel across the full valley width has occurred over several decades, it is also clear from the more recent aerial photographs and field observations that significant lateral shifts in the channel and intertidal deposits also occur over shorter time-scales. Aerial photographs were not available throughout the 1996–2001 field survey period, but oblique photographs from the high ground of the southern margin supplemented topographical surveys to provide sufficient information to delineate the ebb channel, and inter- and supratidal sediment bodies for comparable inlet configuration analysis. The configurations presented (Fig. 5) document the extension of the intertidal barrier and southward migration of the ebb channel from the 1994 mid-estuary position.

Further definition was afforded by the field surveys, allowing the identification of barred morphology within the Maghera foreshore region (illustrated in Fig. 5 as variable shading of the intertidal deposits). It is clear from the short-term evolution of the basic configuration that the bar features on the foreshore are integral to the development of the intertidal barrier by maintaining a surplus of sediment in the nearshore region of the inlet. While the channel exits in a mid-inlet position, the Maghera foreshore is characterised by bar-trough morphology. These features are not present on the foreshore when the configuration changes to the full width intertidal barrier and southerly channel exit position. Morphology associated with this latter configuration comprises a planar upper foreshore, with break of slope near the neap low-water mark to a lower foreshore of shallower gradient that continues toward the channel.

Over an annual scale, the development of the intertidal barrier at the mouth appears to be relatively progressive, but the actual shift from barred foreshore/mid-inlet channel configuration to intertidal barrier/south-margin channel configuration occurred between November 1997 and March 1998 (Fig. 6). The configuration in January 1998 presented a transitional form: the intertidal barrier was not fully developed, and the foreshore lacked the established bar/trough system. The ebb channel had divided, and although the majority of tidal flow was still contained within the mid-inlet channel, some ebb tidal flow was accommodated by the smaller channel toward the southern margin of the inlet. The intertidal shoal, seaward of this small channel, had a broad low relief bar form, which was distinctly different from the bars typically present on the Maghera foreshore.

Throughout the summers (~ March–November) of 1996 and 1997, beach profile surveys documented continuous onshore migration of the intertidal bar and trough system that characterised the foreshore (Burningham, 1999). During both subsequent winters, the barred system was destroyed or moved to an extreme lower foreshore position (i.e. it migrated offshore), and it is via this process that the change in configuration between November 1997 and March 1998 occurred. The significant difference between the 1996/97 and 1997/98 redistribution of sediment is in the subsequent onshore transport. During 1997, the Maghera foreshore retained the surplus sediment in the form of migrating onshore bars; whereas in 1998, the sediment was re-deposited seaward of the ebb channel, thereby adding to and extending the longshore intertidal barrier.

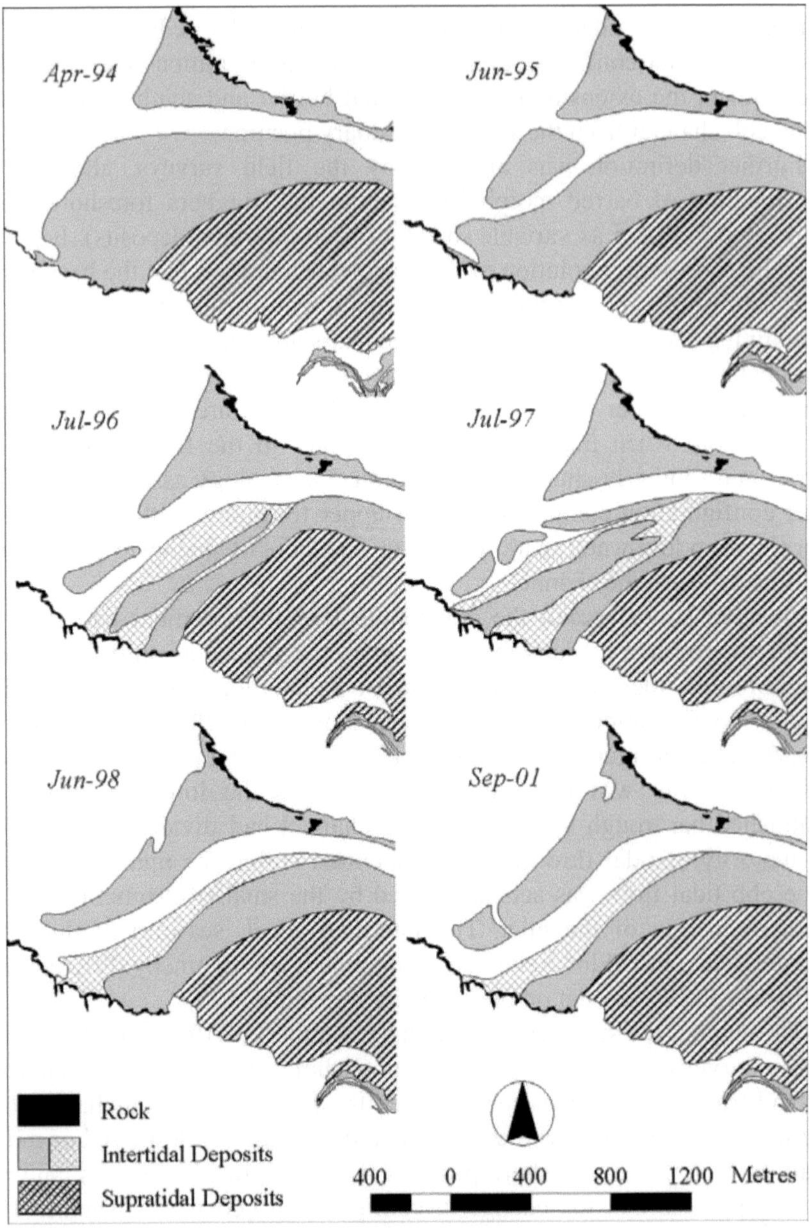

Figure 5. Short-term changes in configuration of the Loughros Beg inlet (1994-2001), derived from aerial/ground photographs, field observations and surveys. Variable shading of intertidal deposits delineates bar/trough features or upper/lower foreshore definition.

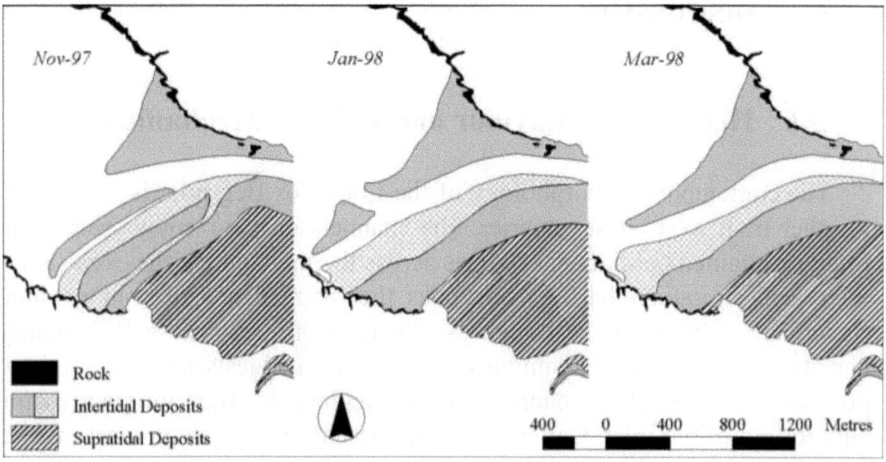

Figure 6. Definitive change from the barred foreshore/mid-mouth channel configuration to intertidal barrier/south-margin channel configuration: November 1997 – March 1998.

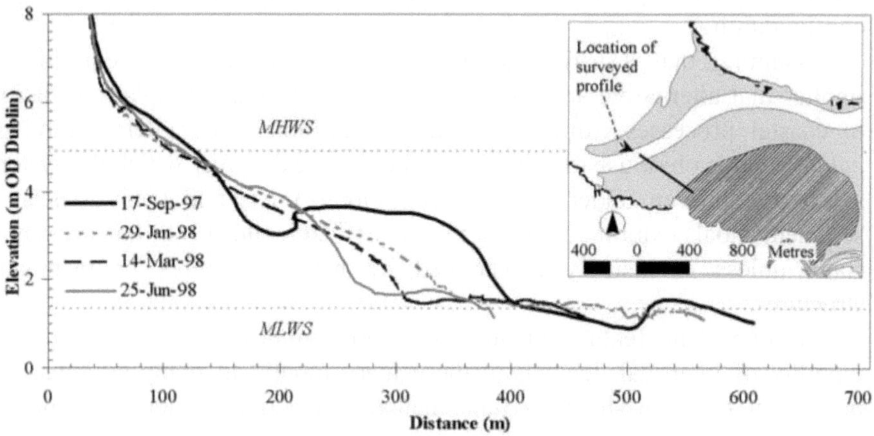

Figure 7. Topographical evolution of the Maghera foreshore: September 1997 – June 1998.

The behaviour of the Maghera foreshore following the change in inlet configuration is not dissimilar to its previous summer behaviour. Throughout 1998, the ebb channel, intertidal barrier and foreshore moved shoreward (Fig. 7), suggesting that the intertidal barrier was responding to the same processes and behaving in an equivalent manner to the 1996 and 1997 foreshore bar-trough systems.

5. DISCUSSION

5.1 Historical Behaviour and Driving Mechanisms

The contemporary morphology of the Loughros Beg inlet is comparable to other Irish west coast estuaries: ebb channels in these sandy systems are often maintained on diagonal courses across the inlet, where the extension of inter- and supratidal barriers precludes the maintenance of a more direct route. The stability of the Loughros Beg inlet throughout the 19th century, and its recent return to a similar configuration, suggests that the southerly position of the ebb channel mouth represents the morphodynamic equilibrium of the inlet. Carter (1988) suggested that the south-west trend of ebb channels is related to the Coriolis effect, whereby the flood tide is deflected to the right in the northern hemisphere, corresponding to a southerly direction for west-flowing estuaries (and east-flowing flood tides). This may explain the tendency for the ebb channel, within the Loughros Beg inlet, to ostensibly favour the southerly exit position and associated configuration.

Assuming that a particular configuration is evidence of equilibrium implies that alternative arrangements are induced by a significant deviation in coastal processes from the 'normal' conditions. The interpretation of *equilibrium* is obviously dependent on the time scales over which the system is observed. Even under 'normal' conditions, inlets and their associated sediment shoals exhibit a range of mobility (FitzGerald, 1988) and are rarely static unless extensively constrained by natural features or manmade structures.

An important feature of inlet morphodynamics is the position and mobility of intertidal deposits at the estuary mouth. In many systems, the movement of these shoals is fundamental to the overall behaviour of the inlet, and can contribute to cyclical morphological changes (Robinson, 1975) and progressive shifts in inlet position (Hayes, 1972).

On high wave energy shorelines, or those where tides are less important, wave processes frequently preclude the formation of extensive ebb tidal deposits. Form and behaviour of inlet sediment shoals is often attributed to the longshore transport of littoral sediments (FitzGerald, 1988) or fluvial erosion and deposition within the estuarine system (Cooper, 1994), in comparison to flood and ebb driven sediment dynamics in tide-dominated inlets. Marine-driven inlet closure is a common occurrence in response to storm events and associated large waves (Webb *et al.,* 1991). Fluvial flooding can then act to breach wave-formed mouth deposits, thereby changing inlet configuration (Van Heerden, 1986). In river-dominated systems, floods can contribute large quantities of sediment to the inlet region, which are then subsequently reworked within the inlet (Cooper, 1994).

In the case of Loughros Beg, the inlet system is notably constrained and exists within a longshore-drift – limited bedrock cell. Despite this, the swathe of the ebb channel position within the 1 km-wide valley has occupied the majority of this width at some point over the last 160 years (Fig. 8). In association with this channel movement, the intertidal shoals at the inlet have exhibited considerable temporal variability in form and position. Whilst it is clear that the short-term movement of these deposits has a significant control on the configuration of the inlet, it is less clear what the overall driving mechanism is.

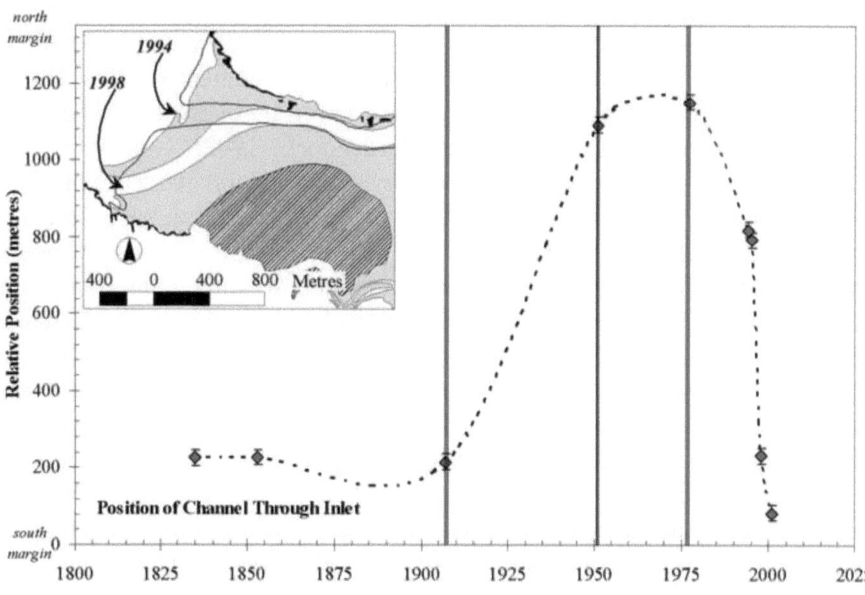

Figure 8. Change in the relative position of the main ebb channel within the Loughros Beg inlet. [Size of symbols corresponds to error margin. Relative Position: 0 m = south margin of bedrock valley, 1350 m = north margin. Years 1907, 1951 and 1977 are highlighted for cross-reference with Fig. 9.]

The inlet does not exhibit classic cyclical behaviour, and does not appear to be significantly influenced by fluvial processes or longshore sediment supply. The dominant processes are clearly marine: wave and tidal processes are important, and within the inlet region, the interplay of these is complex. Webb *et al.* (1991) showed that inlet closure on a wave-dominated coastline required the occurrence of storm waves following a period of neap tides to ensure a sufficient supply of sediment. At Loughros Beg, the 1907–1951 shift in inlet configuration suggests an equivalent concurrent sequence of processes. It is unclear from the long-term morphological changes whether this change was progressive or abrupt.

In most published examples, the shortening of the ebb channel within the inlet region is associated with breaching of an inlet barrier (e.g. FitzGerald, 1988). The switch in position generally occurs over a short time-scale as a response to either hydraulic inefficiency driven by excessive extension of a longshore barrier, or storm-driven breaching by waves or floods. The original stability of the intertidal barrier across the Loughros Beg inlet (i.e. the morphological constancy from 1835 to 1907) suggests that hydraulic inefficiency was not the cause of the changing configuration. Storm-related sediment redistribution is thought to be the cause of change in inlet configuration. What is less certain is the actual time period over which this occurred, and the morphodynamic behaviour associated with the change.

The impact of storm conditions on coasts is complex and not entirely predictable (Delaney and Devoy, 1995). In most documented cases, storm-related marine breaching of inlet barriers is driven by overwash and erosion by storm waves (e.g. Hume and Herdendorf, 1992; Conley, 1999). Morphological records of breaching events rely on concurrent field observations or surveys: in the absence of records relating specific storm events to their effects in the coastal zone, it is difficult to identify the actual driving mechanisms associated with the switch in inlet configuration at Loughros Beg. The climate of the west coast of Ireland is notoriously windy and susceptible to large scale storms: the region presents the first land-fall of large scale depressions that build up in the mid-Atlantic and move north-eastwards toward Europe (Rohan, 1986). Analysis of storm tracks over recent years has provided specific accounts of tropical systems which have had notable effects on the Irish coastline (Cooper and Orford, 1998). Re-assessment of past storm tracks has also pointed to a common course toward this north-west European coastline (Fernandez-Partagas and Diaz, 1996).

Assessment of the historical wind climate of Donegal is limited due to the lack of pre-1950s data. However, several accounts of storms exist (e.g. Burt, 1987; Shields and Fitzgerald, 1989; Lamb, 1991) which detail specific episodic climatic events that have affected the west coast of Ireland. Large storms in 1839 and 1961 stand out as being particularly destructive on the west coast, and both have been linked to large erosive events in south-west Ireland (Orford *et al.*, 1999). A storm in 1903, considered to be the second largest in recorded Irish history (Anon, 1903; Lamb, 1991), may have resulted in the deterioration of the Maghera dune system (Knowles, 1901; D'Evelyn, 1933), but this is clearly speculation. Other extreme storms throughout the early 20th century (e.g. 1927 (Lamb, 1991) and 1933 (Burt, 1987)) could be the initiators of the 1907–1951 shift in ebb channel position, but the lack of specific references detracts from the potential importance of these observations.

Although identification of the exact timing of the 1907–1951 change in inlet configuration is speculative, it is interesting to note that during this period, extensive changes in the coastal systems too place throughout west Donegal, particularly in terms of inlet intertidal shoals and dune systems (Burningham, 1999). Furthermore, Pye and Neal (1994) found that windier than normal conditions in the early 20th century may have contributed to erosional trends on the north-west coast of England. The winter storm climate is known to be a significant mechanism behind the contemporary annual changes in beach and dune morphology and sediment volumes on the west Donegal coast (Burningham, 1999), and is thought to be a key factor in the 1997–1998 shift in inlet configuration. The North Atlantic Oscillation (NAO) also plays a significant role in the storm climate of the north-east Atlantic and can provide an extension to the historical wind record (Hurrell, 1995).

The NAO refers to the latitudinal pressure gradient between the Arctic and subtropical extents of the North Atlantic, and is based on the difference in normalised sea level pressures between Iceland and the Azores (Hurrell, 1995). The Azores pressure record dates back to 1865, although extension of the NAO back to the early 1820s can be achieved through the use of data from Gibraltar (Jones *et al.*, 1997).

The winter NAO index, which refers to specific winter months (e.g. November–March), is thought to be related to weather conditions in the northern hemisphere: a sustained positive NAO index relates to wetter, stormier conditions over Europe and negative NAO index values relate to drier, calmer weather (Hurrell, 1995). This winter index, records of which extend back to 1823 (Jones *et al.,* 1997), can therefore provide a measure of climate variability in Europe, particularly the increased storm track activity associated with positive NAO phases (Ulbrich and Christoph, 1999), over a time-scale corresponding to the morphological changes presented here.

The November–March (NDJFM) index (Fig. 9a) shows evidence of the recent positive NAO trend that has been linked to increasing storminess over the last 25 years (Dawson *et al.,* 2002), but also shows a notable positive phase throughout the early 20th century. The markedly strong positive phase between ~ 1900 and ~ 1930, which was associated with strong westerly winds onto Europe (Hurrell and Van Loon, 1997), coincides with the period of major change in inlet configuration at Loughros Beg. Furthermore, Bouws *et al.* (1996) suggest a correlation between increases in observed wave heights in the north-east Atlantic and the recent positive phase of the NAO. This would suggest that the positive NAO index of the early 20th century was also accompanied by increased wave heights, a factor that could be associated with inlet barrier breaching and shifts in channel position.

Examination of the winter NAO index with respect to the historical storm record and the post-1956 wind climate reveals significant associations (Fig. 9). Timing of specific storm records, identified by several authors as low pressure and/or high energy wind events impacting the north-west of Ireland (Anon, 1903; Joyce, 1912; Burt, 1987; Shields and Fitzgerald, 1989; Lamb, 1991; Cooper and Orford, 1998), appear to be clustered during periods of sustained positive winter NAO index (Fig. 9a). Furthermore, storminess within the Malin Head hourly wind record (1956-1998) exhibits a notable similarity to the post-1956 winter NAO index (Figs. 9b,c). Storminess is delineated through the identification of sustained high energy wind 'events', defined by hourly wind speeds sustained at 22 knots (Force 6) for more than 24 hours: the 'event' period continues until the threshold of 22 knots is not sustained for a further consecutive 24 hours. This separation of the wind climate, and indication of 'event' duration and peak wind speed (Fig. 9c), provides a measure of storm character by recognising extreme endurance of windy conditions and their maximum energy. Although the wind climate exhibits a high degree of variability, it displays a similar temporal trend to that of the winter NAO index over the last 50 years. The strengthening of the positive NAO phase toward the 1990s is marked by an increasing occurrence and duration of high-energy wind events.

A further breaching mechanism, associated with storm conditions, is the elevated water level of a storm surge. Extreme surges in the past have resulted in extensive coastal flooding and erosion (e.g. North Sea, 1953) and the depositional record in saltmarshes at the tidal limit of Loughros Beg shows evidence of storm surge sedimentation (Orford *et al.*, 1996; Wheeler *et al.*, 1999). Furthermore, Wheeler *et al.* (1999) found peaks in saltmarsh deposition rates in the late-1930s to early-1940s, early-1960s and mid-1980s, which correlate well with periods of significant change in inlet configuration. Although fluvial flooding and breaching is unlikely in Loughros Beg, it links to the process of surges and tidal flooding. If a storm and associated surge arrive at the coast at high tide, wave conditions and water levels are elevated. In extreme conditions within a narrow inlet such as Loughros Beg, a surge of this type coincident with high tide would result in the generation of non-fluvial 'flood' conditions capable of breaching a sand-dominated inlet barrier.

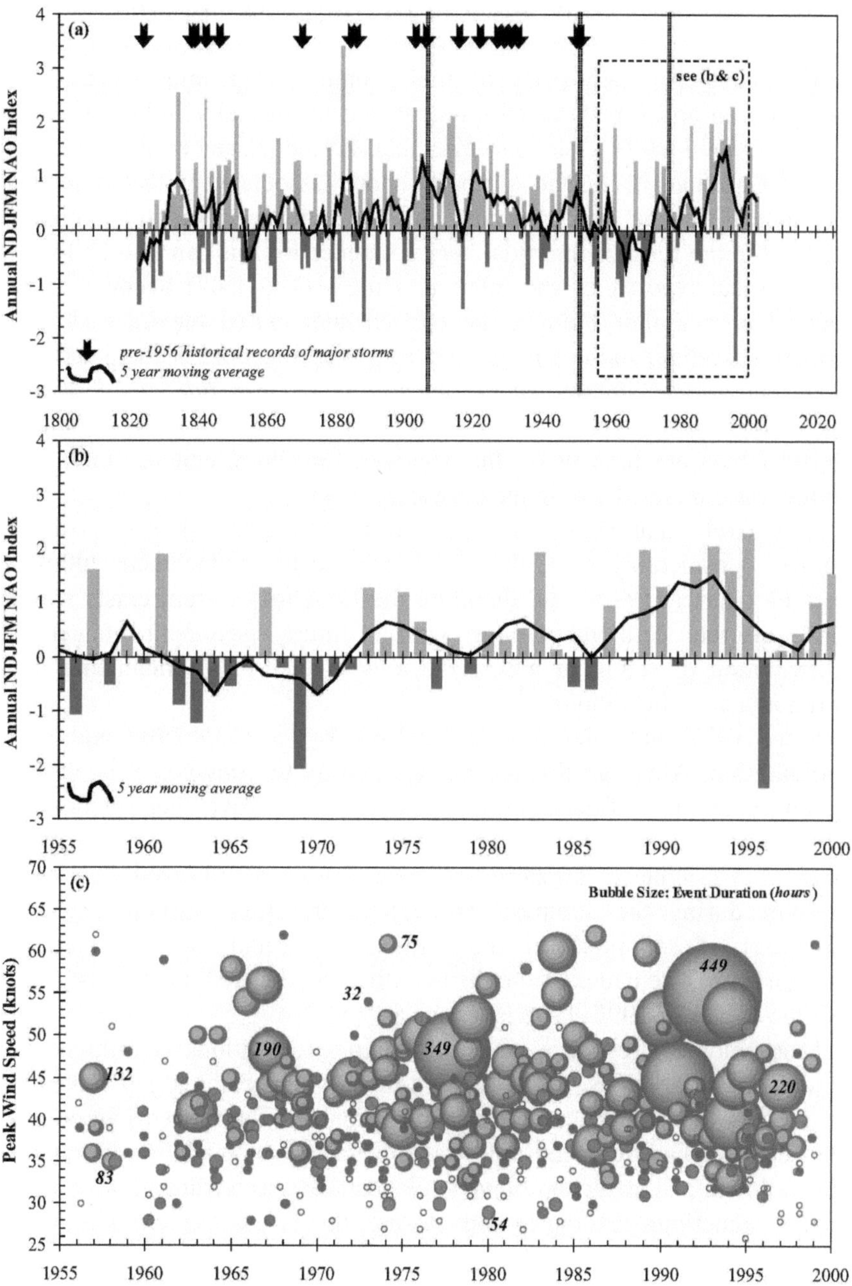

*Figure 9. (**a**) North Atlantic Oscillation winter (NDJFM) index (1823-2000) [Source: Jones, 2002] and dates of historical storms (years 1907, 1951 and 1977 are highlighted for cross-reference with Fig. 8). (**b**) Subset of NAO winter index (1955-2000). (**c**) Storminess at Malin Head (1956-1998). See text for further explanation.*

5.2 Recent Behaviour and Driving Mechanisms

The short-term behaviour of the Loughros Beg inlet indicates that configuration and foreshore morphology are intrinsically linked. It is clear from the 1996–1998 field surveys, that in the absence of a longshore intertidal barrier attached to the north-shore (associated with the southerly exit course for the ebb channel), the foreshore comprises a surplus of sediment in the form of intertidal bars. Evidence for this can also be found in the historical records of the inlet: in King's (1965: 42) account of west Donegal coastal morphology, the ebb channel exited the inlet along the northern shoreline, and it is noted that the lower foreshore is characterised by "*very low ridges...situated at the levels of low neap tide and low spring tide*". Conversely, when the intertidal barrier extends across the inlet, no intertidal bars are present on the Maghera foreshore, and the bulk of the sediment is contained within the seaward deposit.

It is likely that sediment supply from glacigenic shelf deposits has declined considerably since the initial large scale emplacement 4000-6000 years BP (Carter, 1990), and therefore the Loughros system exists within a small, discrete cell, functioning with a limited sediment budget. The redistribution of sediment within the inlet is therefore fundamental to its morphodynamic behaviour.

Since 1977, the inlet has tended back to its 1835–1907 equilibrium configuration. Although this has been generally progressive, it is clear that the key switch in configuration occurred during the November 1997 – March 1998 period. The annual behaviour of the inlet is thought to follow a cycle of winter storm-induced offshore sediment transport followed by onshore transport during the summer. Throughout the field survey period this movement of sediment, in the form of intertidal bars, defined inlet configuration. The cause of the distinct difference in sediment redistribution over the winter of 1997–1998 is less clear.

Morphological response of inlets to storm conditions is influenced by stage of the tidal cycle, which dictates the position of breaking waves over inlet shoals and the magnitude and direction of tidal currents through the inlet region. Neap tides can reduce the efficiency of tidal flow through the inlet, allowing storm conditions to deposit excess sediment at estuarine mouths, sometimes leading to inlet closure; this is less likely in the event of spring tides (Webb *et al.*, 1991). On the Portuguese coast, Morris *et al.* (2001) found that spring tides allowed storm waves to break closer to inlet regions, thereby resulting in focused intensive erosion: in comparison, lower tide levels caused waves to be refracted and dissipated over the ebb tidal shoals, causing less focused and less intensive erosion.

The temporal scale of the topographical surveys at Loughros Beg precludes the identification of the exact storm event associated with specific morphological change. The north-west Ireland winter wind climate (November–March) is generally stormy. Wind records from Belmullet and Malin Head show that several periods of 'stormy' weather occurred during the winters of both 1996–1997 and 1997–1998 (Figure 10: delineated by dashed boxes). Wind speeds greater than 22 knots, corresponding to Beaufort scale Force 6, at which *large waves begin to form*, occur almost daily throughout the winter, and certainly cover high, low, spring and neap elements of the tidal cycle.

Wind speeds of at least Force 8 (34 knots), the threshold that Orford *et al.* (1999) use to signify storm wave potential, also occur relatively frequently; there are extended periods of notably strong winds. Again, these strong winds occur often enough to coincide with most stages within the tidal cycle, suggesting that tide level is not significant in instigating changes in inlet morphodynamics.

Both winters can be characterised by 2 main storm periods. The first storm period occurs in late October/early November in the 1996–1997 winter, but not until mid to late December in the 1997–1998 winter. Late January to mid-March is remarkably similar during the 2 winters, whereby a short lull is followed by ~ 4 weeks of strong winds (> 22 knots), predominately from a south-westerly direction. The distinction between winters, with respect to the first storm period, is important in terms of timing, duration and discrete differences in storm character. The short November storm in 1996 is followed by ~ 1 month of moderate wind speeds, and a further month of considerably lower wind speeds. This results in a time period that is sufficient for post-storm recovery of the foreshore. During the 1997–1998 winter, storm conditions prevail throughout mid-December to early January, and the following 2 weeks in January are characterised by moderate winds. One week of calm conditions is then followed by the second February–March main storm period.

The successive nature of stormy conditions in the 1997–1998 winter is a possible mechanism driving the changing configuration of the Loughros Beg inlet. Post-winter, fairweather periods allow the onshore migration of sediment in the form of intertidal bars: it is probable that the period between the storms of November 1996 and February 1997 allowed a degree of recovery on the Maghera foreshore through the onshore transport of sediment. The storm periods during the 1997–1998 winter are, however, unlikely to have allowed such recovery, and offshore sediment transport may have been accentuated. Hence, the inlet in March 1998 comprised a larger seaward accumulation of sediment, which was sufficient to redefine the position of the ebb channel and configuration of the inlet.

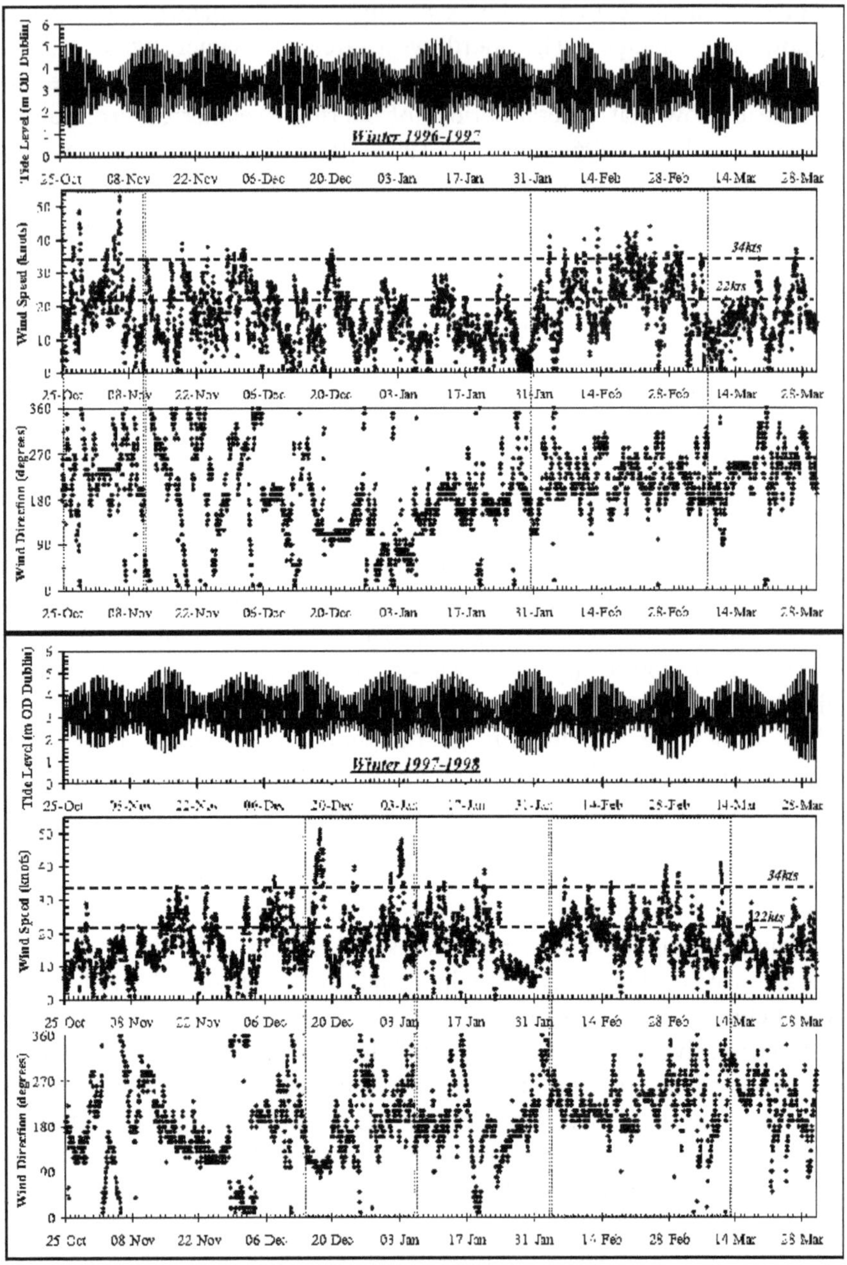

Figure 10. Tidal curve and wind climate of west Donegal for the winters of 1996–1997 and 1997–1998 (Malin Head wind records shown, although Belmullet exhibits a similar temporal character). Storm periods discussed in the text are defined by dashed boxes.

5.3 Scales of Coastal Monitoring

Inlet configuration has remained stable since March 1998, although there is clear evidence that the seaward intertidal barrier underwent a similar onshore migration to the 1996 and 1997 foreshore bar-trough systems. Field surveys in September 2001 indicated that the intertidal barrier and ebb channel had migrated further landward than the position observed during the previous survey in June 1998. The lack of intermediate surveys between 1998 and 2001 prevents a more detailed understanding of the morphodynamics over this period, but raises the issue of scales of monitoring. Considering a coastal system over the historical time scale is an important and valuable approach to morphodynamic analysis and understanding. The ability to then examine the same system over much shorter time-scales is beneficial and provides the opportunity to research the process-response characteristics of the system, particularly those identified over the longer time-scale.

Examination of the morphological behaviour of the Loughros Beg inlet is inherently difficult due to the variable accuracy of the diverse array of data sources and monitoring techniques. There is no temporal overlap between the map surveys, aerial images and field observations/surveys. Temporal continuity is crucial to the analysis of such data. Aerial photographs provide an instantaneous representation of a specific environment; tidal, wave and wind states can all contribute to errors of interpretation. The first and second editions of OS maps are the cumulative result of field mapping surveys that would have been conducted over several weeks: this time scale is potentially larger than that exhibited by the changing configuration of the Loughros Beg inlet. The exact extent to which detailed revisions were made is dubious, and the prioritised focus on land boundaries (Andrews, 1993) raises questions about the attention given to coastal areas in the revised first and second editions. It is clear that the use of early maps in the detailed examination of morphological changes can be problematic, particularly in the true meaning of variable annotations, for example the changes in the shading style used to denote dune environments. As Baily and Nowell (1995) noted, the worth of early maps lies in the more fundamental record of specific features and delineation of high- and low-water marks, and these are characteristics that should be transferable across the various survey techniques.

The lack of long-term, high-resolution climatological and morphological data on the Donegal coast hinders direct correlation between climatic events, coastline development and coastal morphodynamics. Without pre- and post-storm morphologies, specific understanding of process and morphological response is lost, and speculative discussions are unavoidable. The ability to carry out surveys in response to changing climatic events is dependent on numerous factors, the least obstructive of which is the storm event itself. Difficulties in capturing morphological changes associated with episodic events are not unique to this study, particularly when shifts occur over hours or days (Webb *et al.,* 1991).

The role of fixed point, ground photography is highlighted here as an additional and effective method of obtaining high-resolution morphological data, to supplement conventional topographical field surveys. In this study, photography from high ground on the southern margin of the inlet provided greater spatial coverage of morphological information, and also linked to interpretations obtained from standard aerial photography. The recent work of Morris *et al.* (2001) is a testament to the validity and worth of this form of data. The work is pioneering in that, although high-resolution records of coastal climates and dynamics are commonplace, it is the first published study using oblique digital image acquisition and processing techniques to provide high-(temporal)-resolution morphological data. Used in conjunction with the aforementioned climate/dynamics data, specific correlations between process and response can be made.

6. CONCLUSIONS

The Loughros Beg inlet on the west coast of Ireland presents various scales of change, historically and over the short term. The inlet has been relatively stable for periods of at least 70 years, but has also exhibited significant shifts over periods of months. The coastal systems on this part of the Donegal coast are unusual in the fact that human interference and management are negligible. It can be stated therefore, with relative certainty, that the morphological behaviour observed is a consequence of the changes in, and general regime of, the north-east Atlantic climate.

The significant shift in inlet configuration experienced in the early 20th century is tentatively linked to storminess, and accounts of several major storms during that period support this argument. The strong, sustained positive phase of the winter NAO at that time, which implies a significantly stormier period than throughout the preceding century or during the mid-20th century, supports this notion. There is a distinct correlation between periods of inlet configuration stability (1835–1907 and 1951–1977) and negative or weakly-positive phases of the winter NAO, and an equally clear association between periods of changing configuration (1907–1951 and 1977–1998) and strong positive phases.

Over a shorter time-scale, storm conditions were found to be an important control on the location and behaviour of intertidal deposits at the mouth of Loughros Beg, which were intrinsic to the overall configuration of the inlet. Winter storms invariably resulted in the offshore transport of inlet shoals; this was followed, during summer months, by the progressive onshore transport of sediment in the form of intertidal bars. When the ebb channel was located toward the middle or north margin of the inlet, the foreshore comprised intertidal bar-trough systems: when the ebb channel course cut diagonally across the inlet to exit close to the southern margin, the sediment was contained within a seaward intertidal barrier.

As discussed above, combining diverse sources and scales of morphological and climatological data can be problematic. The historical approach to coastal morphodynamic understanding is potentially flawed due to the discontinuous nature of datasets. The interpretation of system behaviour relies on identifying the morphological extremes, as well as the interludes, and it is impossible to achieve this temporal coverage from random maps and aerial photographs. The meso-scale understanding of coastal environments is fundamental to their present day and future management. To ensure that management strategies evolve away from the conventional reactive approach, it is essential that monitoring in support of conceptual, hydrodynamic and morphodynamic modelling, is encouraged.

7. ACKNOWLEDGEMENTS

This work benefited from a DENI Distinction Award studentship while the author studied at the University of Ulster, and more recently the UCL Faculty of Social & Historical Sciences Deans Travel Fund.

Thank you to Joanne Millington and Jon French, who commented on early drafts and the reviewers, John McKenna, Andy Wheeler and Jasper Knight, for their helpful suggestions.

8. REFERENCES

Andrews, J.H. 1993. *History of the Ordnance Survey*. Ordnance Survey, Dublin, 63pp.

Anon, 1903. Ireland and the recent gale. *The Irish Builder and Engineer*, 40, 1656-1658.

Bacon, S. and Carter, J.T. 1991. Wave climate changes in the North Atlantic and North Sea. *International Journal of Climatology*, 11, 545-558.

Baily, B. and Nowell, D. 1995. Techniques for monitoring coastal changes: a review and case study. *Ocean & Coastal Management*, 32, 85-95.

Bouws, E., Jannick, D. and Koman, G.J. 1996. The increasing wave height in the North Atlantic Ocean. *Bulletin of the American Meteorological Society*, 77, 2275-2277.

Buller, A.T., McManus J. and Green, C.D. 1975. Dynamics and sedimentation: The Tay in comparison with other estuaries. In: Hails, J. and Carr, A. (Eds) *Nearshore Sediment Dynamics and Sedimentation*. John Wiley & Sons, London, 201-250.

Burningham, H. 1999. *Morphodynamics of West Donegal Estuaries*. Unpublished DPhil Thesis, University of Ulster, 281 pp.

Burt, S.D. 1987. A new North Atlantic low pressure record. *Weather*, 42, 53-56.

Carter, R.W.G. 1982. Sea level changes in Northern Ireland. *Proceedings of the Geologists' Association*, 93, 7-23.

Carter, R.W.G. 1988. *Coastal Environments*. Academic Press, London, 617pp.

Carter, R.W.G. 1990. Coastal processes in relation to geographic setting, with special reference to Europe. *Senckenbergiana Maritima*, 21, 1-23.

Ciavola, P. 1997. Coastal dynamics and impact of coastal protection works on the Spurn Head spit (UK). *Catena*, 30, 369-389.

Conley, D.C. 1999. Observations on the impact of a developing inlet in a bar built estuary. *Continental Shelf Research*, 19, 1733-1754.

Cooper, J.A.G. 1994. Sedimentary processes in the river-dominated Mvoti estuary, South Africa. *Geomorphology*, 9, 271-300.

Cooper, J.A.G. and Orford, J.D. 1998. Hurricanes as agents of mesoscale coastal changes in western Britain and Ireland. *Journal of Coastal Research*, SI 26, 123-128.

Dawson, A.G., Hickey, K., Holt, T., Elliott, L., Dawson, S., Foster, I.D.L., Wadhams, P., Jonsdottir, I., Wilkinson, J., McKenna, J., Davis, N.R. and Smith, D.E. 2002. Complex North Atlantic Oscillation (NAO) Index signal of historic North Atlantic storm-track changes. *The Holocene*, 12, 363-369.

Delaney, C. and Devoy, R. 1995. Evidence from sites in Western Ireland of late Holocene changes in coastal environments. *Marine Geology*, 124, 273-287.

D'Evelyn, A.M. 1933. A sandhill settlement, Maghera, Co. Donegal. *Journal of the Royal Society of Antiquaries of Ireland*, 63, 88-100.

Fernandez-Partagas, J. and Diaz, H. F. 1996. Atlantic hurricanes in the second half of the Nineteenth Century. *Bulletin of the American Meteorological Society*, 77, 2899-2906.

FitzGerald, D.M. 1988. Shoreline erosional-depositional processes associated with tidal inlets. In: Aubrey, D.G. and Weishar, L. (Eds) *Hydrodynamics and Sediment Dynamics of Tidal Inlets*. Springer-Verlag Inc., New York, 29, 186-225.

Hayes, M.O. 1972. Forms of Sediment Accumulation in the Beach Zone. In: Meyer, R.E. (Ed) *Waves on Beaches and Resulting Sediment Transport*. Academic Press, New York, 297-356.

Haynes, J. and Dobson, M. 1969. Physiography, foraminifera and sedimentation in the Dover estuary (Wales). *Geological Journal*, 6, 217-256.

Hume, T.M. and Herdendorf, C.E. 1992. Factors controlling tidal inlet characteristics on low drift coasts. *Journal of Coastal Research*, 8, 355-375.

Hurrell, J.W. 1995: Decadal trends in the North Atlantic Oscillation and relationships to regional temperature and precipitation. *Science*, 269, 676-679.

Hurrell, J.W. and Van Loon, H. 1997. Decadal variations in climate associated with the North Atlantic Oscillation. *Climatic Change*, 36, 301-326.

Jago, C.F. 1980. Contemporary accumulation of marine sand in a macrotidal estuary, southwest Wales. *Sedimentary Geology*, 26, 21-49.

Joyce, R.W. 1912. Recollection of two great storms. *Catholic Bulletin*, 2, 15-19.

Jones, P.D. 2002. North Atlantic Oscillation Index Data: UEA Climate Research Unit http://www.cru.uea.ac.uk/cru/data/nao.htm (accessed June 2002).

Jones, P.D., Jónsson, T. and Wheeler, D. 1997: Extension to the North Atlantic Oscillation using early instrumental pressure observations from Gibraltar and South-West Iceland. *International Journal of Climatology*. 17, 1433-1450.

King, C.A.M. 1965. Some observations on the beaches of the west coast of County Donegal. *Irish Geography*, 2, 40-50.

Knowles, W.J. 1901. Prehistoric remains from the sandhills of the coast of Ireland. *Proceedings of the Royal Irish Academy*, 22, 331-389.

Lamb, H.H. 1991. *Historic Storms of the North Sea, British Isles and North-West Europe*. Cambridge University Press, Cambridge.

MacClenahan, P., McKenna, J., Cooper, J.A.G. and O'Kane, B. 2001. Identification of highest magnitude coastal storm events over western Ireland on the basis of wind speed and duration thresholds. *International Journal of Climatology*, 21, 829-842.

Morris, B.D., Davidson, M.A. and Huntley, D.A. 2001. Measurements of a coastal inlet using video monitoring techniques. *Marine Geology*, 175, 251-272.

Orford, J.D., Wheeler, A.J., McCloskey, J., Dardis, O., Doherty, J. and Gallagher, K.A. 1996. Variations in climate forcing of coastal processes and the coastal response along the European Atlantic shoreline. *EU Environmental Programme - Phase II: Climate Change and Coastal Evolution in Europe.*

Orford, J.D., Cooper, J.A.G. and McKenna, J. 1999. Mesoscale temporal changes to foredunes at Inch Spit, south-west Ireland. *Zeitschrift für Geomorphologie*, 43, 439-461.

PSMSL. 2002. Permanent Service for Mean Sea Level: revised local reference sea level data http://www.psmsl.ac.uk/gloss (accessed June 2002).

Pugh, D.T. 1982. A comparison of recent and historical tides and mean sea-levels of Ireland. *Geophysical Journal of the Royal Astronomical Society*, 71, 809-815.

Pye, K. 1996. Evolution of the shoreline of the Dee estuary, United Kingdom. In: Nordstrom, K.F. and Roman, C.T. (Eds) *Estuarine Shores: Evolution, Environments and Human Alterations*. John Wiley & Sons, Chichester, 15-37.

Pye, K. and Neal, A. 1994. Coastal dune erosion at Formby Point, north Merseyside, England: Causes and mechanisms. *Marine Geology*, 119, 39-56.

Robinson, A.H.W. 1960. Ebb-flood channel systems in sandy bays and estuaries. *Geography*, 45, 183-199.

Robinson, A.H.W. 1975. Cyclical Changes in Shoreline Development at the Entrance of Teignmouth Harbour, Devon, England. In: Hails, J. and Carr, A. (Eds) *Nearshore Sediment Dynamics and Sedimentation*. John Wiley & Sons, London, 181-200.

Rohan, P.K. 1986. *The Climate of Ireland*. The Stationery Office, Dublin, 146 pp.

Scott, B. 1996. *The morphodynamics of coarse clastic beaches: Examples from North Donegal, Ireland.* Unpublished DPhil Thesis, University of Ulster.

Shaw, J. and Carter, R.W.G. 1994. Coastal peats from northwest Ireland: implications for late-Holocene relative sea-level change and shoreline evolution. *Boreas*, 23, 74-91.

Shields, L. and Fitzgerald, D. 1989. The 'Night of the Big Wind' in Ireland, 6-7 January 1839. *Irish Geography*, 22, 31-43.

Taylor, R.B., Carter, R.W.G., Forbes, D.L. and Orford, J.D. 1986. Beach sedimentation in Ireland: contrasts and similarities with Atlantic Canada. *Current Research, Part A, Geological Survey of Canada*, Paper 86-1A, 55-64.

Ulbrich, U. and Christoph, M. 1999. A shift of the NAO and increasing storm track activity over Europe due to anthropogenic greenhouse gas forcing. *Climate Dynamics*, 15, 551-559.

Van Der Spek, A.J.F. 1997. Tidal asymmetry and long-term evolution of Holocene tidal basins in The Netherlands: simulation of palaeo-tides in the Schelde estuary. *Marine Geology*, 141, 71-90.

Van Heerden, I.L. 1986. Fluvial sedimentation in the ebb-dominated Orange Estuary. *South African Journal of Science*, 82, 141-147.

Webb, C.K., Stow, D.A. and Howard, H.C. 1991. Morphodynamics of southern California inlets. *Journal of Coastal Research*, 7, 167-187.

Wheeler, A.J., Orford, J.D. and Dardis, O. 1999. Saltmarsh deposition and its relationship to coastal forcing over the last century on the north-west coast of Ireland. *Geologie en Mijnbouw*, 77, 295-310.

Chapter 12

COMPLEX MORPHO-HYDRODYNAMIC RESPONSE OF ESTUARIES AND BAYS TO WINTER STORMS: NORTH-CENTRAL GULF OF MEXICO, USA

Gregory W. Stone[1,2], B. Prasad Kumar[1], A. Sheremet[1,2] and Dana Watzke [2]
[1]*Coastal Studies Institute, Louisiana State University, Baton Rouge, Louisiana, USA*

[2]*Department of Oceanography and Coastal Sciences, Louisiana State University, Baton Rouge, Louisiana, USA*

1. INTRODUCTION

1.1 Background

Concepts pertaining to our understanding of estuarine dynamics have been heavily influenced by work carried out on the east and west coasts of the United States and western Europe (Pritchard, 1967). Antecedent geological controls have played an important role in predetermining the dominant type of estuaries along these coasts, namely drowned river valleys on coastal plains and fjord type systems tuned to moderate/high tidal regimes. Along the northern Gulf of Mexico (Fig. 1), however, estuaries are predominantly bar-built where the latest Holocene "stillstand" in sea level has permitted waves to build barrier islands/spits/beaches supplied by sediment from updrift and offshore sand sources (Stone *et al.*, 1992; Stapor and Stone, 2004). Tides in the Gulf of Mexico are microtidal (0-0.3 m), predominantly diurnal and mixed (Marmer, 1954). Characteristically broad regions of low bathymetric relief result in minimal bathymetric steering of the otherwise low-frequency flow (Schroeder and Wiseman, 1999). Due to a incidence of tropical cyclones in the northern Gulf (Stone *et al.*, 1997; Muller and Stone, 2001), low profile barriers are susceptible to multiple

D.M. FitzGerald and J. Knight (eds.),
High Resolution Morphodynamics and Sedimentary Evolution of Estuaries, 243-267.

breaches and inlet development. Such occurrences play an important role in estuarine circulation patterns due to phase lags in tidally driven waves. These interlinkages have, however, yet to be fully explored (Schroeder and Wiseman, 1999).

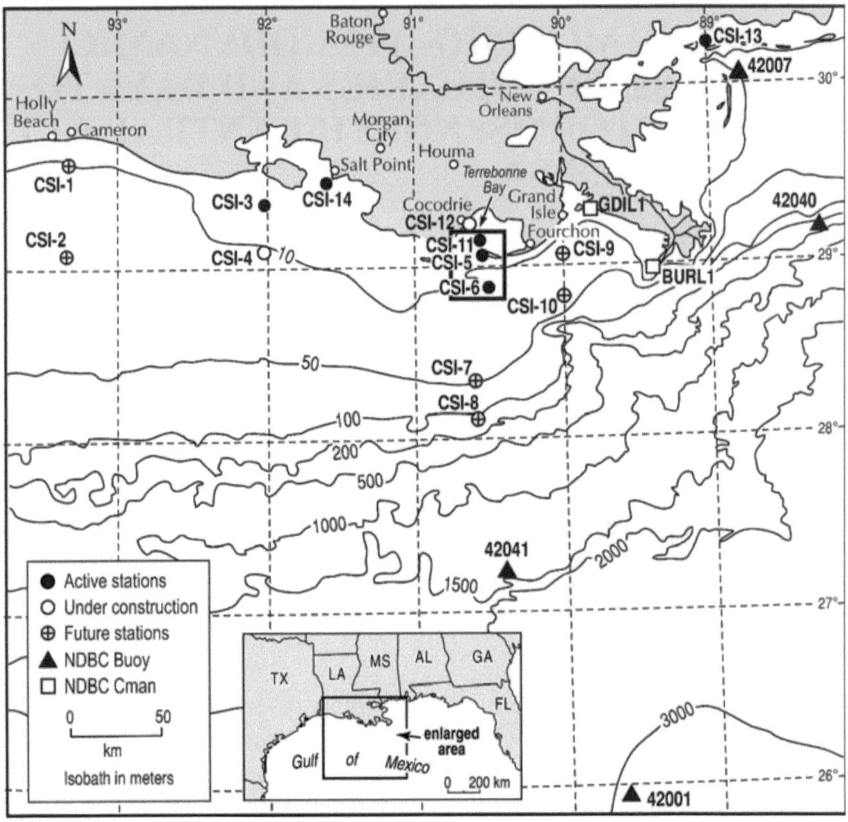

Figure. 1. Study area showing distribution of WAVCIS stations and NDBC buoys off the Louisiana coast. The rectangle shows the grid location used for numerical modeling and is enlarged in figure 4.

A land-sea breeze effect is apparent along the northern Gulf all year long, but becomes particularly pronounced during the summer months. Much of the synoptic variability in the wind field, however, is due to the impact of cold air outbreaks, or cold fronts. Approximately 30 fronts per year pass over the northern Gulf every 3-10 days in winter (Chaney, 1999); their strength and frequency decrease during summer (DiMego *et al.*, 1976). Analysis of wind data indicates that the fronts generally have a southwest-northeast orientation and can exert significant control of coastal processes given the east-west alignment of the northern Gulf of Mexico coast (Fernandez-Partaga and Mooers, 1975; Chaney, 1999; Pepper and Stone, 2002).

1.2 Objectives

Although it is becoming apparent that cold fronts play a significant role in the short-term evolution of bays and estuaries along the northern Gulf (e.g. Roberts *et al.*, this volume), detailed evaluations of their morpho-hydrodynamic response to the complete cold front cycle have not been carried out to date. Thus, it is the objective of this paper to accomplish this goal using *in situ* observations coupled to a wave model along the mid and upper shoreface fronting a substantial bay system, Terrebonne Bay, Louisiana (USA). A review of the salient morphologic response of bay shorelines to frontal forcing is also provided.

2. FRONTAL IMPACTS ON COASTAL PROCESSES

2.1 Hydrodynamic Effects

Pre-frontal winds blow from the south and generate energetic waves along the Gulf-facing beaches (Sheremet and Stone, 2003). Post frontal winds blow from the north and generate high frequency, energetic waves in the adjacent estuaries and bays where the fetch is long enough. Data recorded by WAVCIS Station CSI 11 located in Terrebonne Bay on the 3.5 m isobath (Fig. 1) are presented as an example of frontal effects on coastal processes. In figure 2 time series of sustained wind speed, wind direction, significant wave he t, wave direction, and wave spectral evolution are presented for a one month period in January 2004.

Four strong fronts crossed the site during January 2004. Sustained wind speeds approximate 10-12 ms^{-1} during the peak of each event when winds were from the north. Prior to the arrival of each event winds from the south show a persistent veering to the west and north as the frontal boundary nears the coast. North winds associated with the post-frontal phase usually persist for 1-2 days, gradually decreasing in speed and veering to the south. Maximum significant wave height approximates 0.5 m for each event and is coincident with peaks in sustained wind speed. Wind and wave direction are strongly correlated since peak energy is associated with peak wind speed. Shifts in energy to higher frequencies are also apparent and are coincident with wind veering to the north. Assuming that a cutoff frequency of 0.2 Hz separates swell from seas (Sheremet and Stone, 2003), frontal energy dominates the short wave band and waves propagate south towards the flanking barrier islands along the southern portion of the bay. In the intervening periods between fronts, wave energy is negligible, even during pre-frontal winds, due to the sheltering effect of the barrier islands.

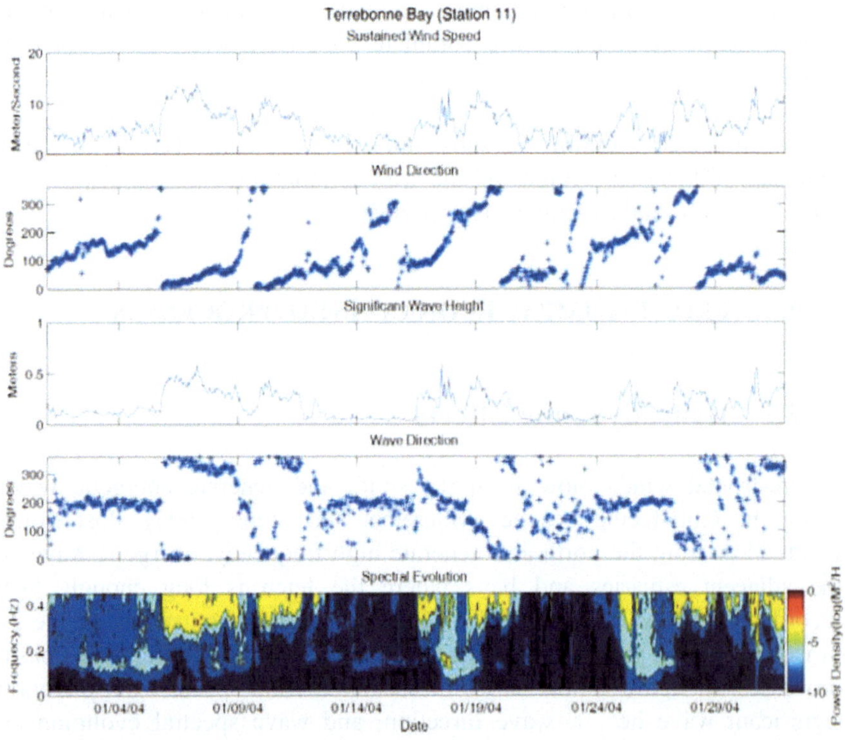

Figure. 2. Time series of sustained wind speed and direction, significant wave height, direction and spectral evolution during January 2004. Four strong cold fronts impacted the area during this period. Data were obtained from CSI 11.

2.2 Morphological Effects

Within time scales of years to decades, cold fronts play an important role in the morphologic evolution of the foredune-beach nearshore system in estuarine/bay environments along the northern Gulf. This is not only because of their high frequency of occurrence, but also the east-west orientation of the coast which is generally perpendicular to the front (Stone *et al.*, in press). Chronic beach erosion along estuarine/bay beaches of the northern Gulf over an expansive reach of coast spanning the Florida panhandle to Louisiana (> 300 km) was attributed to cold front passages by Stone *et al.* (1999; in press). Their data support the contention that winter cold fronts play the key role in beach erosion along north-facing barriers flanking the southern rim of estuaries and bays over short time scales. Their work centered on evaluating a high resolution time series of bathymetric/topographic change over a 6.5 year period. Their data show that after significant morphological change to a barrier island in Florida (Santa Rosa Island) during Hurricane Opal (1995), the Gulf beach-nearshore and associated dunes began recovery approximately two years after landfall. Throughout the entire monitoring period, however, the bay beach continued to lose sediment through wave erosion (Fig. 3). The prevalence of fronts over the Florida site explains the continued loss of sediment from the bay beach; however, what is less clear is the actual pathway of sand transport during these events. Transport and ultimate fate of material eroded is not yet understood. In addition, the relative significance of longshore and offshore transport is not clear. The presence of oblique transverse bars, observed at numerous locations along the northern Gulf (Zapel, 1984) and other estuarine beaches (cf. at Fire Island, New York, Nordstrom *et al.*, 1996), indicates a combination of longshore and cross-shore transport. The strike of the oblique transverse bars cannot be explained by one mechanism alone and Stone *et al.* (in press) hypothesize a genetic relationship between the sum of longshore and cross-shore transport vectors. Based on findings from a field experiment carried out along the Florida panhandle during frontal passages, Stone *et al.* (in press) discussed the possibility of near-bottom bayward flow. A bottom boundary layer tripod was deployed at the 1 m isobath and captured two fronts over a nine-day period. Two intervals are apparent when offshore currents occur as the significant wave height increases with wind speed and northward veering. Between fronts when wave energy is low, onshore currents occur but as wave energy increases with the second post-frontal period, offshore currents are apparent.

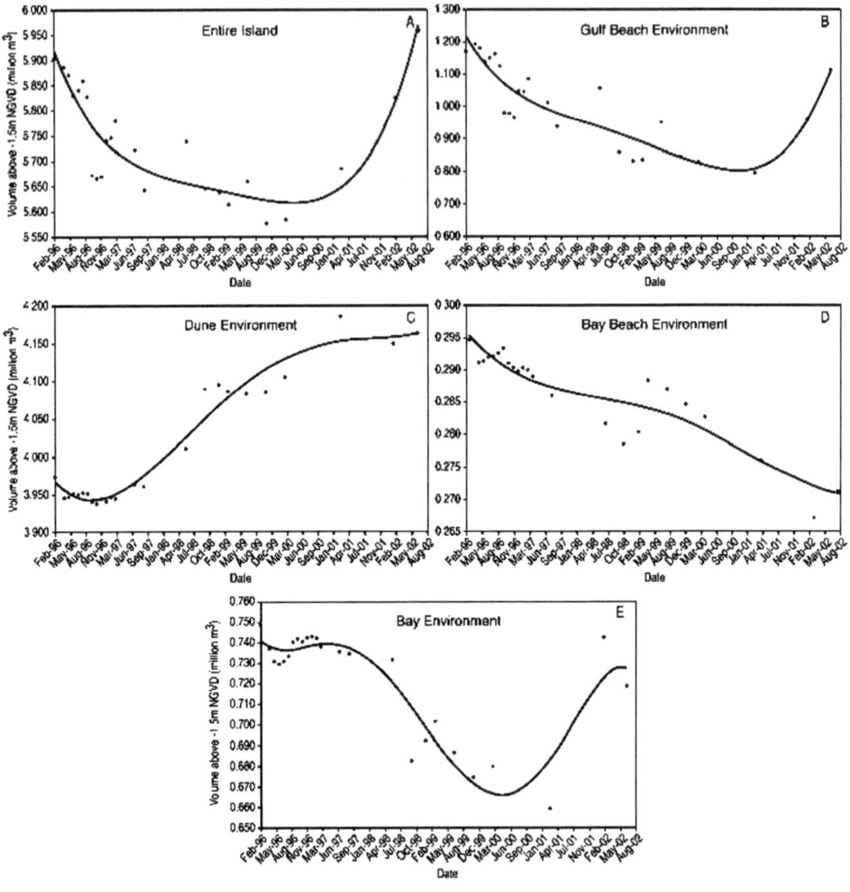

Figure 3. Time series of volumetric change from 2/96-8/02 along Santa Rosa Island, Florida, for the entire section of the study site (A), Gulf beach (B), dune system (C), bay beach (D) and bay platform (E) (modified from Stone et al., in press).

Similar observations have been made elsewhere along the northern Gulf where, in Mississippi Sound, the north-facing beaches of West Ship Island have eroded at rates of 1.6 m yr^{-1} over a four year period (Chaney and Stone, 1996). Ongoing research focuses on the short-term (5 years) response of an island in Terrebonne Bay to frontal activity. Preliminary data show 1.5 m yr^{-1} of erosion along the north-facing flank of the island, largely attributable to post-frontal wind-wave forcing.

3. EXPERIMENTAL SETTING

A recent study shows that along with Key West, Florida, south-central Louisiana ranked the highest over a 100 year period (1900-2000) in frequency of major storm strikes (category 3 and above) for an area extending from Texas to North Carolina (Muller and Stone, 2001). The incidence of tropical cyclones remains high throughout the northeastern Gulf and the late Holocene barrier islands that dominate the coast are exceptionally vulnerable to storm impacts (see reviews in Stone and Finkl, 1995; Stone *et al.*, 1997; in press). With a frequency of ~ 30/year, the most common meteorological events in coastal Louisiana are, however, cold fronts. Many authors have examined relationships between nearshore waves, currents, sediment concentrations and erosion/deposition patterns during frontal passages (e.g. Davis and Fox, 1975; Dingler *et al.*, 1993; Chaney and Stone, 1996; Addad and Matrins-Neto, 2000; Perez *et al.*, 2000; Keen, 2002). Although waves and currents during cold fronts are weaker compared to extreme events, their high frequency of occurrence has proven a significant factor in their role in low energy coastal morphodynamics in the Gulf of Mexico (Roberts *et al.*, 1987; Moeller *et al.*, 1993; Stone *et al.*, 1999; in press; Huh *et al.*, 2001).

Synoptic scale classification systems have also been applied to the meteorology of the northern Gulf of Mexico. Notably, Muller (1977) subdivided New Orleans' weather into eight synoptic types that included both storms and fair weather. Roberts *et al.* (1987) identified two end member types of extratropical storms in coastal Louisiana: the migrating cyclone, characterized by the passage of a cold front aligned oblique to the coast; and the Arctic surge, in which a front is aligned parallel to the coast. Chaney (1999) subdivided characteristic synoptic weather patterns responsible for extratropical storms over the northern Gulf of Mexico into seven categories. Different synoptic types were shown to be associated with unique meteorological conditions capable of generating a range of hydrodynamic responses. The primary and secondary fronts categorized in the synoptic weather pattern account for approximately 90% of storm activity along the northern Gulf of Mexico. The cold front cycle has commonly been used to characterize the sequence of events that accompanies a typical extratropical storm passage (Roberts *et al.*, 1987, 1989; Armbruster *et al.*, 1995; Chaney, 1999). The initial pre-frontal phase includes strong, warm moist winds that blow from the southerly direction. The ensuing frontal phase is characterized by a sudden drop in air pressure, erratic winds with short life, but occasionally intense squalls. Finally, the post-frontal phase occurs during which temperature and humidity drop, air pressure rises and winds are strong and northeasterly to northwesterly. A summary of wave field response on the inner shelf off western Louisiana is provided in Roberts *et al.* (this volume).

4. NUMERICAL MODELS AND *IN SITU* OBSERVATIONS

4.1 Outline

Due to recent advances in technology, many wave buoys and shallow water gauges deployed along the coasts of the U.S. measure directional wave spectra. Here, circulation and wave forecast models are used in operational and semi-operational modes under the responsibilities of state and federal agencies. Notable examples are along the east coast (Aikman *et al.*, 1996), the west coast (Clancy *et al.*, 1996), Great Lakes (Schwab and Bedford, 1994), Tampa Bay (Vincent *et al.*, 2000) and Galveston Bay in the Gulf of Mexico (Schmatz, 2000). Apart from these the U.S. Navy disseminates information on waves and currents on global and regional domains through operational models (Horton *et al.*, 1992). Operational third generation models, e.g. WAM (WAMDI Group, 1988) and WaveWatch-III (Tolman, 1997, 1999), are used for global forecasts and regional applications worldwide (Clancy *et al.*, 1986; Burgers, 1990; Cavaleri *et al.*, 1991; Dell'Osso *et al.*, 1992; Bauer *et al.*, 1992; Monaldo and Beal, 1998; Prasad *et al.*, 2003). Numerous experiments reveal satisfactory performance of these models for deep water applications. However, they cannot be realistically applied to coastal regions with horizontal scales less than 20-30 km and water depths less than 20-30 m. This limitation also pertains to coasts characterized by the presence of estuaries, tidal inlets, barrier islands, tidal flats, channels etc. The third generation model SWAN (Booij *et al.*, 1999) is a more appropriate model to study wave transformation in these coastal areas and nearshore regions. SWAN is being used by scientists worldwide and has been extensively validated for regional scale applications under different environmental forcing (Christopoulus, 1997; Ris *et al.*, 1999; Rogers *et al.*, 2002; Sheremet and Stone, 2003; Cerqueiro *et al.*, 2003). This paper examines an application of SWAN (Version 40.11) to study wave transformation due to a cold front outbreak across the inner shelf south of Terrebonne Bay in coastal Louisiana during March 2003 (Fig. 1). The area is characterized by a gentle seaward slope broken up by ridges up to 2 m in relief (Fig. 4) over short spatial scales. *In situ* observations provided by the WAVCIS (Wave Current Surge Information System, accessible through http://wavcis.csi.lsu.edu) array at three shallow water locations are used as a benchmark to study the evolution of wave transformation across the inner shelf and into Terrebonne Bay.

4.2 Model Formulation

Input to the model consists of bathymetry, water level change and wind fields. The model has the capability to incorporate deep water wave forcing at the open boundaries. It calculates refraction, wave breaking, dissipation, wave-wave interaction, and local wind generation. The model does not compute diffraction and it should not be used when wave heights are expected to vary over a few wavelengths. Thus, the wave field is not generally accurate within the immediate vicinity of obstacles. Dissipation of wave energy is computed for whitecapping, bottom friction, and depth-induced wave breaking. SWAN uses whitecapping formulations as adapted by the WAMDI Group (1988). The depth-induced dissipation formulation in the model is based on the JONSWAP bottom friction formulation with a friction coefficient of 0.067 m^2s^{-3} (Hasselmann *et al.*, 1973).

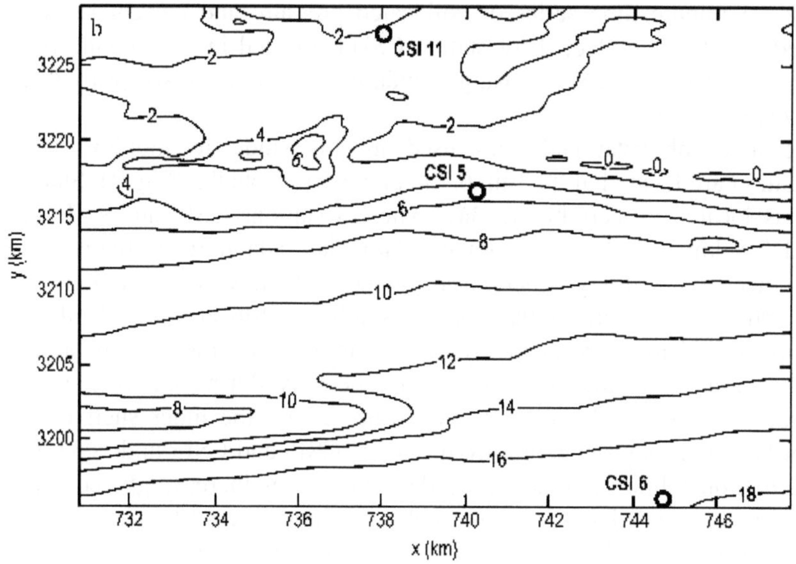

Figure. 4. Fine scale bathymetric grid for SWAN run. Time varying JONSWAP spectra was provided as boundary condition at station CSI6 to study wave propagation along CSI5 and CSI11. Location of this grid is shown in Figure 1.

4.3 *In situ* Observations

The WAVCIS array has provided a stable and comprehensive source of metocean field observations for the Louisiana inner shelf region. The program which transmits online realtime data has provided unprecedented opportunities to observe coastal processes under a wide range of metocean conditions (Stone *et al.*, 2003; Sheremet and Stone, 2003). WAVCIS is an array of observational networked stations distributed off the Louisiana coast, typically designed to withstand tropical cyclone conditions in the Gulf of Mexico. Stations used in this paper are shown in figure 1. In 2002, the system withstood and operated successfully during four tropical cyclones, one of which, Hurricane Lili, was category 4 while in the central Gulf (Stone *et al.*, 2003). The WAVCIS project has six active stations, with three additional ones under construction. The data measured include directional waves, currents, water level water temperature, wind speed and direction, air temperature, barometric pressure, humidity and visibility. Wave measurements consist of hourly, 17.08 minute time series of collocated pressure and current velocity sampled at 2 Hz for stations CSI5 and CSI6. Hourly 8.5 minute time series of collocated pressure and current velocity sampled at 4 Hz is collected for station CSI11, located in Terrebonne Bay. Time series were processed using standard spectral analysis procedures (Earle *et al.*, 1995).

The three observational sites used in this study to monitor wave propagation are located at 20.3 m, 6.7 m and 3.5 m isobaths (Fig. 4). Station CSI11 is located in Terrebonne Bay, whereas stations CSI5 and CSI6 are located 2.5 km and 20 km south off Timbalier Island respectively. The nearest deep water buoys encompassing these shallow water stations are NDBC buoys 42041 and 42001 located south of CSI6 at water depths of 1435 m and 3246 m respectively. The symmetry of these buoy arrays provides valuable metocean information especially during extreme events from the mid Gulf of Mexico in deep water across the inner shelf to the interior bay. Waves propagating from stations CSI6 and CSI5 undergo rapid transformation while approaching CSI11 due to an increasingly complex bathymetry associated with an inlet ebb and flood tide system connecting Terrebonne Bay and the Gulf. We investigate wave evolution and decay at CSI11 in Terrebonne Bay using appropriate boundary condition at CSI6 coupled with meteorological forcing from CSI5 and CSI6 during the cold front episode.

4.4 Case Study: March 2003 Cold Front

In this study, the SWAN model is used to perform a local application in south Terrebonne Bay (Fig. 1) with a geographic domain of 18 km in the east-west direction and 30 km in the north-south direction. The geographical area covers all three WAVCIS stations (CSI6, CSI5 and CSI11) with imposed time varying boundary condition at CSI6, the southern boundary of the computational grid. During the period 28-31 March, 2003, a cold front passed over the study site and the respective wave and wind field signatures were captured by the array. Model simulations were performed during this event which covered the pre-frontal and post-frontal phase.

4.5 Bathymetry

Bathymetry was obtained from three different sources; (1) the United States Geological Survey; (2) digitized National Oceanic and Atmospheric Administration high resolution navigation chart data for the entire Louisiana coast, and (3) deeper water data obtained from Geophysical Data System by the National Ocean Service. The bathymetric grid was created using the triangulated irregular network (TIN) with a system of contiguous non-overlapping triangles. The modeling grid can be converted from the TIN model to the desired resolution for purposes of running SWAN. The numerical grid for the coarse run consisted of 114 points in the X-direction (parallel to the equator) and 61 points in the Y-direction (perpendicular to the equator) with a uniform spatial resolution of 1000 m. The derived nested grid which is a subset of the coarse grid, consisted of 87 points in the X-direction and 167 points in the Y-direction at a spatial resolution of 200 m along both X and Y directions. The isobaths are gently uniform sloped with low gradients amidst stations CSI6 and CSI5. The bathymetric features immediately surrounding CSI11, however, are more complex with multiple sand bars and the presence of barrier islands oriented in southwest-northeast and southeast-northwest directions. The barriers impeded incident waves approaching this station from offshore.

4.6 Boundary conditions

Two types of boundaries are distinguished, viz; coastlines and open sea boundaries. Both boundaries are fu absorbing (no reflection) for wave energy that leaves the computational domain. At the open sea boundaries wave energy can enter the computational area. Boundary conditions were specified along the 10 m isobath encompassing station CSI6. For the present study, the incoming wave components at the up-wave boundaries with

SWAN prescribes a time varying JONSWAP spectrum, the information being provided from a message file of observations from CSI6.

5. RESULTS AND DISCUSSION

5.1 Observations

During pre-frontal conditions wind speeds were observed higher at all three stations with a mean wind speed of 3.5 ms^{-1} blowing from the south (Figs. 5, 6). The pre-frontal period experienced higher air temperatures of approximately 22.5°C observed at stations CSI5 and CSI6. The air temperature at station CSI11 is lower by approximately 2°C compared to the mean temperature recorded at CSI5 and CSI6 during this episode. In the later stage of the pre-frontal phase (03Z March 29, 2003), wind speeds at all three stations fell rapidly. Thereafter, over a five hour time period wind speed steadily increased at all three stations (with no observed phase lag) attaining a maximum sustained wind speed of 14.7 ms^{-1} at CSI5 (maximum wind gust = 17.4 ms^{-1}) during the later part of 29 March. During the onset of the front, relative wind speeds were higher at stations CSI5 and CSI11 compared to CSI6. Dominant wind direction (Fig. 6) was observed predominantly in the northern/northwest quadrant during the post-frontal period as initial wind veering was nearing completion.

As the front moved out over the northern Gulf, wind speeds were seen to increase and then steadily decrease at station CSI6, unlike that observed at the other two stations. This observation provides a unique opportunity to examine the interaction between swell waves in the vicinity of stations CSI5 and CSI11 and locally generated wind waves at CSI6 during the waning stage of the front. The post-frontal phase shows a sharp gradient in decreasing wind speeds at all three stations.

Significant wave he t is shown in figure 7 for the three stations during the cold front event. The pre-frontal period depicts wave growth at CSI6. Although wind speeds show almost similar trends at all three stations, wave growth at CSI5 and CSI11 is not pronounced when compared to CSI6. At CSI 5 we attribute this observation to depth- induced breaking; at CSI11 we attribute it to the sheltering effect attributable to very shallow bathymetry and the presence of barrier islands south of the station. The mean wave direction (Fig. 8) shows waves reverting to northwest/south-southwest at CSI6 whereas, at CSI5 waves were observed propagating south/south-southwest during the initial phase (pre-frontal period) of the cold front. No data on mean wave direction were available from CSI11 during this event.

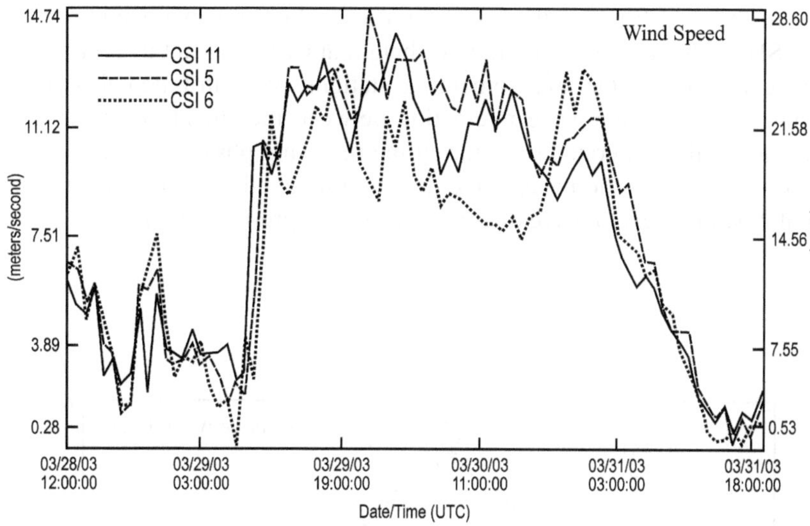

Figure. 5. Time series of wind speed for CSI11, 5 and 6 during the cold front event 28-31 May, 2003.

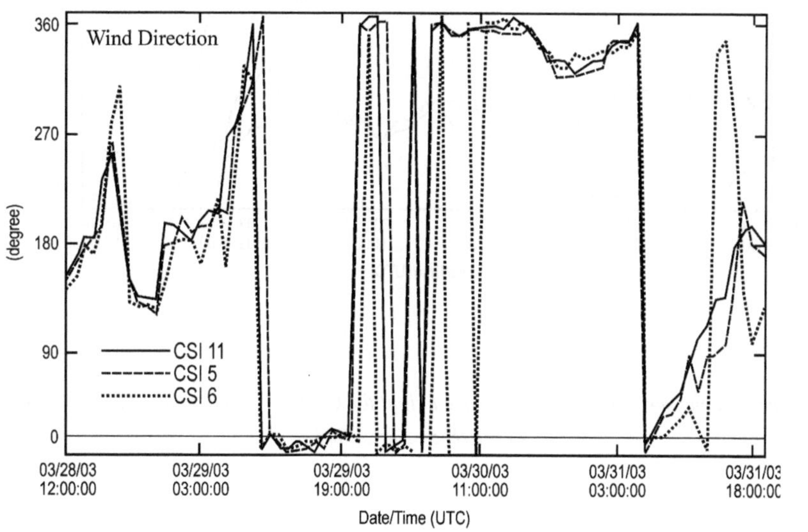

Figure. 6. Time series of wind direction for CSI11, 5 and 6 during the cold front event 28-31 May, 2003.

The frontal activity linked with higher wind speeds consistently from a northerly direction (sustained for ~ 48 hours) resulted in wave propagation to the northeast at CSI5 and CSI6. Significant wave heights (Fig. 7) show a maximum of 1.43 m during the peak frontal activity at CSI6. The decay stage of the cold front resulted in a significant drop in wave height at CSI6 and CSI11. The rate of decay in wave height at CSI5 was not as pronounced when compared to the two remaining stations. We attributed this to the strong nonlinear wave-wave interaction between the already existent wave field and locally generated seas during the reversing phase of winds. During post-frontal activity the mean wave direction reverted back to southerly and was almost identical to wave direction during the pre-frontal phase.

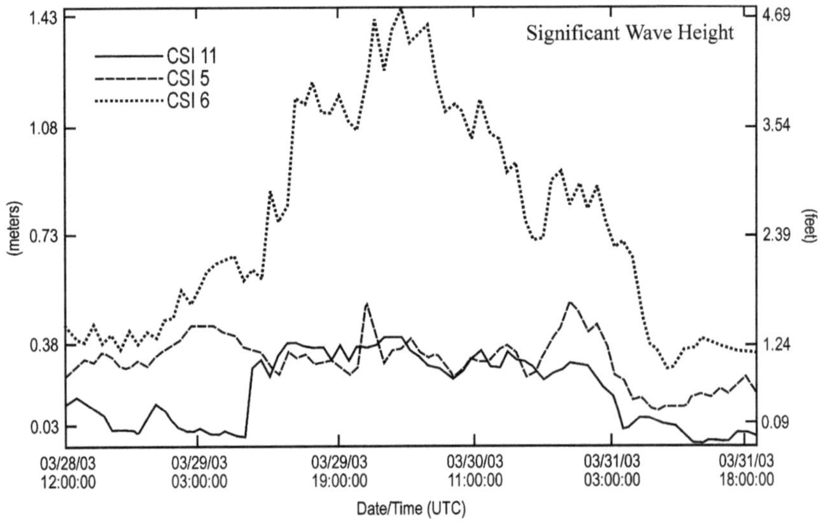

Figure. 7. Time series of significant wave height for CSI11, 5 and 6 during the cold front event 28-31 May, 2003.

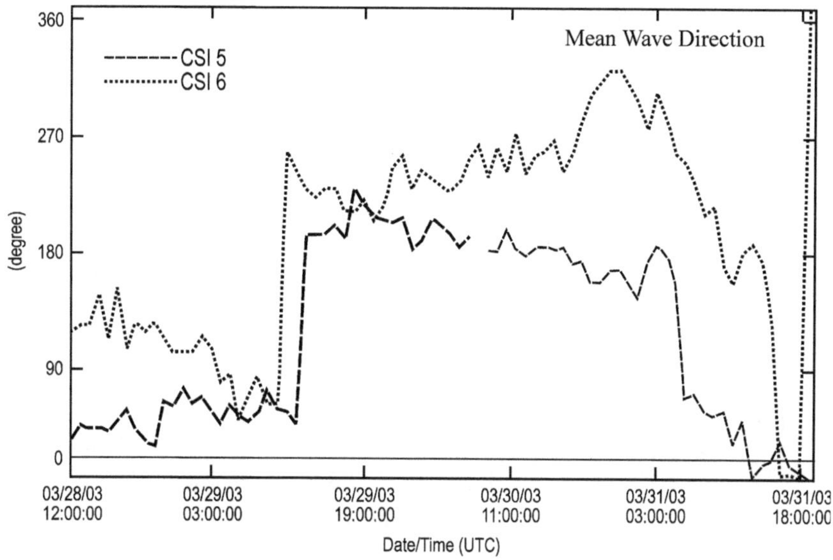

Figure. 8. Time series of mean wave direction for CSI 5 and 6 during the cold front event 28-31 May, 2003.

Mean wave periods at CSI5, CSI6 and CSI11 are shown in figure 9. Long period, short-crested waves are observed at all three stations during the pre-frontal phase and sustained until mid 29 March. An increase in wind speed resulted in short period, high wind waves (Fig. 7). During the decay phase of the front, wave periods were found increasing (primarily dominated by long period waves) in the study domain. Interestingly, the mean wave period at CSI5 is higher during the post-frontal phase when compared to CSI6 which is located in considerably deeper water.

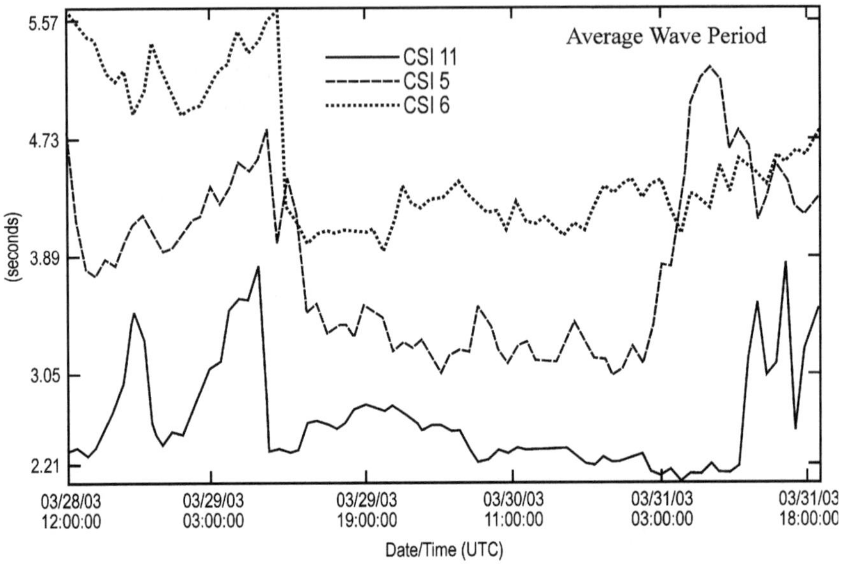

Figure. 9. Time series of mean wave period for CSI11, 5 and 6 during the cold front event 28-31 May, 2003.

5.2 Spectral Evolution

SWAN simulations of the spectral evolution of wave energy at CSI5 and CSI11 are shown in figure 10 (a-b) and corresponding wind speed at all three stations in figure 10 (c). The spectral evolution of wave energy at CSI5 is shown in figure 10 (a) which depicts the presence of high frequency waves. This is seen during the transition phase when the wind direction shifted after the pre-frontal phase. Although it was evident that the fully developed stage was achieved during the frontal period at station CSI5, the slope of the high energy band reveals the dominance of wind generated waves during the later stage of the frontal activity. The spectra during the post-frontal period however, notably show no presence of isolated swells. The relative shift in the energy band to higher frequencies is linked with strong winds (Fig. 10c) during the peak frontal activity.

At CSI11 (Fig. 10b) spectral peaks corresponding to 0.5 Hz and 0.46 Hz were noticed both during the pre-frontal and post-frontal periods. The period corresponding to the shift in wind direction (later stage of pre-frontal activity) show the presence of low frequency low energy waves centered in the band around 0.2 Hz. As was noted at CSI5 (Fig. 10a) the post-frontal period shows a dominance of high frequency waves with no signatures in the low frequency part of the spectrum.

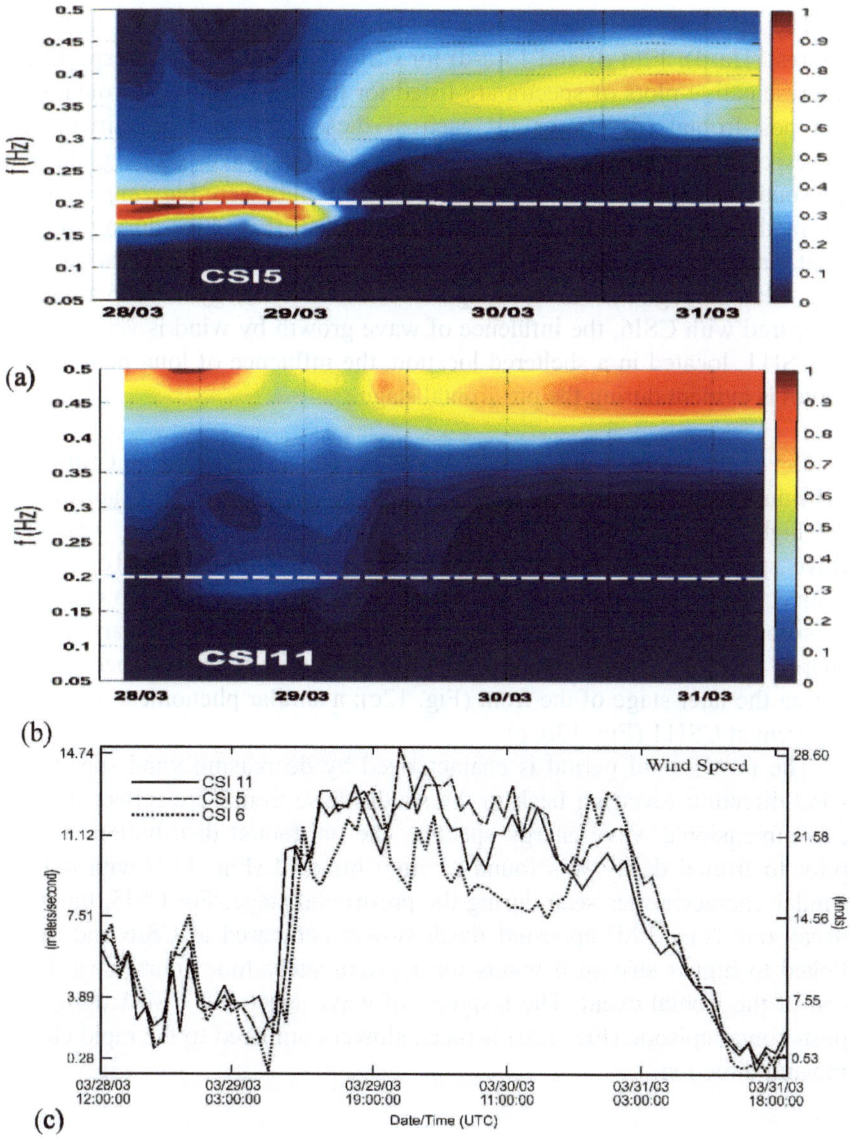

Figure. 10. Temporal evolution of wave spectrum at (a) CSI5, (b) CSI11 and (c) wind speed for the cold front period 28-31 March, 2003

5.3 Two dimensional Directional Spectra

The spectral evolution of two dimensional wave energy is shown in figures 11(a-d), 12(a-d) and 13(a-d) for CSI6, CSI5 and CSI11 respectively. Representative plots of spectra are listed for the pre-frontal (a), frontal (b-c) and post-frontal (d) phases for each of these stations. The initial phase corresponding to the pre-frontal conditions signifies waves propagating from the south, the initial direction from where the winds were blowing. A strong uni-modal spectral distribution is evident at CSI6 (Fig. 11a) during the pre-frontal activity centered at 0.2 Hz with a higher directional spread at CSI5 (Fig. 12a). Though the relative magnitudes of wave energy at CSI5 are lower compared with CSI6, the influence of wave growth by wind is very evident. At CSI11, located in a sheltered location, the influence of long period swell waves is evident during the pre-frontal stage.

Development of wind-induced wave growth is apparent at CSI6 (Fig. 11b) during the onset of the front when wave energy growth is apparent in the high frequency tail; this occurs concurrently at CSI5 (Fig. 12a). The wave spectra during the later stage of frontal activity (Fig. 11c) show a distinct sharp spectral peak associated with strong winds blowing in a persistent direction. At CSI5 (Fig. 12b), the energy spectrum clearly reveals energy build-up at high frequency and its shift with a narrow spectral peak during the later stage of the front (Fig. 12c); a similar phenomenon was also apparent at CSI11 (Fig. 13b, c).

The post-frontal period is characterized by decreasing wind speeds and wind direction reverting back to the south, these trends are reflected in the two dimensional wave energy spectra. The uni-modal distribution at CSI6 prior to frontal decay was found to align bimodal (Fig. 11d) with notably similar characteristics seen during the pre-frontal stage. For CSI5, the wave decay rate (Fig. 12d) appeared much slower compared to CSI6 and this is linked to higher sustained winds for approximately four hours towards the end of the frontal event. The response of wave spectra at CSI11 during the post-frontal episode (Fig. 12d) is much slower compared to the rapid change in wind direction.

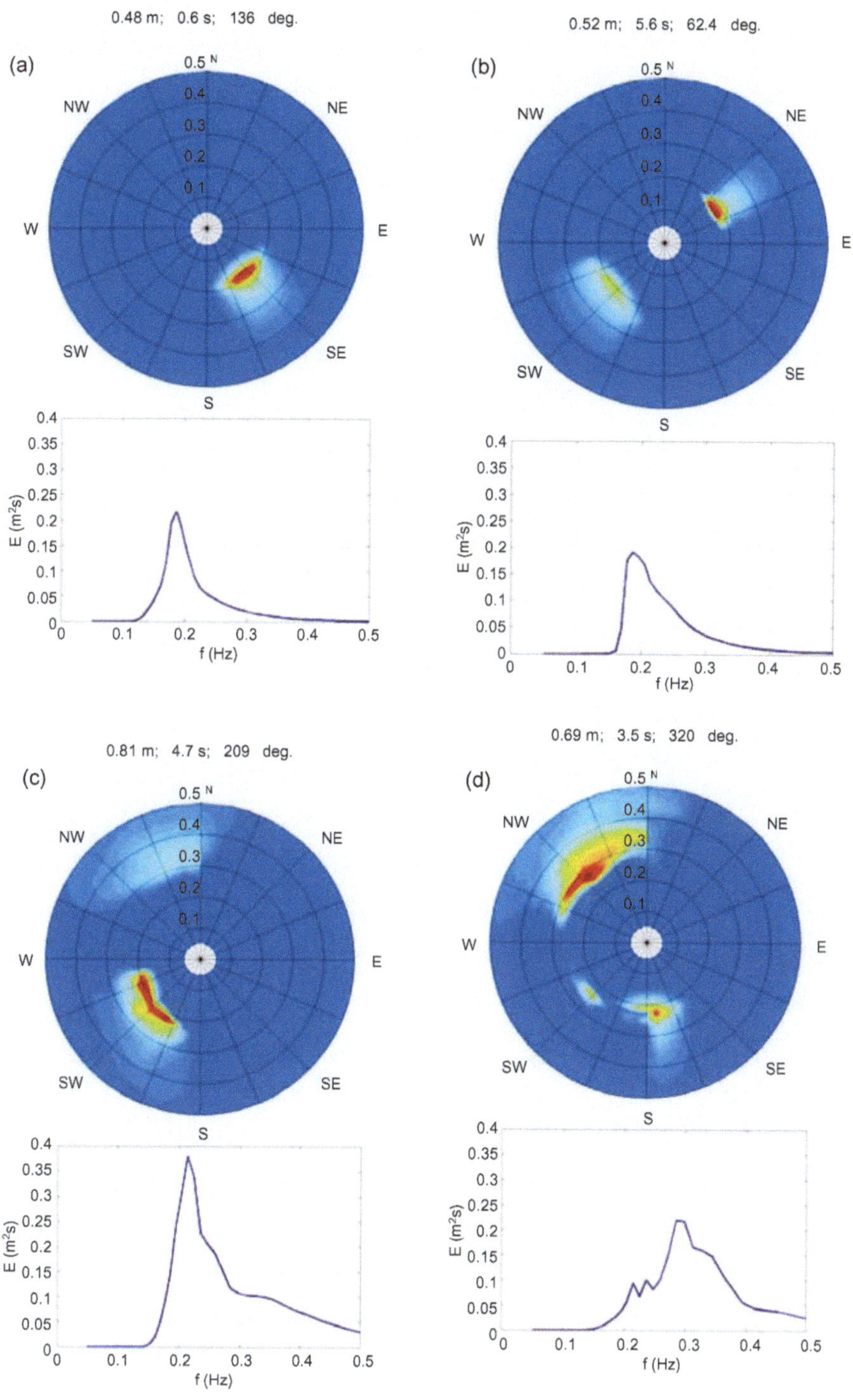

Figure. 11. Spectral evolution of wave energy at CSI6 during (a) pre-frontal stage, (b) and (c) frontal and (d) post-frontal stage

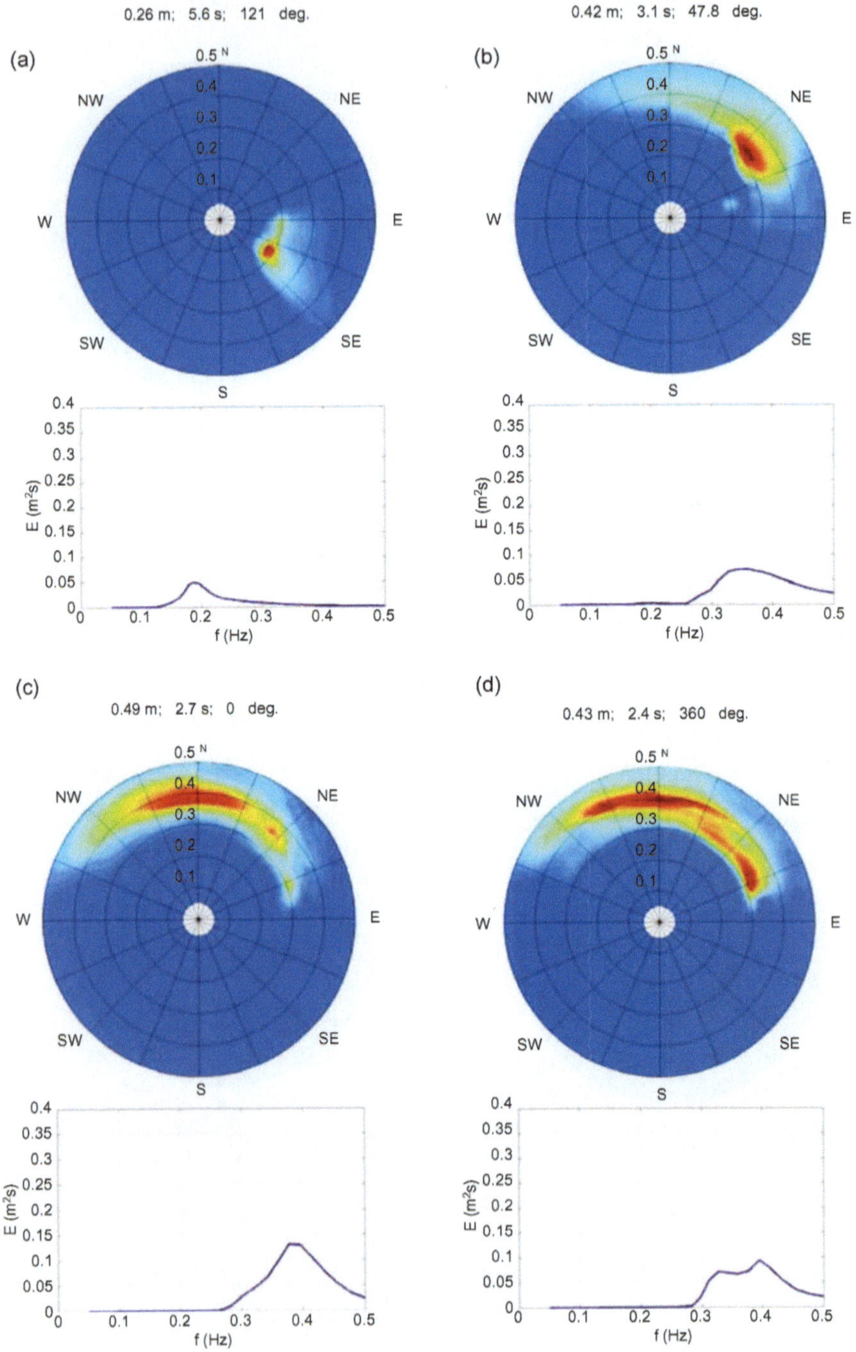

Figure. 12. Spectral evolution of wave energy at CS I5 during (a) pre-frontal stage, (b) and (c) frontal and (d) post-frontal stage.

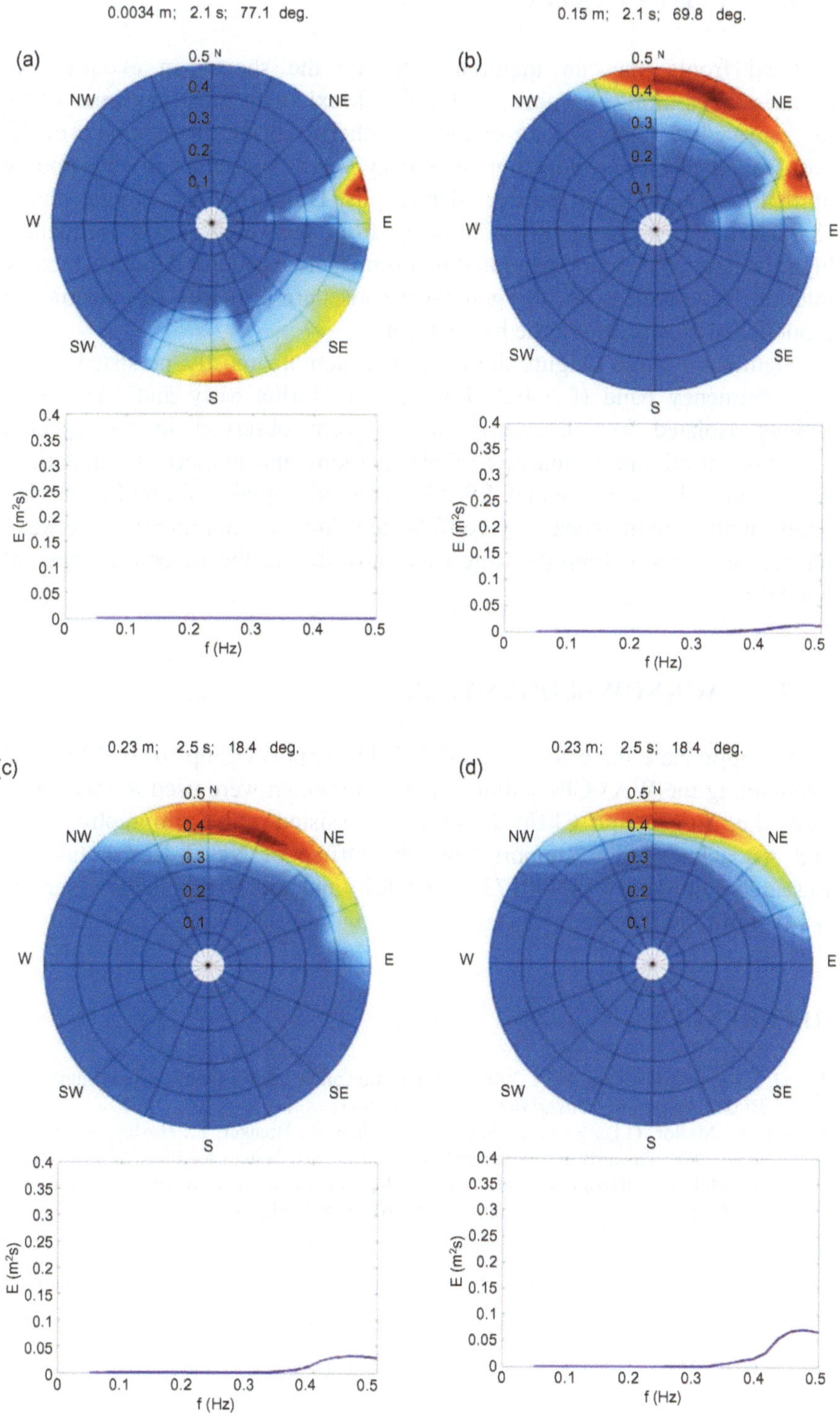

Figure. 13. Spectral evolution of wave energy at CS I11 during (a) pre-frontal stage, (b) and (c) frontal and (d) post- frontal stage

6. CONCLUSIONS

Cold fronts play an important role on the short-term evolution of estuaries/bays along the northern Gulf of Mexico. In most systems, where the estuary/bay fetch is long enough, northerly winds characteristic of the post-frontal phase develop high frequency waves that are actively eroding north facing beaches of barrier islands/spits. During the pre-frontal phase, characterized by strong winds blowing from the south, lower frequency, higher energy waves are generated that propagate through inlets into the bay interior. Nevertheless, locally generated waves are primarily responsible for pronounced erosion along the bay margins.

Significant wave heights during frontal activity mostly persisted in the high frequency band ($f > 0.2$ Hz), whereas during early and late frontal activity isolated low frequency waves were observed in the spectral evolution. Preliminary analysis of observations and numerical simulations signify that the wave model SWAN captured signals of rapidly varying winds with a high degree of confidence. Complex nonlinear wave-wave interaction plays a dominant role especially during the reversing phase of wind direction.

7. ACKNOWLEDGEMENTS

We appreciate the efforts of CSI's Field Support Group for building and maintaining the WAVCIS stations, data from which were used in this paper. Mary Lee Eggart and Clifford Duplechin assisted with cartography. GWS and AS acknowledge support from the Office of Naval Research/Naval Research Laboratory (#N00173-03-1-6907) and the National Park Service (#1443CA532097010).

REFERENCES

Addad, J. and Matrins-Neto, M.A. 2000. Deforestation and coastal erosion: A case study from Brazil. *Journal of Coastal Research*, 16, 423-431.

Aikman, F., Mellor, G.L., Ezer, T., Sheinin, D., Chen, P., Breaker, L., Bosley, K and Rao D.B. 1996. Towards an operational nowcast/forecast system for the U.S. East Coast. In: Malanotte-Rizolli, P. (ed) *Modern Approaches to data assimilation in Ocean Modeling*. Elsevier Oceanographic Series 61, New York. 347-376.

Armbruster, C.K., Stone, G.W. and Xu, J.P. 1995. Episodic atmospheric forcing and Bayside foreshore erosion: Santa Rosa Island, Florida. *Gulf Coast Association of Geological Societies, Transactions*, 45, 31-37.

Bauer, E., Hasselmann, S., Hasselmann, K and Graber, H.C. 1992. Validation and assimilation of Seasat altimeter wave heights using the WAM wave model. *Journal of Geophysical Research*, 97 (C8), 12671-12682.

Booij, N., Ris, R.C. and Holthujsen, L.H. 1999. A third generation wave model for coastal regions. Part I: Model description and validation. *Journal of Geophysical Research*, 104 (C4), 7649-7666.

Burgers, G. 1990. *A guide to the Nedwam wave model*. Scientific Report WR-90-04, KNMI, Netherlands, De Bilt.

Cavaleri, L., Bertotti, L. and Lionello, P. 1991. Wind wave cast in the Mediterranean Sea. *Journal of Geophysical Research*, 96 (C6), 10739-10764.

Cerqueiro, D., Gomez, L.M., Gomez-Gesteira, M. and Carretero, J.C. 2003. Sensitivity of the SWAN model in a local application to the Artabro Gulf (NW Spain). *Thalassas*, 19, 33-43.

Chaney, P.L. 1999. *Extratropical storms of the Gulf of Mexico and their effects along the Northern Coast of a Barrier Island: West Ship Island, Mississippi.* Unpublished PhD Dissertation, Louisiana State University, 211pp.

Chaney, P.L. and Stone, G.W. 1996. Soundside erosion of a nourished beach and implications for winter cold front forcing: West Ship Island, Mississippi. *Shore and Beach*, 64, 27-33.

Christopoulus, S. 1997. Wind wave modeling aspects within complicate topography. *Annales Geophysicae*, 15, 1340-1353.

Clancy, R.M., Kaitala, J.E and Zambresky, L.F. 1986. The Fleet Numerical Oceanography Center Global Spectral Ocean Wave Model. *Bulletin of the American Meteorological Society*, 67, 498-512.

Clancy, R.M., DeWitt., P.W., May, P. and Ko, D.S. 1996. Implementation of a coastal ocean circulation model for the west coast of the United States. *Proceedings of the American Meterological Society, Conference on Oceanic and Atmospheric Prediction*, 72-75.

Davis, R.A. and Fox, W.T. 1975. Process-response patterns in beach and nearshore sedimentation: 1 Mustang Island, Texas. *Journal of Sedimentary Petrology*, 45, 852-865.

Dell'Osso, L., Bertoti, L. and Cavaleri, L. 1992. The Gorbush Storm in the Mediterranean Sea: Atmospheric and wave simulation. *Monthly Weather Review*, 120, 77-90.

DiMego, G.J., Bosart, L.F. and Endersen, G.W. 1976. An examination of the frequency and mean conditions surrounding frontal incursions into the Gulf of Mexico and Caribbean Sea. *Monthly Weather Review*, 104, 709-718.

Dingler, J.R., Reiss, T.E. and Plant N.G. 1993. Erosional patterns of the Isles Derniers, Louisiana, in relation to meteorological influences. *Journal of Coastal Research*, 9, 112-125.

Earle., M.D., McGehee, D. and Tubman, M. 1995. *Field Wave Gaging Program, Wave data analysis standard.* U.S. Army Corps of Engineers Instructions Report CERC-95-1, 33pp.

Fernandez-Partegas, J. and Mooers, C.N.K. 1975. *Some front characteristics over the eastern Gulf of Mexico and surrounding land areas*. Final report to the Bureau of Land Management under contract 08550-CT4-L6.

Hasselmann, K., Barnett, T.P., Bouws, E., Carlson, H., Cartwright, D.E., Enke, K., Ewing, J.A., Gienapp, H., Hasselmann, D.E., Kruseman, P., Meerburg, A., Muller, P., Olbers, D.J., Ritcher, K., Sell, W. and Walden, H. 1973. Measurements of wind wave growth and swell decay during the Joint North Sea Wave project (JONSWAP). *Dtsch. Hydrogh. Z. Suppl. A*, 8, 12, 95pp.

Horton, C., Clifford, M, Cole, D., Schmitz, J. and Kantha, L. 1992. Operational modeling: semi-enclosed basin modeling at the Naval oceanographic Office. *Oceanography*, 5, 69-72.

Huh, O.K., Walker, N.D. and Moeller, C. 2001. Sedimentation along the eastern Chenier Plain coast: Down drift impact of a delta complex shift. *Journal of Coastal Research*, 17, 72-81.

Keen, T.R. 2002. Waves and currents during a winter cold front in the Mississippi Bight, Gulf of Mexico: Implications for barrier island erosion. *Journal of Coastal Research*, 18, 622-636.

Marmer, H.A. 1954. Gulf of Mexico: Its Origin, Waters and Marine Life. Tides and sea level in the Gulf of Mexico. *Galtsoff, P. S. Fish. Bull. 89. Wild. Serv.* 101-103 55.

Moeller, C.C., Huh, O.K., Roberts, H.H., Gumley, L.E. and Menzel, W.P. 1993. Response of Louisiana coastal environments to a cold front passage. *Journal of Coastal Research*, 9, 434-447.

Monaldo, F.M. and Beal, R.C. 1998. Comparison of SIR-C SAR wavenumber spectra with WAM model predictions. *Journal of Geophysical Research*, 103 (C9), 18815-18825.

Muller, R.A. 1977. A synoptic climatology for environmental baseline analysis: New Orleans. *Journal of Applied Meteorology*, 16, 20-33.

Muller, R.A. and Stone, G.W. 2001. A climatology of tropical storm and hurricane strikes to enhance vulnerability prediction for the southeast U.S. coast. *Journal of Coastal Research*, 17, 949-956.

Nordstrom, K.F., Bauer, B.O., Davidson-Arnott, R.G.D., Gares, P.A., Carter, R.W.G., Jackson, D.W.T. and Sherman, D.J. 1996. Offshore aeolian transport across a beach. Carrickfinn strand, Ireland. *Journal of Coastal Research*, 12, 664-672.

Pepper, D.A. and Stone, G.W. 2002. Atmospheric Forcing of Fine Sand Transport on a Low-Energy Inner Shelf: South-Central Louisiana, USA. *Geo-Marine Letters*, 22, 33-41.

Perez, B.C., Gay, J.W., Rouse, L.J., Shaw, R.F. and Wang, R. 2000. Influence of Atchafalaya River discharge and winter frontal passage of suspended sediment concentration and flux in Four-league Bay, Louisiana. *Estuarine Coastal and Shelf Science*, 50, 271-290.

Prasad Kumar, B., Ig-Chan Pang, Rao, A.D., Kim, T.H., Nam, J.C. and Hong, S.C. 2003. Sea state hindcast for the Korean Seas with a spectral wave model and validation with buoy observations during January 1997. *Journal of the Korean Earth Science Society*, 24, 7-21.

Pritchard, D.W. 1967. What is an estuary: physical viewpoint. In: Lauf, G.H. (ed) *E ries*. AAAS Publication No. 83, Washington, D.C., 3–5.

Ris, R.C., Booij, N. and Holthujsen, L.H. 1999. A third generation wave model for coastal regions. Part II: Verification. *Journal of Geophysical Research*, 104 (C4), 7667-7681.

Roberts, H.H., Huh, O.K., Hsu, S.A., Rouse, L.J., and Rickman, D. 1987. Impact of cold-front passage on geomorphic evolution and sediment dynamics of the complex Louisiana coast. *Coastal Sediments '87*. ASCE, New York, 1950-1963.

Roberts, H.H., Huh, O.K., Hsu, S.A., Rouse, L.J. and Rickman, D. 1989. Winter storm impacts on the Chenier plain coast of southwestern Louisiana. *Transactions of the Gulf Coast Association, Geological Society*, 39, 515-522.

Rogers, E.W., Hwang, P.A and Wang, D.W. 2002. Investigation of Wave growth and decay in the SWAN model: Three Regional-Scale Applications. *Journal of Physical Oceanography*, 33, -389

Schmatz, R.A. 2000. Development of a nowcast/forecast system for Galveston Bay. *Proceedings of the 6th Estuarine and Coastal Modeling Conference*. ASCE, New York, 441-455.

Schroeder, W.W. and Wiseman, W.J., Jr. 1999. Geology and hydrodynamics of Gulf of Mexico estuaries. In: Bianchi, T.S., Pennock, J.R. and Twilley, R.R. (eds) *Biogeochemistry of Gulf of Mexico Estuaries*. Wiley, New York, 428.

Schwab, D.J. and Bedford, K.W. 1994. Initial implementation of the Great Lakes Forecasting system: a real time system for predicting lake circulation and thermal structures. *Water Pollution Research Journal of Canada*, 29, 203-220.

Sheremet, A. and Stone, G.W. 2003. Observations of nearshore wave dissipation over muddy sea beds. *Journal of Geophysical Research*, 108 (C11), 1-11.

Stapor, F.W. and Stone, G.W. 2004. Age and Origin of the New Orleans Barrier Complex, Louisiana Coast, U.S.A. *Marine Geology*, 204, 215-234.

Stone, G.W., Stapor, F.W., Jr., May, J.P. and Morgan, J.P. 1992. Multiple sediment sources and a cellular, non-integrated longshore drift system: northwest Florida and southeast Alabama coast, USA. *Marine Geology* 105, 141-154.

Stone, G.W. and Finkl, C.W. 1995. Impacts of Hurricane Andrew on the coastal zones of Louisiana and Florida; August 22-26, 1992. *Journal of Coastal Research, Special Issue* 21.

Stone, G.W., Grymes, J.W., Dingler, J.R. and Pepper, D.A. 1997. Overview and significance of hurricanes on the Louisiana Coast, U.S.A. *Journal of Coastal Research*, 34, 656-669.

Stone, G.W., Wang, P., Pepper, D.A., Grymes, J.M., Roberts, H.H., Zhang, X.P., Hsu, S.A. and Huh, O.K. 1999. Researchers begin to unravel the significance of Hurricanes on the Northern Gulf of Mexico coast. *EOS, Transactions of the American Geophysical Union*, 80, 301-305.

Stone, G.W., Sheremet, A., Zhang, X., He, Q., Lui, B. and Strong, B. 2003. Landfall of two tropical systems seven days apart along Southern Louisiana, USA, paper presented at Coastal Sediments '03, ASCE, Florida.

Tolman, H.L. 1997. *User manual and system documentation of WAVEWATCH-III version 1.15*. NOAA / NWS / NCEP / OMB Technical Note 151, 97pp.

Tolman, H.L. 1999. *User manual and system documentation of WAVEWATCH-III version 1.18*. NOAA / NWS / NCEP / OMB Technical Note 166, 110pp.

Vincent, M., Burwell, D. and Luther, M. 2000. The Tampa Bay nowcast-forecast system. *Proceedings of the 6th Estuarine and Coastal Modeling Conference*. ASCE, New York, 765-780.

WAMDI Group, 1988. The WAM model – A third generation ocean wave prediction model. *Journal of Physical Oceanography*, 18, 1775-1810.

Zapel, C.L. 1984. *Morphology, sedimentary structures and sediment dispersal patterns within a transverse bar field, Horn Island, Mississippi*. Unpublished MS thesis, Louisiana State University, Baton Rouge, Louisiana, 108pp.

Chapter 13

EFFECTS OF COLD FRONTS ON BAYHEAD DELTA DEVELOPMENT: ATCHAFALAYA BAY, LOUISIANA, USA

Harry H. Roberts, Nan D. Walker, Alexandru Sheremet and Gregory W. Stone

Coastal Studies Institute, Department of Oceanography and Coastal Sciences, Louisiana State University, Baton Rouge, LA 70803, USA

1. INTRODUCTION

Delta-building in the Holocene Mississippi River system is characterized by the successive construction and abandonment of delta lobes (Fisk, 1944; Kolb and Van Lopik, 1958; Frazier, 1967). Each major delta-building episode is accompanied by a rather orderly and predictable set of events starting with stream capture followed by filling of an interdistributary basin with lacustrine deltas and swamp deposits, building of a bayhead delta at the coast, and finally construction of a major shelf delta. The process of "delta switching" involves the initiation of a new major delta while the previously active delta is systematically abandoned. These changes associated with shifting fluvial input are commonly referred to as the "delta cycle" (Roberts, 1997). Each major delta lobe in the Mississippi River system is active for about 1000-1500 years.

As a product of diversion of Mississippi River water and sediment down the Atchafalaya River course, the Atchafalaya River discharges an average total sediment load of approximately 75×10^{6} tonnes yr^{-1} into Atchafalaya Bay through two outlets, the natural lower Atchafalaya River Outlet and the man-made Wax Lake Outlet (US Army Corps of Engineers, 2002). The Atchafalaya and Wax Lake deltas building into Atchafalaya Bay are

D.M. FitzGerald and J. Knight (eds.),
High Resolution Morphodynamics and Sedimentary Evolution of Estuaries, 269-298.

evidence of a delta switching event in the making. These bayhead deltas are sedimentologic reminders that the Atchafalaya River is in the process of systematically capturing water and sediment from the modern Mississippi River. Currently, these deltas as geomorphic forms are confined to Atchafalaya Bay (Fig. 1). However, their prodelta deposits have been transported to the adjacent shelf and downdrift coast. The processes of decoupling of the coarse-grained bayhead deltas from their prodelta facies are the subject of this paper.

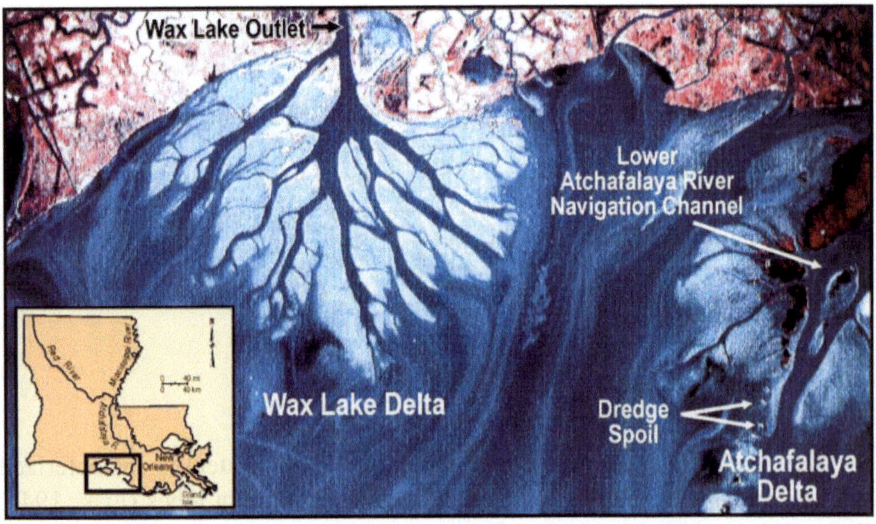

Figure. 1. A high altitude color infrared photograph of the Wax Lake and Atchafalaya deltas taken in December 1990. The inset map shows the location of Atchafalaya Bay along the central Louisiana coast where the deltas are located.

1.1 Diversion History

Diversion of Mississippi River flow down the Atchafalaya River provides a much more efficient route for water and sediment to reach the Gulf of Mexico than down the modern Mississippi River course. From the confluence point at Old River north of Baton Rouge, the distance down the Atchafalaya course is 220 km to the Gulf while it is 520 km down the Mississippi course. This obvious gradient advantage has led to steadily-increasing capture of Mississippi River flow since at least the 1500s when the first European explorers noted that the Atchafalaya River was a distributary of the Mississippi (Fisk, 1952). Initially, sediment captured from the Mississippi River was deposited in a large interdistributary basin located between the levees of the old Teche course of the Mississippi along the western margin of the alluvial valley and the meander belt of the modern Mississippi River to the east. From the 1500s to the beginning of the 20th century, the Atchafalaya River filled this interdistributary basin, the Atchafalaya Basin, with mainly swamp deposits and lacustrine deltas. However, Fisk (1952) noted that the Atchafalaya River steadily captured more and more Mississippi River flow throughout the first half of the 20th century. This trend prompted Congress to appropriate the funding to build a structure at the confluence point at Old River in order to control the diversion down the Atchafalaya to about 30% of the Mississippi River discharge. Although 30% of the Mississippi River discharge is often quoted as the controlled flow down the Atchafalaya, Mossa (1996) emphasized that the actual percentage varies from year to year. Currently, the Atchafalaya carries 30-50% of the Mississippi's discharge at Old River and up to 60% of its suspended load (Mossa and Roberts, 1990). Approximately 5% of the Atchafalaya River discharge can be attributed to the Red River.

1.2 Sediment Transport to Atchafalaya Bay

Until the middle of the 20th century, significant volumes of sediment were not being transported to the coast by the Atchafalaya River. Thompson (1951, 1955) performed the first detailed sedimentological research in Atchafalaya Bay and found a thin layer of brown-to-gray gelatinous clay overlying shelly, gray, old bay-bottom sediments. This surficial unit of thin clay-rich sediment represented the beginning of an increasing trend of Atchafalaya River sedimentation at the coast. Both Thompson (1951) and Cratsley (1975) observed that, between the hydrographic surveys of 1858 and 1935, the bay displayed no significant shoaling or sediment fill. Prior to this time, deposition was taking place in the Atchafalaya Basin. The initial work of Thompson (1951), Morgan *et al.* (1953), and Morgan and Larimore (1957) documented the appearance of mudflats at Chenier au Tigre and along other parts of the eastern Chenier Plain coast downdrift of Atchafalaya Bay. This downdrift coastal accretion along a traditionally retreating shoreline was additional evidence that Atchafalaya sediments were finally bypassing Atchafalaya Basin in sufficient volumes to make a visible impact at the coast.

Before extensive discharge and flood control on the Atchafalaya River, bank sediments became progressively finer downstream, and the channel became narrower and straighter with sand bars becoming more infrequent (Fisk, 1944). Similarly, channel bed sediments in the Mississippi River decrease downstream from sand-dominated deposits near the confluence with the Atchafalaya at Old River to silt-dominated sediments at Head of Passes in the modern "birdfoot" delta (Keown *et al.*, 1986). Consistent with these trends, the coarse fraction presently building the deltas in Atchafalaya Bay is dominated by fine sand and silt.

Between 1935 and 2001, the annual mean discharge of the Atchafalaya River into Atchafalaya Bay through the Wax Lake and lower Atchafalaya River Outlets was 5781 $m^3 s^{-1}$ with a peak discharge of about 20,000 $m^3 s^{-1}$ during the unusually high flood of 1973, measured at Simmesport, Louisiana (US Army Corps of Engineers, 2002). During the period 1981-2000 the average annual discharge increased to 6523 $m^3 s^{-1}$. With about 60% of the Mississippi River's suspended sediment load currently going down the Atchafalaya, huge volumes of fine-grained sediment are delivered to Atchafalaya Bay each year (Mossa and Roberts, 1990). Table I provides estimated suspended sediments entering the bay through both the lower Atchafalaya River Outlet (~ 70% discharge) and the Wax Lake Outlet (~ 30% discharge) over the time period 1980-1994. Because the coarsest sediments being carried by the lower Atchafalaya River are fine sands and silts, during flood discharge these grain sizes can be carried as suspended load. Recent studies of vibracores taken through the Wax Lake delta indicate that it is composed of nearly 70% sand (Roberts *et al.*, 1997) and a similar trend was found earlier for the Atchafalaya delta (van Heerden and Roberts, 1988). In addition to suspended load transport of sands and silts to the bay for delta-building, it is instructive to examine other impacts of major floods. During the years 1972-1975, annual mean Atchafalaya River discharge reached 9948 $m^3 s^{-1}$ and the sediment loads peaked at 143.2 x 10^6 tonnes (McManus, 2002). As a product of the high flood in 1973 and presumably other high floods, the river channel in its lower reaches was eroded and sands stored in the channel were delivered to the bay. Roberts *et al.* (1980) indicated that the bed elevation of the lower Atchafalaya River was lowered 4-5 m below average levels during the 1973 flood. This transfer of channel sand to the bay was largely responsible for emergence of the bayhead deltas in Atchafalaya Bay as subaerial forms. The same process of channel bed erosion was probably true for the Wax Lake Outlet, although no data were available on channel bed response to the major flood of 1973.

Table I. Suspended sediments entering Atchafalaya Bay, 1980-1994

Year	Atchafalaya Bay total suspended sediment (10^6 tonnes)	Wax Lake output contribution (%)	Total Atchafalaya River contribution (%)
1980	58.4	38.0	62.0
1981	64.3	40.9	59.1
1982	90.0	35.4	64.6
1983	119.8	36.3	63.7
1984	88.0	43.3	56.7
1985	70.4	43.2	56.8
1986	53.4	43.5	56.5
1987	71.9	40.0	60.0
1988	63.4	45.3	54.7
1989	75.0	37.4	62.6
1990	89.1	30.4	69.6
1991	55.9	24.7	75.3
1992	69.6	30.0	70.0
1993	89.0	36.1	63.9
1994	60.2	31.4	68.6

2. DELTA-BUILDING

2.1 Subaqueous and Subaerial Growth

Prior to the 1973 flood that forced delta-building into a subaerial phase (Roberts *et al.*, 1980), the work of Thompson (1951, 1955), Morgan *et al.* (1953), Morgan and Larimore (1957), Cratsley (1975), and Shlemon (1975) established that sediments were bypassing Atchafalaya Basin and delta-building was underway in Atchafalaya Bay. Although the early accounts of Atchafalaya River sedimentation in the bay mentioned only clay and silty clay (Thompson 1951, 1955), Cratsley (1975) indicates that some fine sand was being transported to the mouths of both the lower Atchafalaya River and Wax Lake Outlets before the 1973 flood. However, after the unusually high flood of 1973, sand in surface sediments of the bay expanded greatly and both the Atchafalaya River and Wax Lake deltas developed sand-rich bars exposed at low tide (Roberts *et al.*, 1980). As a product of these events, the deltas entered a new phase of subaerial evolution. Once the deltas became subaerial they attracted the interest of biologists and ecologists (Fuller *et al.*, 1985; Shaffer *et al.*, 1992) as well as geologists. In a geologic framework, the emergence of these deltas provided visual evidence of a major "delta switching" event and the initiation of a new delta lobe along the Louisiana coast.

Using satellite imagery, Rouse *et al.* (1978) indicated that by 1976 approximately 32.5 km^2 of new land had formed in Atchafalaya Bay at an average growth rate of 6.5 km^2 yr^{-1}. Later, Majersky *et al.* (1997) estimated the subaerial growth of both deltas based on a terrain model using both bay bathymetry and land elevation data as inputs. The growth estimates are summarized in Table II. Within the time period of the data presented in Table II, the Atchafalaya delta grew at a rate of 3.2 km^2 yr^{-1} while the Wax Lake delta displayed subaerial development at a rate of 3.0 km^2 yr^{-1}.

Table II. Subaerial growth of Wax Lake and Atchafalaya deltas (based on terrain model data, area above -0.6 m NGVD (Majersky et al., 1997)).

Year	Wax Lake (km^2)	Atchafalaya (km^2)
1976	3.8	32.5
1981	19.7	67.3
1989	47.9	85.2
1994	84.2	101.5

Since the beginning of the rapid subaerial expansion phase, which started in 1973, many vibracores have been acquired from both the Atchafalaya and Wax Lake deltas. Studies in the late 1970s and 1980 concentrated on the Atchafalaya delta because it initially grew much faster than its Wax Lake counterpart (Table II). Despite the early observations of Thompson (1951) regarding the initial arrival of fine-grained sediments in the bay, sedimentological studies by Roberts *et al.* (1980), van Heerden (1980, 1983), and van Heerden and Roberts (1980, 1988) demonstrated the sand-rich composition of the emerging delta. These studies also emphasized the lack of a substantial prodelta facies underlying the sand-rich and emergent lobes. As the Atchafalaya delta became more modified by dredging of a navigation channel to the continental shelf, dredge spoil placement, and other human impacts, research was focused on the more natural Wax Lake Outlet delta.

2.2 Sedimentary Architecture

Strike and dip transects of vibracores acquired from the Wax Lake delta demonstrate the coarse nature of the sediments comprising this thin bayhead delta (Fig. 2). Like the delta-front sheet sands described by Fisk (1955), the distributary mouth bars from numerous distributaries merge into a rather continuous sand facies across the delta and many of the channels, especially in a proximal delta location, cut through the delta deposits and into old bay bottom sediments.

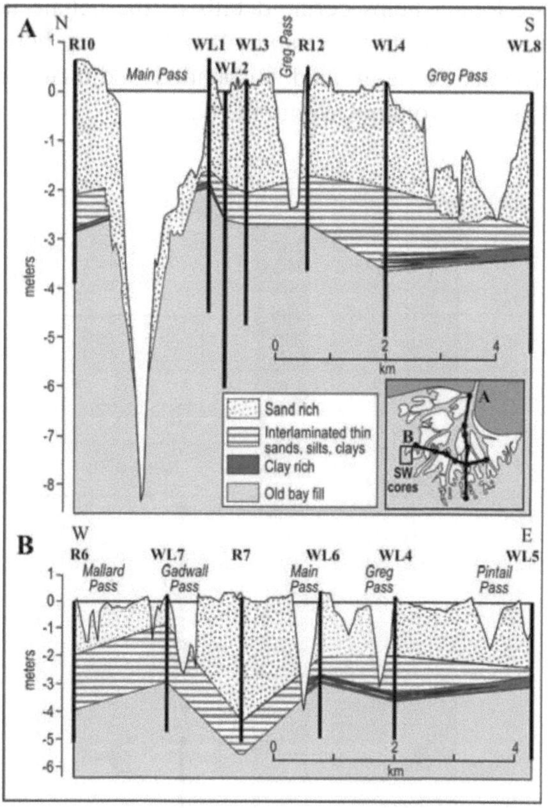

Figure. 2. Sedimentary facies of the Wax Lake delta as determined from (a) dip and (b) strike transects of vibracores. The cores are plotted on 1997 delta topography (modified from Roberts et al., 1997).

Majersky *et al.* (1997) calculated from the vibracores of Figure 2 that the Wax Lake delta averages 2.4 m thick and is composed of 67% sand of distributary mouth bar and subaqueous levee origin. The remainder of the deltaic succession is composed of thin interlaminated sands, silts, and clays. A true prodelta clay is rarely encountered. Majersky *et al.* (1997) and Roberts *et al.* (1997) suggest that the Wax Lake delta has vertically accreted at an average rate of approximately 2.7 cm yr^{-1} since 1981. Since vibracoring has confirmed the sand-rich nature of both the Atchafalaya and Wax Lake deltas, it is curious that the high suspended sediment loads of the Atchafalaya River have not resulted in the deposition of more clay-rich sediments in Atchafalaya Bay. A vibracore acquired from the rapidly developing southwestern part of the delta emphasizes this sedimentary bias toward the sand fraction (Fig. 3). The gamma density profile from a multisensor core logger helps define details of the deltaic facies. It is clear that sediments resting on the clay- and silt-rich old bay bottom deposits are considerably coarser. Fine-grained suspended sediments are clearly not being retained in large volumes within the bay.

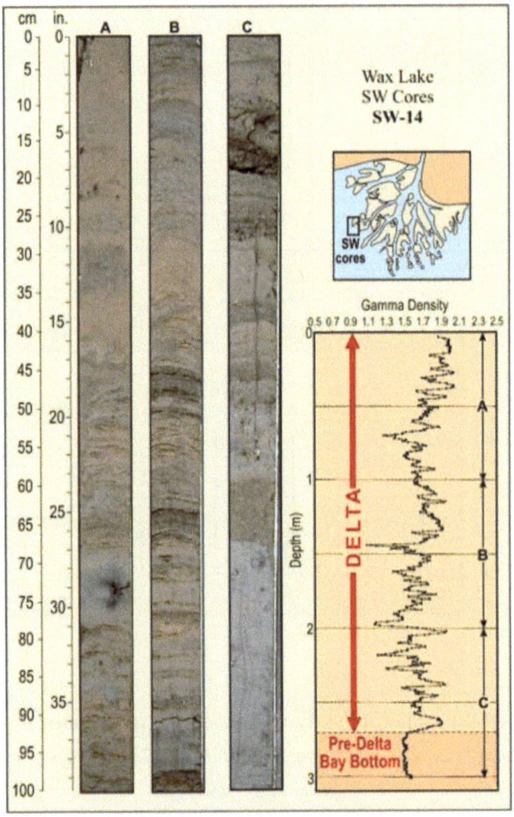

Figure. 3. Photograph of vibracore SW-14 from the Wax Lake delta (see inset map) with associated gamma density profile from a multisensor core logger indicating the sand-rich nature of the delta as compared to sediments of the old bay bottom (clay-rich).

2.3 Prodelta Deposits

The prodelta deposits of the rapidly prograding Atchafalaya and Wax Lake deltas largely reside outside of Atchafalaya Bay, both on the shelf and along the downdrift eastern chenier plain coast. Thompson (1951) first documented deposition of Atchafalaya River–derived fine-grained sediments on the inner shelf opposite Atchafalaya Bay. Recent investigations by Allison *et al.* (2000) and Allison and Neill (2002) have demonstrated that this prodelta facies of the Atchafalaya and Wax Lake deltas forms a wedge-shaped deposit that reaches 2.5 m in thickness on the inner shelf. High-resolution chirp sonar profiles show that the prodelta deposits thin seaward and pinch out against shoals representing the erosional remnants of older Holocene deltaic deposits. Accumulation rates calculated by ^{210}Pb profiles from inner shelf sediment cores show maximum sedimentation rates of 10-20 cm/yr (Allison and Neill, 2002). In areas of most active prodelta sedimentation, the sediments reflect cm-scale interlaminations of silty clays, clayey silts, and silty sands with silty clays being the most common.

In addition to the muds accumulating on the shelf opposite Atchafalaya Bay, fine-grained sediments are advected to the west by the prevailing coastal currents. As noted by Murray (1997), the Atchafalaya sediment plume that normally extends westward along the coast is strongly modulated and sometimes even reversed by the wind cycles associated with the passage of winter cold fronts. However, the westward sediment transport system on the inner shelf has resulted in the deposition of Atchafalaya-Wax Lake delta muds opposite the eastern chenier plain coast (Bentley *et al.*, 2003). Shoreward transport of these fluidized muds has resulted in the mud flat development. Huh *et al.* (2001) has demonstrated that parts of the eastern chenier plain coast are prograding through mudflat accretion at rates as high as 50 m yr^{-1}.

Although the seaward export of fine grained sediments from estuaries and river mouths has been studied from many settings with site-specific physical processes (Nittrouer *et al.*, 1986; Wright and Nittrouer, 1995; Wheatcroft and Borgeld, 2000; Kineke *et al.*, 2000; and others), this study focuses on the export of fine-grained sediments from Atchafalaya Bay by sediment resuspension and seaward advection related to the frequent passage of winter cold fronts (20-30/yr). The following sections of this paper describe the resuspension and advection processes.

3. THE SEDIMENT RESUSPENSION PROCESS

3.1 Wave-Induced Mud Dynamics

A soft, muddy sea-bed has obvious dissipative effects on hydrodynamics. Surface waves, for example, can lose up to 80% of their energy over just three wavelengths (Gade, 1957). Historically, studies of mud-induced wave dissipation have taken an approach based on a rather simple concept, which will be denoted here by the name of "long wave paradigm" (LWP). The basic LWP assumption is that such strong dissipation effects can only result from direct wave-bottom interaction. Therefore, only long-wave motions are affected by sediment fabric, as they "reach" deep enough into the water column and their near-bottom energy is significant. Changes of the seabed and water-column properties due to wave activity are assumed negligible, as are effects involving short waves.

The LWP is a simplistic generalization of an approach developed for sandy beaches, to a type of sedimentary environment characterized, in fact, by completely different physics. Indeed, there is evidence that wave-sediment interaction in muddy environments is more complex than simple wave-bottom friction (for example, mud fluidization 26 m below the sea floor has been recorded during Hurricane Camille (Sterling and Strohbeck, 1975). Even a cursory inspection of routine records of full wave spectrum evolution (not just the long-wave band) will reveal indirect evidence of significant mud reworking by waves and mud-induced dissipative effects on short waves.

Sediment resuspension and subsequent settling can significantly change the properties of the water column. Near-bottom fluid-mud layers can form, changing the character of flow to a multi-phase one, with viscous and even non-Newtonian components (Maa and Mehta, 1991; Chou *et al.*, 1993; Li and Mehta, 2000). New physical aspects could play an important part, such as structural stability of stratified flow under different surface forcing regimes, with different mechanisms for transition to turbulence and mixing. The few systematic observations available (Gade, 1957; Suhayda, 1977; Forristall and Reece, 1985) do not allow for a comprehensive formulation of the sediment-hydrodynamics coupling problem. Moreover, the LWP approach dominates experimental methodology, limiting the scope of the results. Typically, wave attenuation has been studied by comparing wave records from two sensors, one placed in deep, the other in shallow water (312 m and 20 m in SWAMP; Forristall and Reece, 1985), along the predicted wave propagation path. Short waves are not suitably resolved by such an array (they lose coherence quickly between the two locations). Energy input from wind, and nonlinear interactions are neglected, and refraction effects are estimated using linear models. This approach is questionable in highly energetic hurricane conditions (Forristall and Reece, 1985, discuss 8-m waves).

The development and implementation of the WAVCIS array (available from http://www.wavcis.lsu.edu, Fig. 4), operated by researchers in the Coastal Studies Institute (CSI), Louisiana State University, has provided unprecedented means to monitor in great detail sediment-hydrodynamics processes. The system is a comprehensive observation tool, providing continuous, real-time met-ocean data, designed to withstand hurricane-force wind and sea conditions. In the next section, we present some recent results regarding wave evolution over muddy sea beds during frontal passages.

3.2 Indirect Evidence of Sediment Resuspension During Frontal Passages

The basis for LWP is the assumption that significant near-bottom motion is a requisite for effective interaction between waves and the sea bed. Since short wave motions near the bottom are not significant, short-waves should not respond differently to different sedimentary types. To test this hypothesis, Sheremet and Stone (2003) used WAVCIS stations CSI 3 and CSI 5 (Fig. 4) to monitor wave evolution during frontal passages. Except for the sediment type (muddy at CSI 3 and sandy at CSI 5), the two locations are similar (5-m isobath, bathymetry gradient very low, 0.1%).

Wave observations show that strong dissipation mechanisms are active in the short wave band, in contradiction to the LWP. Although atmospheric conditions are almost identical at the two stations, wave records differ significantly, as illustrated in figures 5 and 6, by comparing time series of sea and swell variances at the two stations. Here, the frequency value of 0.2 Hz is taken as separation of sea (short wave) and swell (long wave) frequency bands. Swell variance values at CSI 3 and CSI 5 (Figs. 5a, 6a) differ by about one order of magnitude, regardless of sea state, and exhibit an almost perfect phase match. This behavior is consistent with the LWP and a direct-interaction bottom dissipation mechanism such as bottom friction, which depends on the kinematics of the wave motion (i.e. frequency), rather than the dynamics (wave energy).

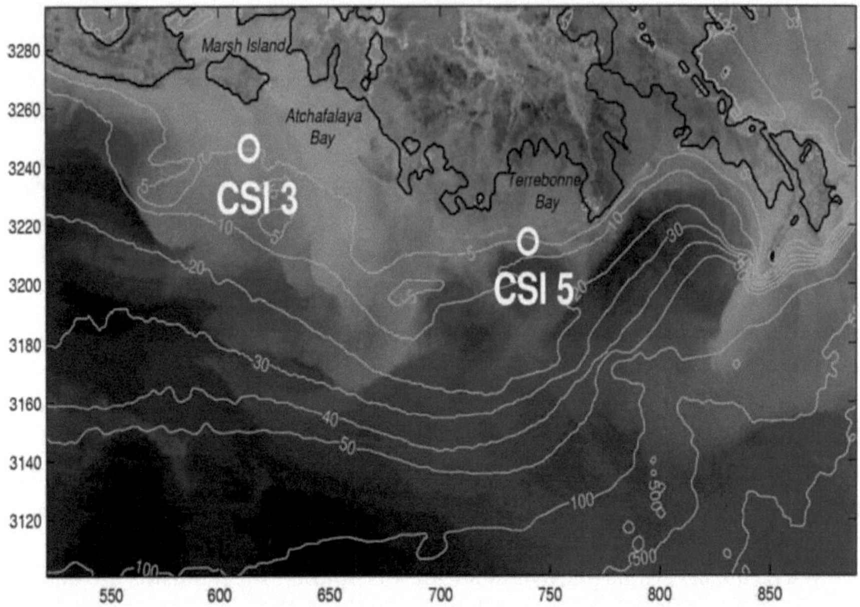

Figure. 4. *TERRA-1 Modis satellite image (gray scale) of the Louisiana coast at 250 m resolution (Earth Scan Laboratory, http://www.esl.lsu.edu), overlain with bathymetry contours at 1 km resolution. Contours are given in meters. Coordinates are given in kilometers with respect to UTM 1983, Zone 15. The two WAVCIS stations used here are represented by circles. The light grey is correlated to high surface sediment concentrations.*

The relationship between short wave energy at the two locations (Figs. 5b, 6b) is more complicated, with energy ratios scattered over two orders of magnitude. For the entire data set, energy ratios cluster around the values of either 0.1 or 1. An analysis of wind records (not shown) suggests that the 0.1 value corresponds to low wind forcing, whereas high winds typically result in similar sea excitations at both locations (e.g. Fig. 5b). High frequency waves respond rapidly to increases in wind speed, reaching comparable energy levels, but falling off by about an order of magnitude at CSI 3 during periods of calm weather. This behavior is not consistent with wave breaking mechanisms, active for energetic waves (i.e. the variance of the wave motion is less than 5% of the surface value anywhere in the first 1 m above the seafloor).

It is worthy to note that advanced stochastic models such as SWAN (Booij *et al.*, 1999), typically developed and tested for sandy environments, reproduce quite accurately the wave propagation characteristics at CSI 5 (the sandy site), but fail to describe correctly short wave dissipation effects observed at CSI 3, even with increased values of the bottom friction coefficients (details are given in Sheremet and Stone, 2003). In a sense this result is not surprising, since the physics of muddy sea beds is different. It is, however, a good indication that short dissipation is not the result of other mechanisms, such as wave breaking, or refractive scattering.

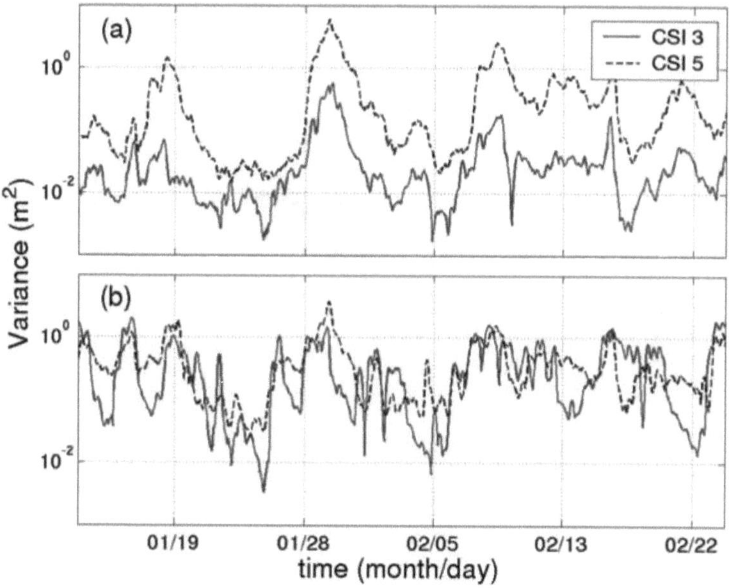

Figure. 5. Variance time series of (a) swell and (b) sea measured at CSI 3 and CSI 5.

The different response of short-wave fields to different sediment fabrics can be explained if we assume that significant reworking of the bottom sediments happens as a result of wave activity. The evolution of suspended sediment concentration during a frontal passage observed by Allison *et al.* (2000) during a cruise along the west Louisiana coast supports this relationship. Recently, Optical Backscatter Sensors (OBS) were deployed at CSI 3, to monitor sediment movement. The new system became operational during summer 2003, and has not, as of November 2003, collected a long enough time series of measurements to allow for a definitive conclusion. However, the array successfully withstood the passage of Hurricane Claudette (2003), providing some remarkable data about sediment and hydrodynamic evolution in energetic sea conditions. We use these data to illustrate the response of a fine-grained sediment sea-bed to wave activity.

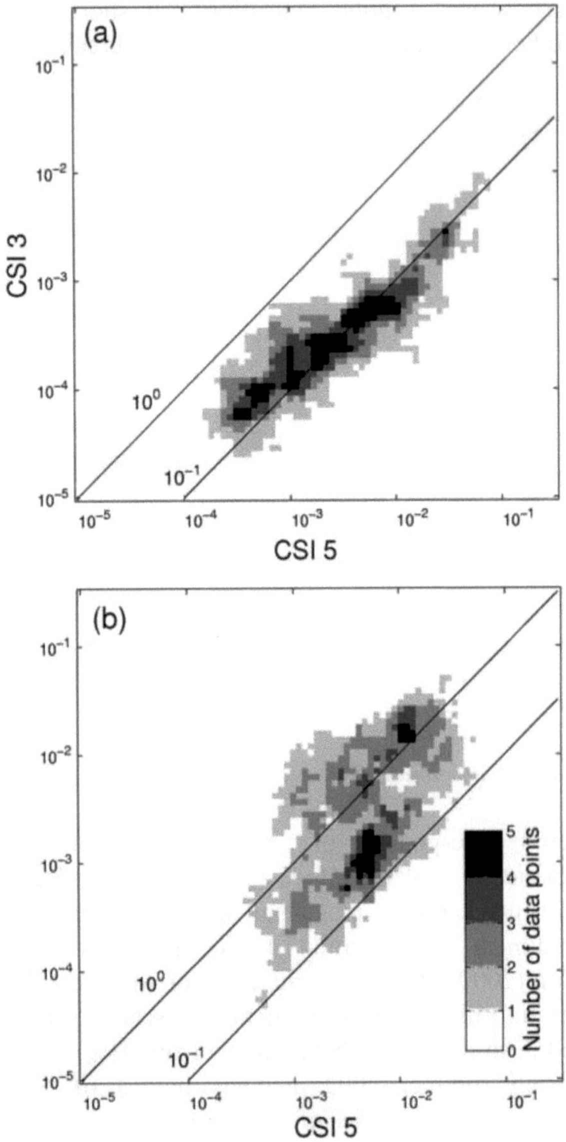

Figure. 6. *Comparison of (a) swell and (b) sea variance measured at CSI 3 and CSI 5.*

3.3 Recent Observations Of Fine-Grained Sediment Dynamics

The data collection system used to study fine-grained sediment dynamics is composed of three OBS, temperature, and salinity sensors, deployed at WAVCIS station CSI 3 at the provisional levels of 1, 2 and 3 m above the seabed. These new data have been integrated into the WAVCIS operational real-time data stream. The first major event to test the array was Hurricane Claudette, a storm which made landfall along the Texas coast on July 15, 2003. The array passed the test successfully, recording throughout the duration of the event. By coincidence, the system had undergone routine maintenance checks and sensor cleaning three days before the arrival of the hurricane; thus, the data it provided is high quality.

A summary of the observations is shown in figure 7. The evolution of high backscatter in the water column is plotted in figure 7d. The data set is rich in evidence of sediment motion, which supports the hypothesis that fine-grained sediments are mobile and respond fast to hydrodynamic forcing. For example, the spikes recorded by the topmost OBS before the storm are typical for a calm period, correlated with diurnal low tides and are related to surface sediment and fresh water influx from the nearby mouth of the Atchafalaya River. This process occurs in the surface layer only and was not recorded by middle or bottom sensors.

As storm waves arrive at the site, backscatter levels show a general increase in sediment resuspension in the first 1 m above the bottom. The substantial increase in turbidity recorded by the bottom sensor two days before the arrival of the storm could be due to advection along the coast, involving sediment resuspension and transport from areas southeast of CSI 3. Sediment resuspension levels at 1 m above the bottom are well correlated with swell; the higher layers show an increase of turbidity only at the peak of the storm, when the signals from the two upper layers are almost identical. The most remarkable element in this plot, however, is the processes taking place in the wake of the storm. The turbidity in the upper layers drops back to normal levels, approximately at the same rate as the decrease in swell energy. In contrast, turbidity in the bottom layer increases in a spectacular fashion, beyond the saturation levels of the bottom sensor (over 1500 NTU, for an approximate 36-hour period). This suggests the formation of a fluid mud layer by settling of suspended sediment, an effect that has been observed before (Allison *et al.*, 2000, with concentrations up to 25 g l^{-1}), but at a much coarser time resolution. The formation of the high-turbidity bottom layer is associated with a marked decrease in both swell and sea energy at CSI 3 (Figs. 7b, 7c). These

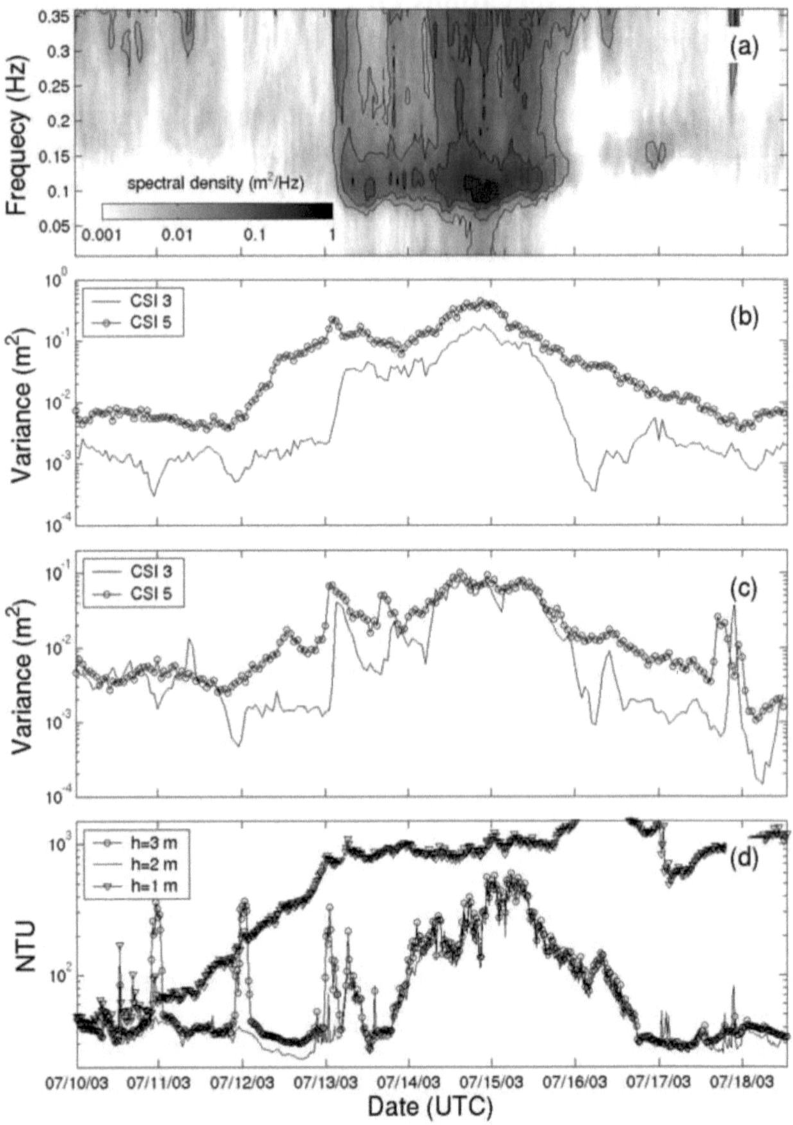

Figure. 7. Wave and turbidity measurements at CSI 3 (in a cohesive sedimentary environment) collected during Hurricane Claudette. (a) Spectral evolution. (b) Long wave (frequency less than 0.2 Hz) variance at CSI 3 and CSI 5. Note the considerable attenuation that occurred at CSI 3 (muddy site). (c) Short wave variance. (d) turbidity time series (OBS) at three levels located at 1, 2 and 3 m from the bottom.

observations support the assumption that short wave attenuation mechanisms are strongly connected to sediment reworking. Overall, observations show that, rather than dealing with one-way effects, such as mud-induced wave dissipation, or wave-induced sediment entrainment, hydrodynamic and sedimentary processes on muddy coasts should be viewed as a single, strongly coupled system, evolving on tightly correlated time and spatial scales. Importing simpler models and paradigms from a different sedimentary environment does not lead to accurate results. Sedimentary and hydrodynamic processes exhibit a strong coupling, which requires reformulating the theoretical approach from the level of the governing equations.

3.4 Bay and Plume Responses To Cold Front Forcing

Satellite remote sensing provides regional synoptic views of near-surface suspended sediment distribution along the Louisiana coast to enable study of plume morphology, sediment resuspension, and sediment transport associated with cold front passages. Figure 8 is a visible band satellite image, obtained by the Terra-1 MODIS sensor, after such an event followed by several days of northerly winds along the Louisiana coastline. The true color enhancement of 21 March 2001 reveals extensive sediment plumes along the coast from the discharge of the Atchafalaya and Mississippi Rivers. The Mississippi plume was discharged into relatively deep water, whereas the Atchafalaya plume was discharged onto a shallow shelf, as revealed by the 10 m isobath (Fig. 8). Resuspension of sediments is a major contributor to the Atchafalaya plume during high wind conditions and particularly during winter storm events (Walker and Hammack, 2000). This MODIS image shows a surface sediment plume extending 75 km seaward from the outer edge of Atchafalaya Bay.

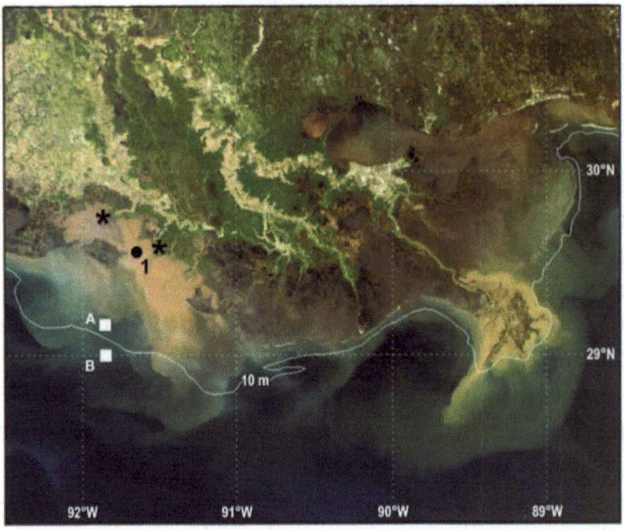

Figure. 8. Visible band satellite image from the Terra-1 MODIS sensor on 21 March 2001 showing the regional distribution of surface suspended sediments on the inner shelf and interior bays after a cold front passage. Site 1, in East Cote Blanche Bay, is depicted with a black dot. The current meter locations on the inner shelf are shown with white squares. Asterisks depict the Cypremort Point meteorological station (west) and the East Cote Blanche Bay water level station (east). The 10 m isobath is depicted by a solid line (image processed at the LSU Earth Scan Laboratory).

Imagery such as this one was used to plan a multi-year field measurement program that began in October 1997. Hourly time-series measurements of current speed and direction, water level, conductivity, temperature, and optical backscatter were obtained at locations to enable the calculation of fluxes in and out of the shallow bays (1-4 m in depth), west of Atchafalaya Bay. Four stations were instrumented to investigate wind-related changes in circulation, sediment resuspension, sediment transport, and salt flux. Water samples were collected concurrently with the time-series measurements at 6-hour intervals to enable "calibration" of the optical backscatter records to total suspended solid (TSS) concentrations. The 6-hour interval was essential to ensure that samples were obtained during high velocity wind events, when concentrations are relatively high. Inorganic sediments were found to comprise at least 90% of the TSS concentrations. Detailed information on field instrumentation and data processing techniques can be found in Walker and Hammack (2000) and Walker (2001).

The most energetic wind events along this coast are winter storms and tropical cyclones. The winter cold-front systems generally moves from the northwest to the southeast with a recurrence interval of 4 to 7 days in winter (Chuang and Wiseman, 1983). With the approach of a cold front, the prevailing easterly winds strengthen and veer to south and west before the frontal passage. Strongest winds generally accompany frontal passage and blow from the northwest or north. Frontal characteristics vary substantially and strong winds (> 10 ms^{-1}) can last for days. Field measurements obtained during a time period in December 1997 and January 1998 are shown in figure 9 to illustrate winter storm impacts on wind speed/direction, water level, TSS concentrations, and sediment flux. Two clear sky NOAA AVHRR images (Fig. 10) enabled an assessment of change in the regional distribution of surface suspended sediments on the adjacent shelf due to the passage of two winter storms between December 26 and 29, 1997. The event under discussion is shaded and the times of image acquisition are shown with black dots on the time series graph (Fig. 9). All satellite data were obtained from the LSU Earth Scan Laboratory (available from http://www.esl.lsu.edu) and processed using atmospheric correction software and algorithms developed especially for this river-influenced region (Walker and Hammack, 2000; Walker, 2001; Myint and Walker, 2002).

On December 25, winds were from a northeast direction with a speed of approximately 5 ms^{-1} (Fig. 9). Satellite imagery revealed that suspended sediment concentrations were relatively low on the shelf and in the interior bays (Fig. 10). Field estimates of TSS also revealed relatively low concentrations at Site 1, situated in the wide channel linking East and West Cote Blanche Bays (Fig. 9).

The strongest north winds (14 ms^{-1}) accompanied the first winter storm on December 26 and 27. Water level fell more than 1 m in less than 24 hours, suspended sediment concentrations increased from 50 to 400 mg l^{-1}, and sediment flux increased substantially with the strong outflow of sediment-laden bay water (Fig. 9). Ebbing currents of 0.65 ms^{-1} were measured at Site 1 during the event. Sediment fluxes were computed using the primary component of the currents and the corresponding hourly estimate of TSS concentrations. A brief period of southerly winds was experienced on December 28 before another cold front crossed the region. On December 29, strong northwest winds (12 ms^{-1}) characterized the second front and a similar time- history of physical processes was experienced during this winter storm at Site 1. Water levels again dropped 1 m, suspended sediment concentrations rose by an order of magnitude, and sediment flux to the southeast was maximized by the strong tidal currents carrying sediment-charged water out of the bays. A clear-sky satellite image obtained on December 29 after the quick succession of the two cold front

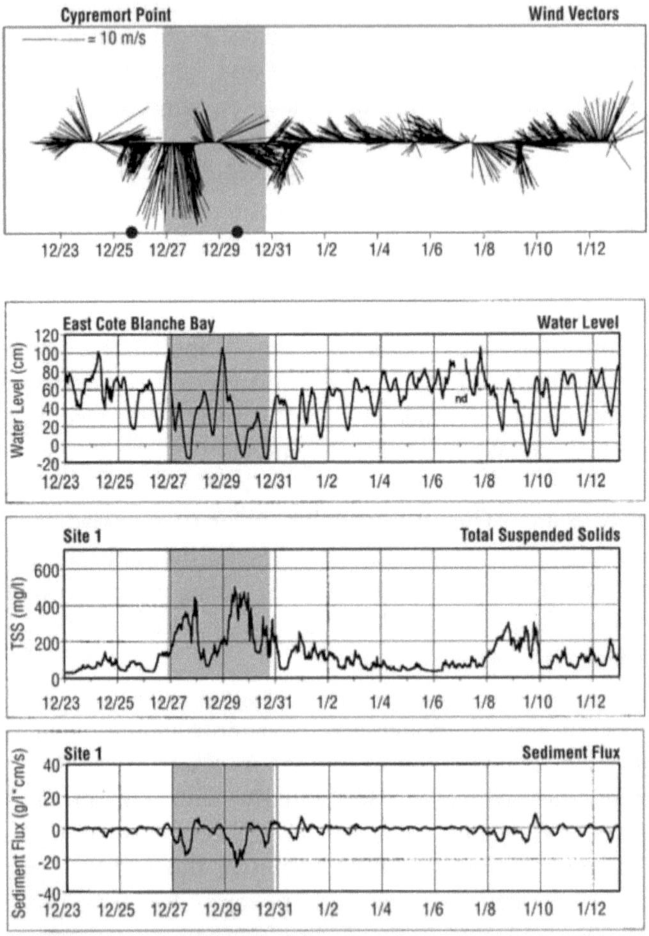

Figure. 9. Field measurements at Site 1 (location given on Fig. 10) from December 23 1997-January 12 1998. The panels from top to bottom are wind speed and direction (Cypremort Point) displayed in stick vector format (winds blowing from the north are represented by wind vectors extending below the horizontal line); water level (East Cote Blanche Bay); total suspended solid concentrations; and sediment flux. Negative values of sediment flux indicate transport out of the bay. The cold front event discussed in the text is shaded. Black dots depict times of satellite imagery. Figure modified from Walker and Hammack (2000).

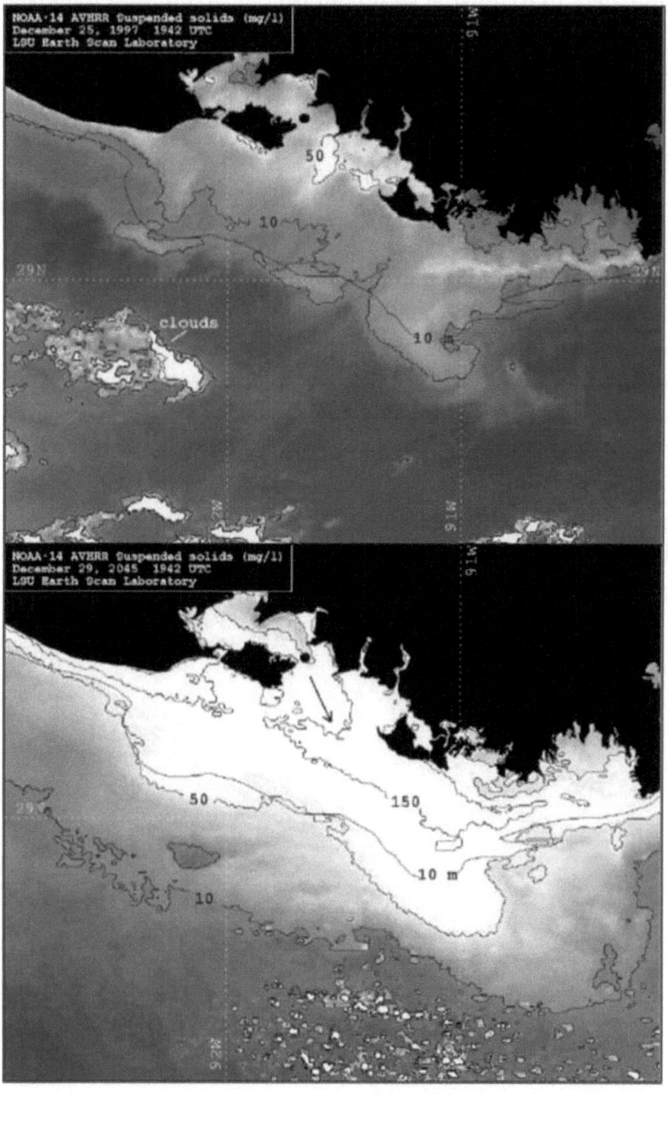

Figure. 10. NOAA AVHRR reflectance images obtained on (top) December 25 1997 and (bottom) December 29 1997. The 10, 50, and 150 mg l^{-1} contours of suspended sediment concentration (estimated using equation of Myint and Walker, 2002) are indicated with black lines. The 10 m isobath is shown with a heavier black line. The arrow in the bottom panel indicates the direction of surface water and suspended sediment movement during the peak of a cold front passage event. Site 1 is shown with solid black dot.

passages revealed an extensive sediment plume on the inner shelf, measuring 180 km along-shelf and 75-90 km cross-shelf (Fig. 10). Sediment concentrations within the plume on the shelf were 10-40 mg l^{-1} during the prefrontal low wind period of December 25. However, after passage of these winter storms, suspended sediment concentrations ranged from 50-150 mg l^{-1} shoreward of the 10 m isobath on the inner shelf (Fig. 10). The areal extent of the Atchafalaya sediment plume increased in size from 1600 km^2 on December 25 to 11,400 km^2 on December 29, based on measurements made in relation to the 25 mg l^{-1} contour.

Current meter data, from locations on the inner shelf, were obtained for a winter storm event in December 1993 (locations given in Fig. 8) to investigate how inner shelf currents seaward of Atchafalaya Bay respond to winter storm events. During this time period, winds from a Burrwood station (180 km to the east) revealed a rapid change in winds from southeasterly to northwesterly early on December 14 (Fig. 11). Unfortunately, wind measurements were not available in the Atchafalaya Bay area until 1996. Inner shelf currents underwent a distinct reversal in current direction at the two current meter moorings (Fig. 11).

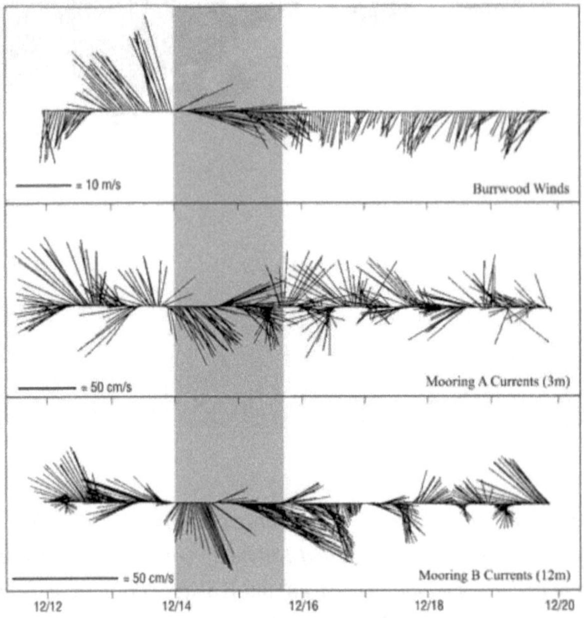

Figure. 11. Time series of wind vectors from Burrwood, LA (180 km to the east of Atchafalaya Bay) and current vectors measured seaward of Atchafalaya Bay for the time period December 11-20 1993. The locations of moorings A and B are shown in figure 8. Current meter data were obtained at mid-depth in water columns of 7 m and 22 m.

Currents at the shallow mooring (A), in 7 m water depth, flowed westward before the wind shift and southeastward after frontal passage. Currents at the deeper mooring (B), in 22 m water depth, demonstrated a similar reversal in flow with the transition to northwesterly winds. Meter locations are shown in figure 8. Both current meters were located mid-water column. Current speeds, associated with the cold-front related current reversals were 0.5 ms^{-1}. Field measurements of currents on the inner shelf opposite Atchafalaya Bay made by Adams *et al.* (1982) showed that shelf currents were directed southeasterly in response to cold front passages and were capable of transporting sand-sized sediment.

The response of the Atchafalaya surface sediment plume to wind forcing was further investigated by compositing clear-sky images representative of the four main wind quadrants. Contour maps of suspended sediment concentrations are shown in figure 12 for northwest, northeast, southwest, and southeast wind conditions.

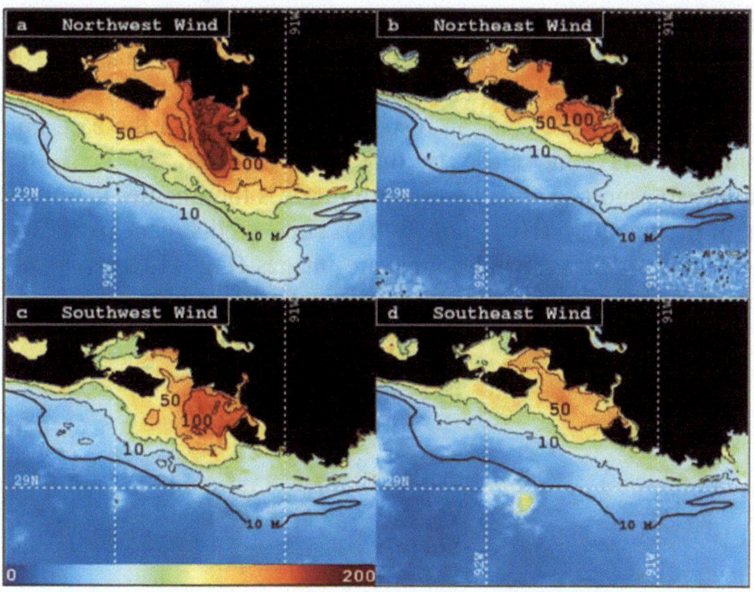

Figure. 12. Satellite image composites of surface suspended sediment concentrations representing each of the four main wind quadrants (modified from Walker and Hammack, 2000). Contours of total suspended solids (10, 25, 50, 100, 150 and 200 mg l^{-1}) are shown as well as a scale relating TSS concentration to color.

Each of these composite images was formed by computing the arithmetic means of suspended sediment concentrations at each pixel from four clear-sky images, during moderate to high river discharge (defined as > 5666 m^3s^{-1}

or 200,000 cfs^{-1}). The composite images revealed highest concentrations of suspended sediments within Atchafalaya Bay, where the two delta lobes are located. Active river discharge and resuspension of unconsolidated sediments along the delta front regions were the most likely sources of sediment. Both side-scan sonar and chirp sonar data support the observation that erosion is occurring on the delta front (Fig. 13). Plume area and surface suspended sediment concentrations within the Atchafalaya sediment plume were highest in association with winds from the northwest. The plumes for the northwest wind case averaged 4400 km^2 in area with maximum surface sediment concentrations exceeding 200 mg l^{-1} (Fig. 12a). The composite plume extended beyond the 10 m isobath. These plume measurements were based on a minimum concentration of 10 mg l^{-1} and did not include the interior bay areas. The second largest plume resulted from southwest winds with an average area of 1925 km^2 (Fig. 12c). Plume orientation during west winds indicated surface sediment transport to the east. Winds from the west caused offshore near-surface flow due to Ekman dynamics, increasing the seaward extent of the plumes and their areas. In contrast, during northeast and southeast wind events, the sediment plumes were smaller (1060 and 1134 km^2, respectively) and closer to the coast (Figs. 12b, d). Water level set-up occurs along the coast in response to winds from the east, confining the plumes to a zone close to the coast (Walker, 2001). East winds effectively transport sediment westward towards the Chenier Plain, an active region of progradation (Huh *et al.*, 2001). This "mud-stream" is shown in the northeast and southeast wind composite images (Figs. 12b, d) as relatively high concentrations extending from Atchafalaya Bay along the coast to the west.

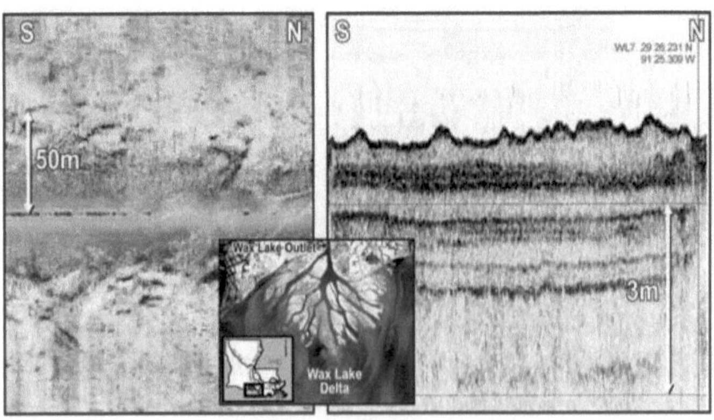

Figure. 13. A side-scan sonar image (left) and chirp sonar subbottom profile (right) of a delta front area along the southeastern flank of the Wax Lake delta. Note the irregular bottom. This bottom type is interpreted as the product of wave and current erosion associated with winter cold front passages.

In situ measurements and satellite imagery demonstrate distinct hydrodynamic and sediment responses to cold front passage within the interior bays and on the shallow shelf seaward of these bays. Winds, water levels, and currents combine to maximize sediment resuspension and transport in both areas. As presented earlier, it appears probable that prefrontal wind-waves and swell play a role in the initiation of the resuspension of bottom sediments, a process that is then maximized by the lowering of coastal and bay water levels during west and north wind periods. Field measurements demonstrate that water column sediment concentrations can increase by an order of magnitude (50 to 500 mg l^{-1}) during the storm. Although wave measurements were not available during the data collection periods shown in figure 9, the close correlation between wind speed and suspended sediment concentrations indicates that resuspension processes are maximized by strong winds and associated waves. The rapid and substantial (> 1 m) decrease in water levels that occurs with northwest and north wind events dramatically increases the resuspension potential of the wind-waves in these shallow bays. The boundary layer turbulence created by the intrusion of cold air over warmer water may enhance wind-wave generation processes. Walker and Hammack (2000) showed that water level changes are maximized when northwest winds blow down the axis of this bay system, in synchrony with the lowering of coastal water levels. Sediment flux from the bays onto the shelf is maximized by the strong ebbing currents carrying sediment-laden water. Additional satellite imagery and field measurements indicate that the case studies presented are typical of winter storm events (Walker and Hammack, 1999; Walker and Hammack, 2000). Based on the Site 1 measurements, Walker and Hammack (2000) estimated that 400,000 metric tons of sediment may be transported from the 1500 km^2 bay system onto the shelf during an average winter storm. Using this estimate, about 106 tonnes of sediment can be transported to the shelf during the year, as approximately 20-30 cold-front events occur annually. This value is about 12% of the yearly average sediment discharge of the Atchafalaya River. Cold-front related erosion of sediments from the bays is a process that reduces the rate of subaerial delta development (van Heerden and Roberts, 1980) and also slows the rate of infilling in the shallow bays to the east and west of Atchafalaya Bay.

On the inner shelf, current direction, sediment resuspension, and sediment transport are closely controlled by wind forcing. An analysis of satellite imagery reveals that surface sediment concentrations increase substantially and the surface expression of the Atchafalaya sediment plume reaches maximum dimensions during the northwest and north wind periods, associated with winter storms. Sediment plumes measuring 75 km in the cross-shelf and 150 km in the along-shelf direction are common after winter storm events. Walker and Hammack (2000) estimated that 25% of the sediment plume on the inner shelf results from sediment fluxes out of the interior bays, whereas the largest fraction (75%) is most likely resuspended from the bottom on the inner shelf and transported seaward by strong wind-driven currents. Walker (2001) demonstrated that large surface sediment plumes can also result from tropical storms and hurricanes, when the storm centers pass to the east of Atchafalaya Bay. One example was Hurricane Georges, that made landfall along the Mississippi coast (300 km to the east) on September 28 1998. It created a large sediment plume seaward of Atchafalaya Bay because strong north-northwest winds, similar in strength to those of a winter storm, impacted the Atchafalaya area for a few days.

4. SUMMARY

The combination of high suspended load fluvial input to Atchafalaya Bay coupled with vigorous sediment flux out of the bay through processes related to the impacts of repeated winter storms has led to residual delta deposition skewed toward the coarsest sediments delivered to the bay, fine sand and coarse silt. Even though the Atchafalaya River delivers approximately 40-50% of the Mississippi River's suspended load to Atchafalaya Bay (Mossa and Roberts, 1990), the bay is filling with sand-rich deposits whereas most of the fine fraction of the sediment load is exported to the adjacent continental shelf and nearshore regions of the downdrift chenier plain (Huh *et al.*, 2001; Roberts *et al.*, 2003). Sediment bypassing of the bay was first well-documented by Adams *et al.* (1982) when their near bottom current meter data illustrated that as winter storms (cold fronts) crossed the Louisiana coast from a northwesterly direction, large volumes of sediment were exported from the bay. They showed that the normal flow field on the inner shelf is a tidally dominated regime susperimposed on a slower wind-driven westward drift. This pattern is frequently interrupted by brief periods of intense cross-shelf flow. Calculated shear stresses and suspended shelf sands in the water column suggested that sand-sized and silt-sized sediment may be preferentially transported southeastward and offshore of Atchafalaya Bay while the fine sediment fraction is carried offshore and downcurrent (westward) with the mean flow once normal non-cold front conditions were resumed.

More recent research by Walker and Hammack (2000) utilizing both satellite imagery and *in situ* measurements of suspended sediments and physical processes provides further support and documentation of the relationships between storms and the resuspension and transport of fine-grained sediments from Atchafalaya Bay. Elevation of bay water levels occurs by wind and wave set-up along the coast as a result of strong southerly winds associated with winter storms. Water levels may be elevated as much as 50 cm above the tidal signal. Because of bay sediment resuspension due to increased wave activity in the delta front regions of the Wax Lake and Atchafalaya deltas, suspended sediment concentrations frequently increase by an order of magnitude or more before and during the frontal passage across the coast. When winds relax or are quickly reversed as a cold front passes, suspended sediments are transported out of the bay and onto the adjacent continental shelf. Water level set-down accompanied by increased wind and wave activity exposes the delta front to substantial sediment resuspension as the wind shifts to the northerly quadrant. Figure 13 illustrates the irregular delta front bay bottom of the Wax Lake delta as observed in side-scan sonar and chirp sonar data. This irregular and erosional bay bottom topography is attributed to cold front associated waves and currents. Although hurricanes and other storm types may also play a role in transporting sediments out of Atchafalaya Bay, the cumulative impacts of the cyclic passage of winter cold fronts (20-30 per year) is thought to far outweigh the effects of other storms. Walker and Hammack (2000) note that turbid plumes from cold front-related events extend as much as 75 km offshore as measured from satellite images. Allison and Neill (2002) illustrate through high resolution subbottom profile data from the continental shelf that the well defined prodelta mud facies from Atchafalaya Bay extends offshore distances that are consistent with turbid plumes observed on satellite-derived images.

In conclusion, the coarse nature of both the Wax Lake and Atchafalaya River bayhead deltas, each nearly 70% sand, is largely a product of their winter storm-related physical process setting. Prior to a cold front moving across the coast, strong easterly and southerly winds cause water level set-up in the bay. Waves erode the delta front of both bayhead deltas causing sediment resuspension and charging of the water column with a high suspended sediment load. As a cold front passes the coast, rapid wind reversal from southerly-to-northerly accompanied by water level set-down amplifies that resuspension process and energetically forces water out of the bay. Large volumes of fine-grained sediment are transported to the adjacent continental shelf. By these processes, the prodelta facies is largely decoupled and transported seaward from the sand-rich deltas that are rapidly filling available accommodation space within the bay.

5. ACKNOWLEDGEMENTS

The data on which this paper is based were collected under the support of several different funding sources including the US Army Corps of Engineers (DACW 29-M-1664, DACW 39-96-K-0032, and noncurrent previous contracts), US Geological Survey (14-08-0001-23411), Louisiana Board of Regents, LEQSF (2002-5-RD-A-10), ONR (N00014-03-1-02000), and proceeds from the J.P. Morgan Professorship in Coastal Studies Institute, Louisiana Sea Grant, and Basin Research Institute.

REFERENCES

Adams, C.E., Jr., Wells, J.T. and Coleman, J.M. 1982. Sediment transport on the central Louisiana continental shelf: Implications for the developing Atchafalaya River delta. *Contributions to Marine Science*, 25, 133-143.

Allison, M.A., Kineke, G.C., Gordon, E.S. and Goni, M.A. 2000. Development and reworking of a seasonal flood deposit on the inner continental shelf off the Atchafalaya River. *Continental Shelf Research*, 20, 2267-2294.

Allison, M.A. and Neill, C.F. 2002. Accumulation rates and stratigraphic character of the modern Atchafalaya River prodelta, Louisiana. *Transactions Gulf Coast Association of Geological Societies*, 52, 1031-1040.

Bentley, S.J., Roberts, H.H. and Rotondo, K. 2003. The sedimentology of muddy coastal systems: The research legacy and new perspectives from the Coastal Studies Institute. *Transactions Gulf Coast Association of Geological Societies*, 53, 52-63.

Booij, N., Ris, R.C. and Holthuijsen, L.H. 1999. A third generation wave model for coastal regions: Part I: model description and validation. *Journal of Geophysical Research*, 104, 7649-7666.

Chou, H.-T., Foda, M.A. and Hunt, J.R. 1993. Rheological response of cohesive sediments to oscillatory forcing. In Mehta, A.J. (ed) *Nearshore and Estuarine Cohesive Sediment Transport*. Coastal and Estuarine Studies Series, 42, 126-148. AGU, Washington, DC.

Cratsley, D.W. 1975. *Recent deltaic sedimentation, Atchafalaya Bay, Louisiana*. Unpublished MS thesis, Louisiana State University, Baton Rouge, 142pp.

Fisk, H.N. 1944. *Geological investigation of the alluvial valley of the Lower Mississippi River*. US Army Corps of Enginering, Vicksburg, Mississippi, 78pp.

Fisk, H.N. 1952. *Geologic investigation of the Atchafalaya Basin and the problem of Mississippi River diversion*. US Corps of Engineers, Mississippi River Commission, Vicksburg, Mississippi, 1, 145pp.

Fisk, H.N. 1955. Sand facies of recent Mississippi delta deposits. *Proceedings 4th World Petroleum Congress, Rome, Italy*, Section 1-6, 377-398.

Forristall, G.Z. and Reece, A.M. 1985. Measurements of wave attenuation due to a soft bottom: the SWAMP experiment. *Journal of Geophysical Research*, 90, 3367-3380.

Frazier, D.E. 1967. Recent deltaic deposits of the Mississippi River, their development and chronology. *Transactions Gulf Coast Association of Geological Societies*, 17, 287-315.

Fuller, D.A., Sasser, C.E., Johnson, W.B. and Gosselink, J.G. 1985. The effects of herbivory on vegetation on islands in Atchafalaya Bay, Louisiana. *Wetland*, 4, 105-114.

Gade, H.G. 1957. *Effects of a non-rigid impermeable bottom on plane surface waves in shallow water*. Unpublished PhD thesis, Texas A&M University, 35pp.

Huh, O.K., Moeller, C.C. and Walker, N.D. 2001. Sedimentation along the eastern Chenier Plain coast: Down drift impact of a delta complex shift. *Journal of Coastal Research*, 17, 72-81.

Li, Y. and Mehta, A.J. 2000. Fluid mud in wave dominated environment revisited. *In* McAnally, W.H. and Mehta, A.J. (eds) *Coastal and Estuarine Fine Sediment Processes*. Proceedings of Marine Science, 3, 79-93.

Keown, M.P., Dardeau, Jr., E.A. and Causey, E.M. 1986. Historic trends in the sediment flow regime of the Mississippi River. *Water Resources Research*, 22, 1555-1564.

Kineke, G.C., Woolfe, K.J, Kuehl, S.A., Milliman, J.D., Dellapenna, T.M. and Purdon, R.G. 2000. Sediment export from the Sepik River, Papua New Guinea: evidence for a divergent plume. *Continental Shelf Research*, 20, 2239-2266.

Kolb, C.R. and Van Lopik, J.R. 1958. *Geology of the Mississippi deltaic plain-southeastern Louisiana.* US Army Corps of Engineers, Waterways Experiment Station, Technical Report 2, 482pp.

Maa, P.-Y. and Mehta, A.J. 1991. Soft mud response to water waves. *Journal Waterways, Ports, Coastal and Ocean Engineering*, 116, 634-650.

Majersky, S., Roberts, H.H, Cunningham, R., Kemp, G.P. and John, C.J. 1997. Facies development in the Wax Lake Outlet delta: Present and future trends. *Basin Research Institute Bulletin*, 7, 50-66.

Martinez, J.D. and Haag, W.G. 1987. *The Atchafalaya River and its basin: A field trip.* Guidebook Series No. 4, Louisiana Geological Survey, Baton Rouge, Lousiana, 22pp.

McManus, J. 2002. The history of sediment flux to Atchafalaya Bay, Louisiana. *In* Jones, S.J. and Frostick, L.E. (eds) *Sediment Flux to Basins: Causes, Controls, and Consequences.* Geological Society of London, Special Publication 191, 209-226.

Morgan, J.P., Van Lopik, J.R. and Nichols, J.G. 1953. *Occurrence and development of mudflats along the western Louisiana coast.* Louisiana State University, Coastal Studies Institute Technical Report 2, 34pp.

Morgan, J.P. and Larimore, P.B. 1957. Change in the Louisiana shoreline. *Transactions Gulf Coast Association of Geological Societies*, 7, 303-310.

Mossa, J. and Roberts, H.H. 1990. Synergism of riverine and winter storm-related sediment transport processes in Louisiana's coastal wetlands. *Transactions Gulf Coast Association of Geological Societies*, 40, 635-642.

Mossa, J. 1996. Sediment dynamics in the lowermost Mississippi River. *Engineering Geology*, 45, 457-479.

Murray, S.P. 1997. *An observational study of the Mississippi-Atchafalaya coastal plume.* OCS Study MMS 98-0040, U.S. Department of the Interior, Minerals Management Service, Gulf of Mexico OCS Region, New Orleans, LA, 513pp.

Myint, S. W. and Walker, N.D. 2002. Quantification of surface suspended sediments along a river dominated coast with NOAA AVHRR and SeaWiFS measurements, Louisiana, USA. *International Journal of Remote Sensing*, 23, 3229-3249.

Nittrouer, C.A., Kuehl, S.A., DeMaster, D.J. and Kowsmann, R.O. 1986. The deltaic nature of Amazon shelf sedimentation. *Geological Society of America Bulletin*, 97, 444-458.

Roberts, H.H. 1997. Dynamic changes of the Holocene Mississippi river delta plain: the delta cycle. *Journal of Coastal Research*, 13, 605-627.

Roberts, H.H., Adams, R.D. and Cunningham, R.H.W. 1980. Evolution of sand-dominant subaerial phase, Atchafalaya Delta, Louisiana. *American Association of Petroleum Geologists*, 64, 264-279.

Roberts, H.H., Walker, N.D., Cunningham, R., Kemp, G.P. and Majersky, S. 1997. Evolution of sedimentary architecture and surface morphology: Atchafalaya and Wax Lake deltas, Louisiana. *Transactions Gulf Coast Association of Geologic Societies*, 48, 477-484.

Roberts, H.H., Beaubouef, R.T., Walker, N.D., Stone, G.W., Bentley, S., Sheremet, A. and van Heerden, I. 2003. Sand-rich bayhead deltas in Atchafalaya Bay (Louisiana): Winnowing by cold front forcing. *Coastal Sediments '03*, Clearwater, Florida, 1-15.

Rouse, L.J., Jr., Roberts, H.H. and Cunningham, R.H.W. 1978. Satellite observation of the subaerial growth of Atchafalaya Delta. Louisiana. *Geology*, 6, 405-408.

Shaffer, G.P., Sasser, C.E., Gosselink, J.G. and Rejmanek, M. 1992. Vegetation dynamics of islands in the Atchafalaya River delta, Louisiana. *Journal of Ecology*, 80, 677-687.

Sheremet, A. and Stone, G.W. 2003. Observations of nearshore wave dissipation over muddy sea beds. *Journal of Geophysical Research*, 108 (C11), DOI 10:1029/2003 JC 001885.

Shlemon, R.J. 1975. Subaqueous delta formation-Atchafalaya Bay, Louisiana. *In* Broussard, M.L. (ed) *Delta: Models for Exploration*. Houston Geological Society, 209-221.

Sterling, G.H. and Strohbeck, E.E. 1975. The failure of the South Pass 70"B" platform in Hurricane Camille. *Journal of Petroleum Technology*, 27, 263-268.

Suhayda, J.N. 1977. Surface-waves and bottom sediment response. *Marine Geotechnology*, 2, 135-146.

Thompson, W.C. 1951. *Oceanographic analysis of Atchafalaya Bay, Louisiana and adjacent continental shelf area marine pipeline problems*. Texas A&M Foundation, Department of Oceanography, Section 2, Project 25, 31pp.

Thompson, W.C. 1955. Sandless coastal terrain of the Atchafalaya Bay area, Louisiana. *SEPM Special Publication*, 3, 52-76.

US Army Corps of Engineers, 2002. *New Orleans District Water Control Section*. Available from http://www.mvn.usace.army.mil/eng/edhd/wcontrol/discharg.htm.

van Heerden, I.L.I. 1980. *Sedimentary responses during flood and non-flood conditions, new Atchafalaya delta, Louisiana.* Unpublished MS thesis, Louisiana State University, 76pp.

van Heerden, I.L.I. 1983. *Deltaic sedimentation in eastern Atchafalaya Bay, Louisiana.* Unpublished PhD dissertation, Louisiana State University, 151pp.

van Heerden, I.L. and Roberts, H.H. 1980. The Atchafalaya Delta – Louisiana's new prograding coast. *Transactions Gulf Coast Association of Geological Societies*, 30, 497-506.

van Heerden, I.L.I. and Roberts, H.H. 1988. Facies development of Atchafalaya Delta, Louisiana: a modern bayhead delta. *AAPG Bulletin*, 72, 439-453.

Walker, N.D. 2001. Tropical storm and hurricane wind effects on water level, salinity and sediment transport in the river-influenced Atchafalaya-Vermilion Bay System, Louisiana, USA. *Estuaries*, 24, 498-508.

Walker, N.D. and Hammack, A.B. 1999. *Impacts of river discharge and wind forcing on circulation, sediment distribution, sediment flux and salinity changes: Vermilion/Cote Blanche Bay System, Louisiana.* Final Report, US Army Corps of Engineers, Waterways Experiment Station, Vicksburg, MS, 157pp.

Walker, N.D. and Hammack, A.B. 2000. Impacts of winter storms on circulation and sediment transport: Atchafalaya-Vermilion Bay Region, Louisiana. *Journal of Coastal Research*, 16, 996-1010.

Wheatcroft, R.A. and Borgeld, J.C. 2000. Oceanic flood deposits on the northern California shelf: large-scale distribution and small-scale physical properties. *Continental Shelf Research*, 20, 2163-2190.

Wright, L.D. and Nittrouer, C.A. 1995. Dispersal of river sediments in coastal seas: six contrasting cases. *Estuaries*, 18, 494-508.

Chapter 14

EVOLVING UNDERSTANDING OF THE TAY ESTUARY, SCOTLAND

Exploring the Linkages Between Frontal Systems and Bedforms

R.W. Duck
Department of Geography, University of Dundee, Dundee, DD1 4HN, Scotland

1. INTRODUCTION TO PREVIOUS RESEARCH ON THE TAY ESTUARY

Renowned as the site of the world's most infamous railway accident, the so-called Tay Bridge Disaster of 1879 (Duck and Dow, 1994), the Tay Estuary of eastern Scotland is one of the most widely studied in the country, especially in terms of geological and geomorphological processes. Over the past four decades in particular, numerous studies of the sedimentology and hydrodynamics of this complex, macrotidal system have been undertaken, largely by researchers from the University of Dundee using the boats and equipment of the dedicated Tay Estuary Research Centre. The database (http://www.dundee.ac.uk/crsem/TEF/review.htm_- currently containing over 300 entries) recently compiled by the Tay Estuary Forum (a voluntary partnership established in 2000 "to promote the wise and sustainable use of the Tay Estuary and adjacent coastline"), provides an impressive but still incomplete inventory of the published and unpublished research that has been undertaken on this water body.

In the 1960s and 1970s physical studies were largely focused on the bathymetry and geological evolution of the estuary since the Pleistocene (e.g. McManus, 1970; Buller and McManus, 1971) and it is of interest to note that sub-bottom profiling, in the form of a Sparker survey, was undertaken in the Tay as long ago as the early 1960s (McGuinness *et al.*,

D.M. FitzGerald and J. Knight (eds.),
*High Resolution Morphodynamics and Sedimentary Evolution of Estuaries,*299-315.

1962) as a component of the site investigation along the line of the proposed Tay Road Bridge (Fig. 1; opened in 1966). At this time studies were also directed towards determining the distribution of bottom sediments (e.g. McManus, 1972; Buller and McManus, 1975), salinity and factors controlling water circulation (e.g. Charlton *et al.*, 1975; West, 1972; Williams and West, 1975) and the dynamics of sedimentation (e.g. Buller *et al.*, 1975; Green, 1975).

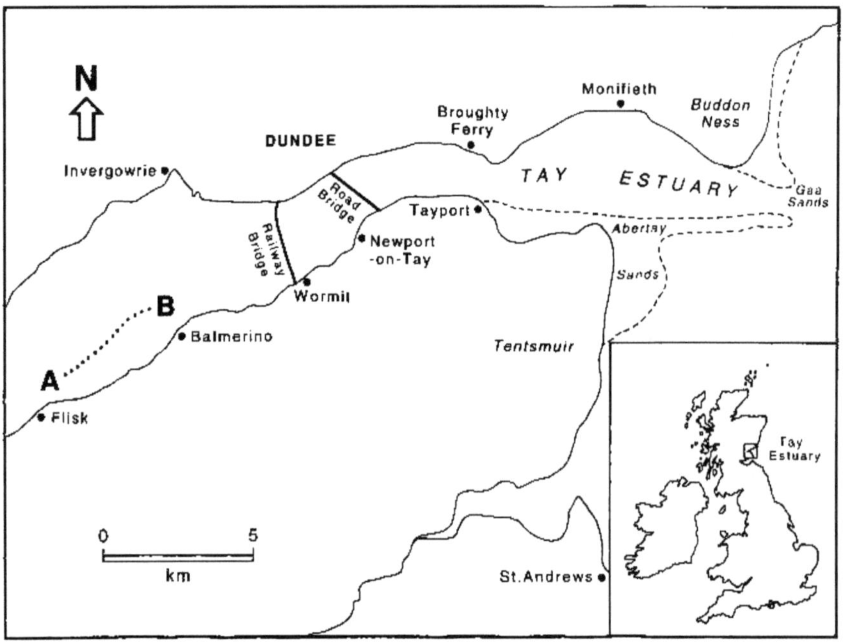

Figure 1. Map of the Tay Estuary showing locations referred to in the text. The dotted line from A to B off Flisk to Balmerino in the upper estuary marks a foam line, delimiting an axially convergent front, that is referred to in studies reported later in the chapter.

By the 1980s attention became focused towards understanding of the generation and migration of turbidity maxima in the Tay Estuary (Dobereiner and McManus, 1983; Weir and McManus, 1987), together with studies estimating sediment inputs to the system from fluvial sources (e.g. Al-Jabbari *et al.*, 1980; McManus, 1986) and the landward migration of marine-derived materials (Al-Dabbas and McManus, 1987). Simultaneously the development of mathematical models to simulate saline intrusion (Nassehi and Williams, 1987), tidal motion and water level (Gunn and Yenigun, 1987; Gunn *et al.*, 1987) was taking place. The 1980s also saw the beginnings of the application of remote sensing studies to the Tay Estuary, initially with the use of Landsat multispectral scanning (MSS) data for monitoring of the changes in positions of emergent sandbanks (Cracknell *et al.*, 1982) and mapping of the intertidal zone (Cracknell *et al.*, 1987).

The acquisition of the synoptic datasets afforded by remote sensing observations has perhaps provided the single most significant advancement in our understanding of the hydrodynamics of the Tay. By the late 1980s remote sensing of the estuary was being carried out using the airborne thematic mapper (Anderson, 1989) which provided the first synoptic observations of foam lines, delimiting frontal systems, at the water surface. Subsequently, in the 1990s, Ferrier and Anderson (1996, 1997a, 1997b) made extensive observations of the temporal and spatial evolution of frontal systems in the Tay through the tidal cycle, at the water surface, using various airborne remote sensing methods linked with water-truth data.

Concurrent with but independent of the airborne observations, studies of the estuary bed using high resolution side-scan sonar were carried out to map subtidal bedform distributions (Wewetzer and Duck, 1996; Wewetzer, *et al.*, 1999a) and to elucidate the relationships between bedform geometries, sediment types and near bottom current velocities (Wewetzer and Duck, 1999; Duck and Wewetzer, 2000). Understanding of the three-dimensional velocity structure within the water column is currently being explored through the use of Acoustic Doppler current profiler (ADCP) measurements which, in particular, are enabling the current systems associated with convergent frontal systems to be better understood (Wewetzer *et al.*, 1999b). Ongoing studies of bedload provenance and transport pathways using magnetic susceptibility measurements, combined with trends in sediment grain size distributions and observations of bedform asymmetry, have for the first time quantified the dominance of marine-derived bedload to the estuary (Duck *et al.*, 2001). Integration of side-scan sonar data (acoustic remote sensing) of the estuary bed with airborne remote sensing observations of the water surface has revealed associations between surface foam lines and bottom sediment facies boundaries (Duck and Wewetzer, 2001) that have implications for both sediment and pollutant dynamics. The latter aspects of the Tay Estuary will be explored further in this chapter.

2. PHYSICAL SETTING OF THE TAY ESTUARY

The Tay Estuary is a major embayment on the east coast of Scotland with a tidal reach of c. 50 km, a maximum width of 5 km and a maximum depth of 30 m. It has a total area of 12128 ha of which 5583 ha are intertidal (Reeves, 1994). The tidal ranges associated with neap, spring and equinoctial tides are 3.5, 5 and 6 m, respectively. One of the cleanest major estuaries in

Europe (McManus, 1998), it receives the drainage from the Rivers Earn and Tay, the latter the foremost British river in terms of discharge, providing a long term mean inflow of c.180 $m^3 s^{-1}$ from a catchment area of c. 6500 km^2. It is, in general terms, a partially mixed estuary (McManus, 1998), in which salinity and the landward penetration of salt water are controlled by the balance between prevailing freshwater discharge and tidal range (Williams and West, 1975; Dobereiner and McManus, 1983). The estuary is of geomorphologically complex origin (Davidson *et al.*, 1991; Paterson *et al.*, 1981), arising from a number of geological constraints; Pleistocene glaciation, river erosion and sea level fluctuation. The bedrock channel in which the estuary is located is, to a large extent, infilled with a varied sequence of Late Glacial – Holocene deposits into which the present day estuarine channels are cut (Buller and McManus, 1971), capped by a veneer of contemporary sediments (Buller and McManus, 1975). The latter, within the sub-tidal zone, are largely of sand and gravel grade.

Here only a brief account of the bed sediments of the Tay Estuary will be given. For a more detailed review see Buller *et al.* (1971), Buller and McManus (1975) and McManus *et al.* (1980). The uppermost fluviatile sector of the estuary is lined with small boulders, pebbles and patches of sand. The bottom sediments of the sub-tidal parts of the upper estuary generally fall within the range of medium to coarse sands with intermixed granules and pebbles. To the north and south of the main channel are inter-tidal areas that are typically characterised by fine sands. These are particularly well developed on the northern shore where they are backed by the largest continuous reed bed (*Phragmites*) in Britain. In this reach of the estuary, from the Tay-Earn confluence to Balmerino (Fig. 1), the sediments of the main channel, which trends along the southern shore, are characterised by the development of flow transverse dune bedforms with wavelengths of between 15 and 50 m and heights of up to 3 m (Grey, 1998).

The bed of the middle estuary is extremely unstable and is characterised by sand, shifting sandbanks and migrating channels. Between the Tay Railway and Road Bridges the dominant sedimentary class is slightly gravelly sand (*sensu* Folk, 1974) which, in sub-tidal areas, is characterised by dunes typically with wavelengths in the range 2-10 m and heights of up to 0.5 m, though larger bedforms are locally developed. The lower middle estuary extends from the Road Bridge to Broughty Ferry (Fig. 1) and is characterised by relatively stable sandbanks, which form the Newcombe Shoal on the south side of the water body. North of the Newcombe Shoal lies the main channel, which is dominated by coarse sands again characterised by the development of dune bedforms. The wide, lower estuary seawards of Broughty Ferry has well developed inter-tidal flats of fine sand on both northern and southern shores, extending as far east as the estuary mouth. The seaward reaches are characterised by the migrating offshore spit complexes of the Gaa and Abertay Sands (Ferentinos and McManus, 1981) and beaches of sand backed by extensive dune systems. Here marine conditions dominate.

3. FRONTAL SYSTEMS IN THE TAY ESTUARY

Like many estuaries world-wide (e.g. Huzzey and Brubaker, 1988; Brown *et al.*, 1991), the Tay is characterised by clearly developed frontal systems. A front, in general terms, is 'a meeting of waters', more specifically 'a region characterized by an anomalous local maximum in the horizontal gradient of some water property (e.g. temperature, salinity, nitrate concentration, chlorophyll concentration)' (Largier, 1993, p. 1). Fronts play important roles in estuarine hydro- and sediment dynamics. In effect they are interfaces, either near vertical or inclined, where discrete water masses converge, diverge or move laterally relative to one another. At the water surface they are typically manifested as bands of foam, floating debris or distinct changes in colour or transparency of the water. However, as Largier (1993) points out, the absence of such features at the water surface in an estuary does not necessarily mean the absence of fronts within the water column. Where developed, fronts serve to compartmentalise the water column, thereby inhibiting complete mixing. In consequence they can cause the occurrence of sharp gradients in turbidity and suspended sediment concentration (Kirby and Parker, 1982; Klemas, 1980; Pinckney and Dustan, 1990; Reeves and Duck, 2001), the concentration of buoyant pollutants in foam lines (Brown *et al.*, 1991; Swift *et al.*, 1996) that may become deposited in inter-tidal areas at low water, and the compartmentalisation of dinoflagellate blooms (Tyler *et al.*, 1982). In the Tay, for example, Ferrier and Anderson (1997b) have reported a 10-fold increase in the concentration of *Escherichia coli* bacteria on the nearshore side of an advective flow front (see Table 19-1) illustrating the inhibiting effect that fronts can have on the mixing and diffusion of effluent into deeper waters.

Their role in inhibiting the transport of fine particulate materials has led Reeves and Duck (2001) to suggest that estuaries, such as the Tay, characterised by fronts, should be considered as acting as 'sieves' rather than filters (cf. UNESCO, 1982; Schubel and Carter, 1984), traps or sinks (cf. UNESCO, 1983; Reeves, 1988). They further suggested that, where characterised by many fronts, an estuary as a whole should most appropriately be considered as a complex of sieves, which collectively create a dynamic "sieve regime" (Reeves and Duck, 2001). The term 'dynamic' alludes to the transient compartmentalisation that fronts create in the estuarine sediment transfer system, according to their mode of formation, subsequent evolution and decay. Within the Tay Estuary, Ferrier and Anderson (1996, 1997a, 1997b) have deduced four principal mechanisms for front formation: tidal intrusion, axial convergence, advective flow and flow separation, the salient features and characteristics of which are summarized in Table I.

As an example, Figure 2 shows the temporally and spatially persistent foam line at the water surface associated with the formation of an axially convergent front in the middle estuary developed on the flood tide in the main channel offshore from Flisk to Balmerino (see Fig. 1 for location). The photograph was taken looking approximately due east with the Tay Road Bridge on the horizon about three hours before high water on a spring tide. This front will be considered further later in the chapter in the context of the bedforms developed on either side of it.

Figure 2. Photograph looking due east of foam line associated with an axially convergent front in the middle Tay Estuary. Location is approximately mid-way between A and B (see Figure 1). Further details are given in the text.

4. STUDIES OF BEDFORMS IN THE TAY ESTUARY

Wewetzer and Duck (1999) undertook the first systematic survey of the bedforms in part of the middle estuary (as delimited by the Tay Railway and

Road Bridges) using high resolution side-scan sonar equipment (Klein Hydroscan, Model 401: 400kHz; Waverley Sonar 3000: 100 kHz) calibrated by bottom sediment sampling. This had the aims of mapping the hitherto unknown spatial distribution of sub-tidal and inter-tidal bedform types and the investigation of relationships among the geometrical parameters of bedforms and water depth. The main bedform types were dunes of various sizes and morphologies (*sensu* Ashley, 1990 as revised by Dalrymple and Rhodes, 1995), with small wavelength dunes occupying most of the channel areas. Dunes of three wavelength classes (small, 0.6-5.0 m; medium, 5.0-10.0 m; large 10.0-100.0 m) were recorded on Middle Bank, a major sandbank that divides the study area into the main Navigation Channel to the south and Queen's Road Channel to the north. Medium height dunes (0.25-0.50 m) were found to be the dominant dune height class in this sector of the estuary, characterising Middle Bank as well as most channel areas. However, a feature of this study was the observation of abrupt changes in sediment facies and bedform geometry, typically without any bathymetric control (Wewetzer *et al.*, 1999a; Duck and Wewetzer, 2001).

Dune dimensions measured from sonographs were analysed in terms of inter-correlations of wavelength, height and corresponding water depth, which vary according to spatial (inter-tidal *versus* sub-tidal) and temporal (ebb tidal *versus* flood tidal) data sub-division. Although several researchers (e.g. Dalrymple *et al.*, 1978; Flemming, 1988; Zarillo, 1982) have found statistically significant correlations between these variables in flume experiments and in field studies of inter-tidal environments, dune height and wavelength were not correlated or weakly correlated with water depth in this study (Wewetzer and Duck, 1999). In consequence, it was suggested that the relationships between these parameters established previously might not be generally applicable in estuarine environments and that the influence of additional variables such as flow strength, sediment textural characteristics and sediment availability should be explored. Flemming (2000), however, criticised this work for not providing explanations as to why the relationships observed in other areas of the world were not apparent in the Tay.

Table I. The four principal mechanisms of front formation in the Tay Estuary and the associated features and characteristics of each type of frontal system (based on Ferrier and Anderson, 1996, 1997a, 1997b and field observations).

Mechanism	Associated Features and Characteristics
Tidal Intrusion	❑ Develop on flood tide ❑ Strong salinity gradients ❑ Surface colour changes ❑ Accumulations of foam and debris ❑ Upstream progressing ❑ V-shaped in plan, apex angle varying according to tidal velocity
Axial Convergence	❑ Result from flow convergence along axis of a tidal flow ❑ Develop on flood tide as bed friction reduces flow velocities at channel margins relative to those in mid-channel ❑ Shear in flood current induces landward advection of higher salinity water ❑ Two-celled lateral surface flow developed (Simpson and Turrell, 1986) ❑ Manifested as foam lines along central axes of channels; can extend for several km. ❑ No colour or temperature variations
Advective Flow	❑ Also known as longitudinal fronts: formation due to intra-tidal and lateral salinity balance ❑ Typically delimited by foam lines, water colour and temperature contrasts ❑ Occur at all stages of the flood-ebb tidal cycle ❑ Orientations are highly variable; closely related to bottom topography ❑ Similar to axially convergent fronts but different mode of formation of density gradient
Flow Separation	❑ Form on both flood and ebb tide downstream from edge of topographic features (e.g. mid-estuary sandbanks) ❑ Also extend outwards from estuary margins ❑ Result from division of tidal flows into discrete sections and subsequent convergence of water masses ❑ Marked by foam lines and sometimes water colour and temperature contrasts; can extend for over 1 km

The study of sub-tidal bedforms in the Tay, using side-scan sonar, echo-sounding (Raytheon Fathometer) and bottom sediment sampling, has subsequently been extended to a large section (c. 1.5 x 13 km) of the upper estuary extending from the Railway Bridge along the main channel to the west of Flisk (Fig. 1). This area is characterised by a major, SW-NE trending channel, lined by deposits of sands, gravelly sands and sandy gravels (*sensu*

Folk, 1974) with emergent banks of finer sediments to the north. Although all of the bedforms observed in this area are various types of dunes, the acoustic surveys have revealed previously unidentified abrupt spatial changes in dune morphology (height, wavelength, sinuosity and superposition). These are similar to those reported from the area of the Tay to the east between the two bridges. In general, the largest dunes observed (large and medium) occur in medium sands (1-2 ϕ) where the water depth is greatest (>5 m). Very few dunes were detected in the coarser, more gravelly deposits, while small dunes form in patches in the finer sands (2-3 ϕ) of the lower inter-tidal flats. In common with observations in the estuary reach between the bridges (Duck and Wewetzer, 2000), dune asymmetry in this area varies temporally and spatially, indicative of the complexity of near-bottom water circulation patterns. In common with the earlier studies, correlations between both dune height and wavelength with water depth are weak and not statistically significant. However, a positive correlation (r^2 = 0.58; significant at the 0.005 level) between dune height and wavelength (cf. Flemming, 1988) was observed for the complete dataset. This is stronger than the correlation between dune wavelength and height recorded by Wewetzer and Duck (1999) in the area to the east (r^2 = 0.32; significant at the 0.01 level), again for the full dataset (i.e. both intertidal and sub-tidal dunes recorded during both ebb and flood tidal conditions). However, it is a weak correlation in comparison with those obtained in the studies of e.g. Flemming (1988) or Zarillo (1982).

5. LINKAGES BETWEEN FRONTAL SYSTEMS AND BEDFORMS

In the Tay Estuary between the Railway and Road Bridges, Duck and Wewetzer (2001) observed, using high resolution side scan sonar, that fronts within the water column may be marked not only by surface foam bands but also by abrupt (i.e. non-gradational) changes in the underlying bedform morphology and/or sediment facies. Sonographs were recorded along traverses at right angles to three foam bands; formed by axial convergence, flow separation and tidal intrusion (see Fig. 3 of Duck and Wewetzer, 2001). In each of seven segments of sonographs crossing the estuary bed beneath these foam lines a clear, abrupt transition in either backscatter levels (as indicated by tonal intensity, which is a function of the density/porosity of the substrate) or bedform geometry was apparent (see Figs. 6 of Duck and Wewetzer, 2001). The sonographs obtained are indicative of differing hydrodynamic conditions on either side of the fronts. Moreover, as a result of this preliminary study, Duck and Wewetzer (2001) suggested that fronts might exert a control not only on surface and intra-water column sediment and pollutant partitioning, but also on the distribution and persistence of bedload transport pathways. Subsequent repetition of the seven sonar traverses has revealed the persistence of the abrupt bedform and sediment facies transitions reported in 2001.

To determine whether this is a phenomenon restricted in its occurrence to the middle estuary, similar investigations have been extended into the upper reaches, particularly in relation to the front shown in Fig. 2, which is developed in a channel characterised by flow transverse bedforms. The surveys incorporated side-scan sonar traverses perpendicular to the front and echo-sounding runs parallel with and crossing the foam line obliquely, with position fixing by Trimble DGPS. As an example, an approximately west to east echogram, crossing the foam line shown in Fig. 2 at an oblique angle, is shown in Fig. 3. To the north of the front (eastern end of echogram) the bedforms developed in the underlying coarse sands are large dunes typically with heights of c. 1.5 m and wavelengths of c. 20 m (Fig. 3). On the southern side of the foam line, however, (western end of echogram) the bedforms developed in the same grade of sediments are much smaller in both height (c. 0.2 m) and wavelength (c. 2 m) and, as such, are classified as small dunes (Fig. 3).

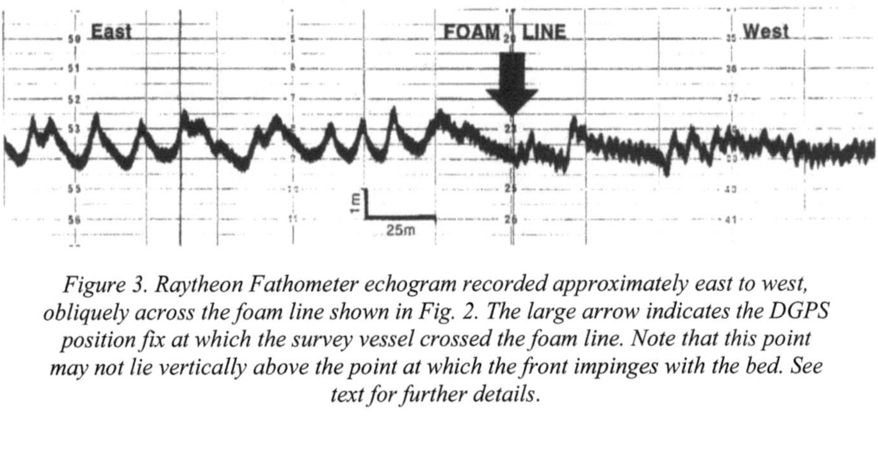

Figure 3. Raytheon Fathometer echogram recorded approximately east to west, obliquely across the foam line shown in Fig. 2. The large arrow indicates the DGPS position fix at which the survey vessel crossed the foam line. Note that this point may not lie vertically above the point at which the front impinges with the bed. See text for further details.

6. CONCLUDING DISCUSSION

The observations made in the upper reaches of the Tay Estuary indicate that the phenomenon of abrupt changes in bedform geometries is not restricted only to association with the frontal systems of the middle estuary. The three-dimensional velocity flow structures within the water column below the foam line of Fig. 2 are currently under investigation by means of ADCP. Clearly, however, there are sharp contrasts in the energy in the water column, giving rise to the very different bedform geometries observed. With reference to detailed direct measurements of the benthic boundary layer, it has been noted that abrupt changes in bathymetry, sediment size and bedform characteristics "indicate that nearbed turbulence changes in space can be equally drastic" (Griffiths *et al.*, 2000, p.3). The Tay Estuary investigations thus reveal an increasing body of evidence to suggest that fronts can exert a control on the distribution and persistence of estuarine bedload transport pathways and bedload segregation (cf. Anthony, 2002).

Previously, in a side-scan sonar study of one of the subtidal channels in the mouth of the Oosterschelde, The Netherlands, Goedheer and Misdorp (1985) reported that the relationships between current velocity parameters and bedform geometries determined in intertidal zones are not applicable. Moreover, sonographs of the floor of this channel (e.g. Goedheer and Misdorp, Fig. 3A) reveal abrupt boundaries (cf. Duck and Wewetzer, 2001) between bedforms of differing geometries. Although Goedheer and Misdorp (1985) did not consider the presence or potential role of frontal systems in the Oosterschelde, it is possible that they may play a role in the abrupt boundaries between bedform types observed. In a recent investigation of flow-transverse bedforms of the northernmost tidal inlet of the Danish Wadden Sea, Bartholdy *et al.* (2002) described the complex evolution of dune dimensions as a function of sediment grain size. This study of the Gradyb inlet, however, also demonstrated no correlation between dune dimensions and water depth. As a consequence of the Tay Estuary observations, it is thought possible that compartmentalisation of the Gradyb water column by frontal systems and associated sharp contrasts in energy levels could be at least partially responsible for this lack of correlation.

The observations of Duck and Wewetzer (2001), together with those reported in this chapter, go some way to explaining the lack of good correlation between dune height, wavelength and water depth in the Tay Estuary. The locally prevailing conditions within the Tay, in particular the partitioning of bedload transport pathways by frontal systems, is suggested to be a reason for the lack of conformity with the correlations reported by others (e.g. Flemming, 2000), which do not take compartmentalisation of the energy levels in the water column and the impacts that these have on the bed into account. Typically, studies of estuarine bedforms take place without consideration of the structure of the overlying water column, whilst studies of fontal systems within the water column are usually carried out without any consideration of the underlying bed. The importance of integrating observations of bedform geometries and dynamics with those of frontal systems has been highlighted in this chapter.

Research in the Tay is continuing with respect to the persistence of bottom sediment transport pathways and the tidal and hydrological conditions under which they function persistently, discontinuously or cease to operate. These aspects are particularly important in terms of not only long-term bottom sediment dynamics but also pollutant transport and accumulation. Despite four decades of intensive research, the Tay Estuary continues to reveal features that are of wider significance to our understanding of estuarine processes. It is, moreover, apparent from the ongoing work reported herein that the Tay Estuary still has much to surrender in terms of our understanding of the complexity of processes in partially mixed systems.

7. ACKNOWLEDGEMENTS

This work would not have been possible without the tireless support, technical skills and seamanship of Ian Lorimer, boatman at the Tay Estuary Research Centre, whose unparalleled knowledge of the Tay has contributed immeasurably to the successful acquisition of field data over the last thirty years. The helpful and constructive comments of three referees, Helene Burningham, Jasper Knight and John McManus, have considerably improved this chapter and are acknowledged with gratitude.

REFERENCES

Al-Dabbas, M.A.M., and McManus, J. 1987. Shell fragments as indicators of bed sediment transport in the Tay Estuary. *Proceedings of the Royal Society of Edinburgh (B)*, 92, 335-344.

Al-Jabbari, M.H., McManus, J. and Al-Ansari, N.A. 1980. Sediment and solute discharge into the Tay Estuary from the river system. *Proceedings of the Royal Society of Edinburgh (B)*, 78, 15-32.

Anderson, J.M. 1989. Remote sensing in the Tay Estuary using the airborne thematic mapper. In: McManus, J. and Elliott, M. (Eds) *Developments in Estuarine and Coastal Study Techniques*, Olsen & Olsen, Fredensborg, Denmark, 15-19.

Anthony, E.J. 2002. Long-term marine bedload segregation, and sandy versus gravelly Holocene shorelines in the eastern English Channel. *Marine Geology*, 187, 221-234.

Ashley, G. 1990. Classification of large-scale subaqueous bedforms: a new look at an old problem. *Journal of Sedimentary Petrology*, 60, 160-172.

Bartholdy, J., Bartholomae, A. and Flemming, B.W. 2002. Grain-size control of large compound flow-transverse bedforms in a tidal inlet of the Danish Wadden Sea. *Marine Geology*, 188, 391-413.

Brown, J., Turrell, W.R. and Simpson, J.H. 1991. Aerial surveys of axial convergent fronts in UK estuaries and the implications for pollution. *Marine Pollution Bulletin*, 22, 397-400.

Buller, A.T., Green, C.D. and McManus, J. 1975. Dynamics and sedimentation: the Tay in comparison with other estuaries. In: Hails, J. and Carr, A.J. (Eds) *Nearshore Sediment Dynamics*, 201-249, John Wiley, London.

Buller, A.T. and McManus, J. 1971. Channel stability in the Tay Estuary: controls by bedrock and unconsolidated Post-Glacial sediment. *Engineering Geology*, 5, 227-237.

Buller, A.T. and McManus, J. 1975. Sediments of the Tay Estuary I. Bottom sediments of the upper and upper middle reaches. *Proceedings of the Royal Society of Edinburgh (B)*, 75, 41-64.

Buller, A.T., McManus, J. and Williams, D.J.A. 1971. Investigations in the estuarine environments of the Tay. *Tay Estuary Research Centre Report*, 1, University of Dundee.

Charlton, J.A., McNicoll, W. and West, J.R. 1975. Tidal and freshwater induced circulation in the Tay Estuary. *Proceedings of the Royal Society of Edinburgh (B)*, 75, 11-27.

Cracknell, A.P., Hayes, L.W.B. and Keltie, G.F. 1987. Remote sensing of the Tay Estuary using visible and near-infrared data: mapping of the inter-tidal zone. *Proceedings of the Royal Society of Edinburgh (B)*, 92, 223-236.

Cracknell, A.P., MacFarlane, N., McMillan, K., Charlton, J.A., McManus, J. and Ulbricht, K.A. 1982. Remote sensing in Scotland using data received from satellites. A study of the Tay Estuary region using Landsat multispectral scanning imagery. *International Journal of Remote Sensing*, 3, 113-137.

Dalrymple, R.W., Knight, R.J and Lambiase, J.J. 1978. Bedforms and their hydraulic stability relationships in a tidal environment, Bay of Fundy, Canada. *Nature*, 275, 100-104.

Dalrymple, R.W. and Rhodes, R.N. 1995. Estuarine dunes and bars. In: Perillo, G.M. E. (Ed) *Geomorphology and Sedimentology of Estuaries*, 359-422, Elsevier, Amsterdam.

Davidson, N.C., Laffoley, D.d'A., Doody, J.P., Way, L, S., Gordon, J., Key, R., Drake, C.M., Peintowski, M.W., Mitchell, R. and Duff, K.L. 1991. *Nature Conservation in Estuaries in Great Britain*, Nature Conservancy Council, Peterborough.

Dobereiner, C. and McManus, J. 1983. Turbidity maximum migration and harbor siltation in the Tay Estuary. *Canadian Journal of Fisheries and Aquatic Sciences*, 40 (Suppl. 1), 117-129.

Duck, R.W. and Dow, W.M. 1994. Side-scan sonar reveals submerged remains of the first Tay Railway Bridge. *Geoarchaeology*, 9, 139-153.

Duck, R.W., Rowan, J.S., Jenkins, P.A. and Youngs, I. 2001. A multi-method study of bedload provenance and transport pathways in an estuarine channel. *Physics and Chemistry of the Earth (B)*, 26, 747-752.

Duck, R.W. and Wewetzer, S.F.K. 2000. Relationship between current measurements and sonographs of subtidal bedforms in the macrotidal Tay Estuary, Scotland. In: Pye, K. and Allen, J.R.L. (Eds) *Coastal and Estuarine Environments: sedimentology, geomorphology and geoarchaeology*. Geological Society Special Publication 175, 31-41.

Duck, R.W. and Wewetzer, S.F.K. 2001. Impact of frontal systems on estuarine sediment and pollutant dynamics. *The Science of the Total Environment*, 266, 23-31.

Ferentinos, G. and McManus, J. 1981. Nearshore processes and shoreline development in St Andrews Bay, Scotland, UK. *Special Publications of the International Association of Sedimentologists*, 5, 161-174.

Ferrier, G. and Anderson, J.M. 1996. The application of remote sensing data in the study of effluent dispersal in the Tay Estuary, Scotland. *International Journal of Remote Sensing*, 17, 3541-3566.

Ferrier, G. and Anderson, J.M. 1997a. The application of remotely sensed data in the study of frontal systems in the Tay Estuary, Scotland. *International Journal of Remote Sensing*, 18, 2035-2065.

Ferrier, G. and Anderson, J.M. 1997b. A multi-disciplinary study of frontal systems in the Tay Estuary, Scotland. *Estuarine, Coastal and Shelf Science*, 45, 317-336.

Flemming, B.W. 1988. Zur Klassifikation subaquatischer, strömungstransversaler Transportkörper. *Bochumer Geologische und Geotechnische Arbeiten*, 29, 44-47.

Flemming, B.W. 2000. The role of grain size, water depth and flow velocity as scaling factors controlling the size of subaqueous dunes.

http://mozart.shom.fr/sci/sedim/MSD/Tpflemprem.html (accessed 17.08.00).

Folk, R.L. 1974. *Petrology of Sedimentary Rocks*, 182p, Hemphill, Austin, Texas.

Goedheer, G.J. and Misdorp, R. 1985. Spatial variability and variations in bedload transport direction in a subtidal channel as indicated by sonographs. *Earth Surface Processes and Landforms*, 10, 375-386.

Green, C.D. 1975. A study of hydraulics and bedforms at the mouth of the Tay Estuary, Scotland. In: Cronin, L.E. (Ed) *Estuarine Research Vol. II*, 323-344, Academic Press, New York.

Grey, W. 1998. *Understanding the Spatial Changes of Bedform Morphology Within the Upper-Middle Reaches of the Tay Estuary Using Acoustic Methods of Remote Sensing*. MSc Thesis, Department of Applied Physics, Electrical and Mechanical Engineering, University of Dundee, unpublished.

Griffiths, G., Fernandes, P.G., Brierley, A.S., Voulgaris, G. and The Autosub Technical Team 2000. Unescorted science missions with the Autosub AUV in the North Sea. http://www.soc.soton.ac.uk/OTD/gxg/NUWC_unescorted_paper.pdf (accessed 30.10.02).

Gunn, D.J., McManus, J and Yenigun, O. 1987. Partial validation of a numerical model for tidal motion in the Tay Estuary. *Proceedings of the Royal Society of Edinburgh (B)*, 92, 275-283.

Gunn, D.J. and Yenigun, O. 1987. A model for tidal motion and level in the Tay Estuary. *Proceedings of the Royal Society of Edinburgh (B)*, 92, 257-273.

Huzzey, L. and Brubaker, J.M. 1988. Formation of longitudinal fronts in a coastal plain estuary. *Journal of Geophysical Research*, 93, 1329-1334.

Kirby, R. and Parker, W.R. 1982. A suspended sediment front in the Severn Estuary. *Nature*, 295, 396-399.

Klemas, V. 1980. Remote sensing of coastal fronts and their effects on oil dispersion. *International Journal of Remote Sensing*, 1, 11-28.

Largier, J.L. 1993. Estuarine fronts: how important are they? *Estuaries*, 16, 1-11.

McGuinness, W.T., Beckmann, W.C. and Officer, C.B. 1962. The application of various geophysical techniques to specialised engineering projects. *Geophysics*, 27, 221-236.

McManus, J. 1970. The geological setting of the bridges of the Lower Tay Estuary with particular reference to the fill of the buried channel. *Quarterly Journal of Engineering Geology*, 3, 197-205.

McManus, J. 1972. Estuarine development and sediment distribution with particular reference to the Tay. *Proceedings of the Royal Society of Edinburgh (B)*, 71, 97-113.

McManus, J. 1986. Land-derived sediment and solute transport to the Forth and Tay Estuaries, Scotland. *Journal of the Geological Society, London*, 143, 927-934.

McManus, J. 1998. Mixing of sediments in estuaries. In: Cracknell, A.P and Rowan, E.S. (Eds) *Physical processes in the Coastal Zone: Computer Modelling and Remote Sensing*, 281-298, SUSSP Publications and Institute of Physics.

McManus, J., Buller, A.T. and Green, C.D. 1980. Sediments of the Tay Estuary VI. Sediments of the lower and outer reaches. *Proceedings of the Royal Society of Edinburgh (B)*, 78, 133-154.

Nassehi, V. and Williams, D.J.A. 1987. A mathematical model for salt intrusion in the Tay Estuary. *Proceedings of the Royal Society of Edinburgh (B)*, 92, 285-297.

Paterson, I.B., Armstrong, M. and Browne, M.A.E. 1981. Quaternary estuarine deposits in the Tay-Earn area, Scotland. *Institute of Geological Sciences Report*, 81/7, HMSO, London.

Pinckney, J. and Dustan, P. 1990. Ebb-tidal fronts in Charleston Harbor, South Carolina: physical and biological characteristics. *Estuaries*, 13, 1-7.

Reeves, A.D. 1988. *The Distribution and Behaviour of Lignin in the Estuarine Environment.* Ph.D. Thesis, University of Liverpool, unpublished.

Reeves, A.D. 1994. Factors influencing water quality in the Tay Estuary. *Tay Estuary Research Centre Report*, 11, 16pp, University of Dundee.

Reeves, A.D. and Duck, R.W. 2001. Density fronts: sieves in the estuarine sediment transfer system? *Physics and Chemistry of the Earth (B)*, 26, 89-92.

Schubel, J.R and Carter, H.H. 1984. The estuary as a filter for fine-grained suspended sediment. In Kennedy, V.S. (Ed) *The Estuary as a Filter*, 81-107, Academic Press, Orlando.

Simpson, J.H. and Turrell, W.R. 1986. Convergent fronts in the circulation of tidal estuaries. In: *Estuarine Variability*, 139-153, Academic Press, Orlando.

Swift, M.R., Fredriksson, D.W. and Celikkol, B. 1996. Structure of an axial convergence zone from Acoustic Doppler current profiler measurements. *Estuarine, Coastal and Shelf Science*, 43, 109-122.

Tyler, M.A., Coats, D.W., and Anderson, D.M. 1982. Encystment in a dynamic environment: deposition of dinoflagellate cysts by a frontal convergence. *Marine Ecology Progress Series*, 7, 163-178.

UNESCO 1982. *Ocean Science for the Year 2000.* Rome, 133p.

Weir, D.J. and McManus, J. 1987. The role of wind in generating turbidity maxima in the Tay Estuar y. *Continental Shelf Research*, 7, 1315-1318.

West, J.R. 1972. Water movements in the Tay Estuary. *Proceedings of the Royal Society of Edinburgh (B)*, 71, 115-129.

Wewetzer, S.F.K. and Duck, R.W. 1996. Side-scan sonograph from the middle reaches of the Tay Estuary, Scotland. *International Journal of Remote Sensing*, 17, 3539-3540.

Wewetzer, S.F.K. and Duck, R.W. 1999. Bedforms of the middle reaches of the Tay Estuary, Scotland. *Special Publications of the International Association of Sedimentologists*, 28, 33-41.

Wewetzer, S.F.K., Duck, R.W. and McManus, J. 1999a. Side-scan sonar mapping of bedforms in the middle Tay Estuary, Scotland. *International Journal of Remote Sensing*, 20, 511-522.

Wewetzer, S.F.K., Duck, R.W. and Anderson, J.M. 1999b. Acoustic Doppler current profiler measurements in coastal and estuarine environments: examples from the Tay Estuary, Scotland. *Geomorphology*, 29, 21-30.

Williams, D.J.A. and West, J.R. 1975. Salinity distribution in the Tay Estuary. *Proceedings of the Royal Society of Edinburgh (B)*, 75, 29-39.

Zarillo, G.A. 1982. Stability of bedforms in a tidal environment. *Marine Geology*, 48, 337-351.

Chapter 15

SEDIMENTOLOGICAL SIGNATURES OF RIVERINE-DOMINATED PHASES IN ESTUARINE AND BARRIER EVOLUTION ALONG AN EMBAYED COASTLINE

ILYA V. BUYNEVICH* and DUNCAN M. FITZGERALD

Department of Earth Sciences, Boston University
685 Commonwealth Avenue, Boston, MA 02215, USA

**Present address:*
Coastal & Marine Geology Program, U.S. Geological Survey, 384 Woods Hole Road, and Geology & Geophysics Department, Woods Hole Oceanographic Institution, MS22, Woods Hole, MA 02543; e-mail: ibuynevich@usgs.gov

1. INTRODUCTION

Embayed coastlines with fluvial bedload contribution are found in many parts of the world. Due to limited longshore sediment transport, fluvial sediment supply, sea-level history and changes in accommodation space are the primary factors controlling the formation and evolution of a variety of coastal accumulation forms (Barnhardt *et al.*, 1997; FitzGerald and van Heteren, 1999; FitzGerald *et al.*, 2000; 2002; Storlazzi and Field, 2000; Ballantyne, 2002). However, few studies have addressed the sedimentological relationships between fluvial systems and associated barrier sequences at millennial time scales, particularly along formerly glaciated coasts (FitzGerald *et al.*, 1994; Forbes and Syvitski, 1994; van Heteren, 1996; Barnhardt *et al.*, 1995; 1997; Buynevich, 2001; Belknap *et al.*, 2002). In areas where the geological record of an initial fluvial vs. inner shelf sediment contribution is ultimately related to a common fluvial source, it may be difficult to establish the link between sediment transport pathways and coastal depocenters. In addition, such records are often confined to the deeper parts of the barrier sequences and require extensive coring efforts.

D.M. FitzGerald and J. Knight (eds.),
High Resolution Morphodynamics and Sedimentary Evolution of Estuaries, 315-334.

The mouth of the Kennebec River, Maine, USA, and the associated Holocene barrier systems of Popham and Seawall Beaches (Fig. 1) provide an ideal setting for evaluating the use of bulk sedimentological properties of recent fluvial-estuarine and nearshore sediments in barrier-stratigraphic research. A recent textural and compositional characterization of modern estuary-mouth deposits by Buynevich and FitzGerald (2003a) supported by an extensive sediment core database along the barriers, offer an opportunity to examine the sedimentary archives of fluvial-coastal interaction along an otherwise sediment starved coastline.

The aims of the study are to: 1) define the riverine-derived lithofacies based on diagnostic sedimentological characteristics of recent sediments; 2) identify and map sedimentary deposits with similar characteristics throughout the barrier lithosome; and 3) present an evolutionary model of fluvially-supplied coastal accumulation forms along an embayed paraglacial coastline.

2. PHYSICAL SETTING

The study region encompasses the mouth of the Kennebec River estuary and the adjoining Popham and Seawall barrier systems located along the indented west-central coast of Maine (Fig. 1). The Kennebec and Androscoggin Rivers join at Merrymeeting Bay about 20 km north of the estuary mouth, and continue toward the Gulf of Maine in a narrow bedrock-carved channel (Fig.1). The combined flow of the two rivers results in a mean annual freshwater discharge of 280 m^3s^{-1} (Nace, 1970), making it the largest river system in Maine. Bedrock in the lower estuarine and coastal region consists of isoclinally folded north-south-trending Proterozoic-Ordovician metasedimentary and metavolcanic rocks, with localized intrusions of Devonian granites and pegmatites (Hussey, 1989). The drainage area of the Androscoggin River is dominated by high-grade metamorphic rocks (schists and gneisses) and granitic plutons, whereas the Kennebec River drains lower-grade slate and phyllite terrain, with occasional granitic intrusions (Osberg *et al.*, 1985). Most of the fluvial sediment is derived from upland glacial outwash deposits (Borns and Hager, 1965; Thompson and Borns, 1985), which inherited their composition from these distinct bedrock lithologies. As a result of the compositional differences of the source areas, Kennebec and Androscoggin River sediments can be distinguished based on both their major mineral composition (Kniskern *et al.*, 1998) and heavy mineral assemblages (Malone

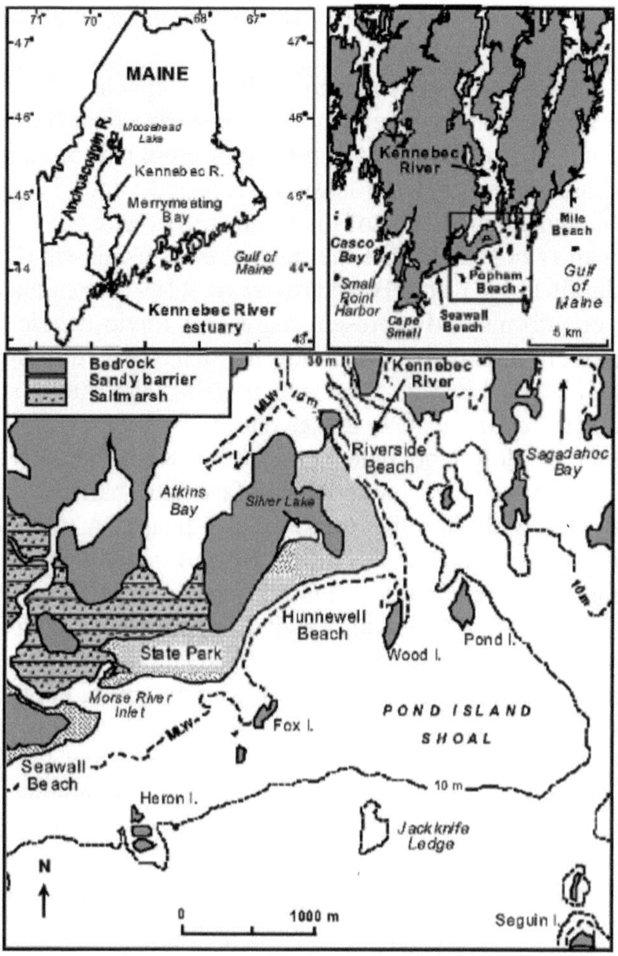

Figure 1. Location of the study area at the mouth of the Kennebec River Estuary, Maine. Note the indented nature of this coastal region and large sandy barriers westward of the estuary mouth.

1997). Today, the Kennebec River estuary seaward of the Merrymeeting Bay receives a mixture of lithologies from the two fluvial systems.

The lower Kennebec River is a partially mixed to stratified mesotidal estuary with seasonal variations in river discharge (Fenster and FitzGerald, 1996). The mean tidal range of 2.6 m (3.5 m during spring conditions) and mean shallow water wave height of 0.5 m (Jensen, 1983) place the estuary mouth region in a mixed energy, tide-dominated coastal classification of Hayes (1979). FitzGerald *et al.* (1989) calculated a mean tidal prism of 1.01×10^8 m^3 for this area, which is 16 times greater than the average freshwater discharge over the same time period. In this study all depth measurements are referenced to mean-high water (MHW) level, which is 1.36 m above the National Geodetic Vertical Datum of 1929 (NGVD-29) for this region (Gehrels *et al.*, 1996).

During the Late Pleistocene and Holocene epochs coastal Maine has experienced a complex sea-level history: 1) a marine inundation of glacially-depressed land; 2) a relative sea-level fall and lowstand (55 m below present sea level) due to postglacial isostatic rebound, and 3) recent marine transgression and submergence of the Kennebec River estuary (Kelley *et al.*, 1987; Belknap *et al.*, 1989). On the inner shelf, seismic-stratigraphic facies D of Barnhardt (1994) and Barnhardt *et al.* (1997) are characterized by large-scale clinoforms and represent regressive fluvio-deltaic deposits (see stage 2, above). Sediment cores indicate that these deposits consist of sedimentologically submature sands and gravels diagnostic of active riverine supply. Belknap *et al.* (1989; 2002) and Barnhardt *et al.* (1995; 1997) proposed that the lowstand paleodelta of the Kennebec River may have supplied sediments to the modern shoreface and barriers in this region through reworking during the transgressive phase. A number of recent studies demonstrate that the Kennebec River estuary continues to supply coarse-grained sediment to the coastal region, especially during spring freshets (Fenster and FitzGerald, 1996; Hannum, 1996; FitzGerald *et al.*, 2000; Fenster *et al.*, 2001; Buynevich and FitzGerald, 2003a).

Situated on the western margin of the estuary mouth, Popham Beach is a 4-km long sandy barrier composed of three segments - Riverside, Hunnewell, and State Park Beaches - anchored to relatively resistant pegmatitic bedrock outcrops (Fig. 1). Intertidal tombolos connect the western and eastern ends of Hunnewell Beach to Fox Island and Wood Island, respectively. Sedimentological signatures of fluvial sediments within the Popham barrier lithosome, as well the eastern portion of the Seawall Barrier, are the subject of the present investigation.

3. MODERN ESTUARY-MOUTH SEDIMENTATION AND FACIES F

In their study of sediment dynamics at the mouth of the Kennebec River, FitzGerald *et al.* (2000) documented a net seaward flux of bedload and described general sediment circulation between the river mouth, nearshore and the adjacent beach. Most recently, Buynevich and FitzGerald (2003a) provided a detailed textural and compositional characterization of the bottom sediments in the same region aimed at strengthening the link between the fluvial-estuarine, nearshore, and barrier environments of deposition. In their study, the area was subdivided into five subenvironments (channel, channel margin, outer bar, shoreface, and offshore), each with a distinct set of sedimentological characteristics. Figure 2A shows the distribution of coarse

(0.0-1.0ϕ) and coarse-medium (1.0-1.5ϕ) sands at the mouth of the estuary, indicating the offshore extension of the main channel and the outer bar complex. These coarse-grained sediments are moderately-well to moderately sorted and are texturally distinguished from shoreface and offshore facies, as well as channel-margin sands found along the Riverside Beach (Fig. 2B). In addition, the average roundness of quartz grains in the channel-derived sands is in the subangular range, compared to sub-rounded to rounded shoreface deposits (Buynevich and FitzGerald, 2003a). The mineralogy of the medium-sand fraction also distinguishes the channel and channel-margin sands as having average mineralogical maturity index (MI = quartz/[feldspar + rock fragments]) of 0.70 and 0.86, respectively, which is lower than that of beachface (1.26), shoreface (1.68), and offshore (2.12) deposits (Buynevich and FitzGerald, 2003a). Similarly, in their study of Yaquina Bay, Oregon, Kulm and Byrne (1967) showed that the amount of rock fragments and weathered grains can be used to distinguish between marine and fluvial sources of the estuarine sediments. These diagnostic sedimentary characteristics of the fluvially-derived sands may be used as lithological fingerprints for similar sediments now preserved within the adjacent barrier system. In the present paper, we define these coarse-grained facies of fluvial origin that have undergone minimal degree of reworking as *lithofacies F* (Fig. 2B). The following sections focus on the occurrence and distribution of facies F in the sedimentary sequence of Popham and eastern Seawall barrier systems.

4. BARRIER-STRATIGRAPHIC RECORD OF DIRECT FLUVIAL CONTRIBUTION

A large sedimentological database from the barriers, consisting of 35 vibracores and 30 pulse-auger cores, is used to examine the spatial distribution and stratigraphic occurrence of lithofacies F. The cores containing sediments that are texturally and compositionally similar to this facies are confined to the landward portion of the Popham and eastern Seawall barriers (Fig. 3). In addition, these facies occur at depth and are not exposed along the present-day barrier systems. A detailed description of barrier lithostratigraphy and the basis for interpreting facies F are presented below.

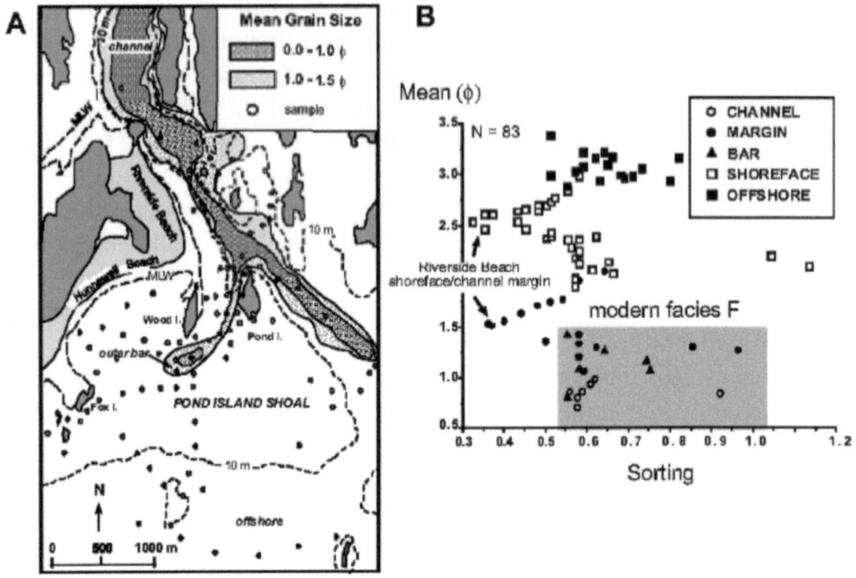

Figure 2. A) Distribution of coarse-medium sands in the lower portion of the estuary and the nearshore region; B) mean grain size vs. sorting of the bottom samples (open circles in Fig. 2A). The textural characteristics of lithofacies F are within the shaded area (modified from Buynevich and FitzGerald, 2003a).

4.1 Correlation of Major Lithostratigraphic Units

Textural and compositional data from pulse-auger cores PO-3, PO-4, and PO-7 provide the most complete record of barrier facies. For interpretation of depositional environments, the textural and compositional characteristics of core samples are compared with those of modern depositional environments (Fig. 2; Buynevich and FitzGerald, 2003a).

Figure 3. Vertical aerial photograph of Popham and eastern Seawall barrier complexes showing the locations of sediment cores. Detailed sedimentology of highlighted cores PO-3, 4, and 7 is shown in Fig. 4. A dashed line indicates the seaward limit of the occurrence of lithofacies F within the barrier sequence.

Figure 4 is a plot of downcore variations in the mean grain size (M_Z), sorting (σ_I), skewness, and mineralogical maturity index (MI) that form the basis for recognizing and correlating the five lithostratigraphic units. The units are numbered based on their overall stratigraphic position and are not represented in all cores.

Core PO-3 is located in a swale between two vegetated beach ridges at the northern part of the Riverside Beach (Fig. 3). The sequence is topped by fine-grained, moderately well-sorted beach sands, with MI = 1.08 at MHW (lithostratigraphic Unit, Fig. 4). It contains a relatively high heavy mineral content (8.4%). This unit is underlain by progressively coarser sands of Unit 4 (2.5 - 6.0 m below MHW), with the coarsest sample at 3.62 m depth being moderately sorted, coarse-grained sand with 6.8% gravel (Fig. 4). This unit has very low mineralogical maturity index (MI<0.5) and is underlain by a coarsening-upward Unit 5 (below 6.0 m), which bottoms at 7.9 m with a moderately well sorted, micaceous medium-to-fine sand (M_Z = 1.84 ϕ). Here sediments have higher mineralogical maturity (MI = 0.69 - 1.61), compared to the overlying strata. Skewness is near-symmetrical to negative for the upper 6.5 m (Units 3 and 4), with a maximum of -1.55 (strongly coarse-skewed) at 2.7 m depth. Below 6.5 m (Unit 5), skewness is positive, increasing to 0.32 (strongly fine-skewed) at the bottom of the core.

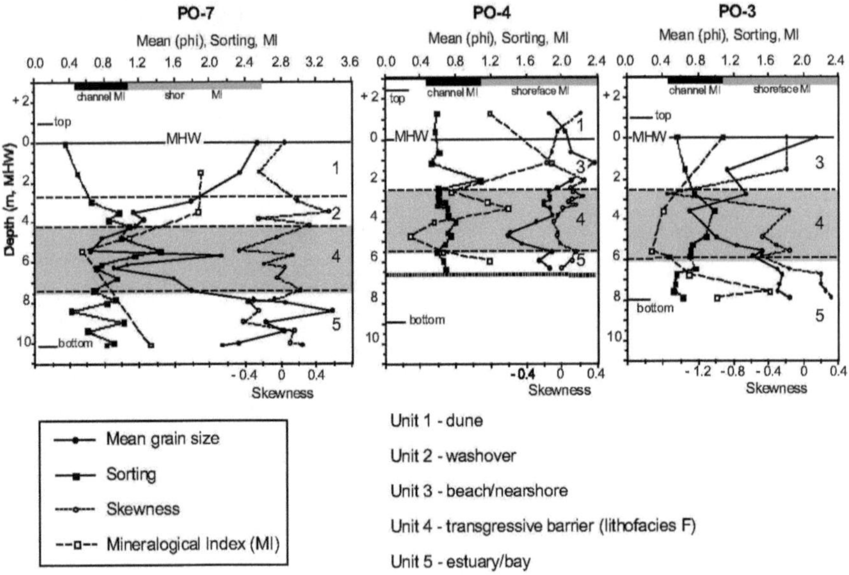

Figure 4. Downcore variations in mean grain size, sorting, skewness, and mineralogical maturity index (MI) in pulse auger cores PO-3, 4, and 7 (see Fig. 3 for core locations). The ranges of mineralogical index values of modern channel and shoreface sands are shown at the top of each panel (MI data based on Buynevich and FitzGerald, 2003a). Shaded interval (Unit 4) corresponds to the stratigraphic range of lithofacies F. MHW- mean high water level.

Core PO-4 is located in a vegetated area of the Hunnewell dunefield (Fig. 3). Here well-sorted, fine sands of Unit 1 are underlain by moderately well-sorted, medium-to-fine sands down to 2.5 m or mean low-tide level (Unit 3, Fig. 4), with finer-grained (M_Z = 2.1 - 2.4 ϕ), mineralogically mature (MI = 1.88) sand at 1.1 m depth. Lithostratigraphic Unit 4 is found between 2.5 and 5.5 m depth. Grain size and sorting vary slightly (M_Z = 1.79 - 1.89; σ_I = 0.59 - 0.72) until 3.7 m, whereas compositional maturity increases downward from 0.75 to 1.39 over this interval. Below 3.7 m depth sediment becomes coarser, reaching M_Z of 1.4 ϕ and MI of 0.28 at 4.7 m below MHW. Moderately well sorted, medium-to-fine sands of Unit 5 (5.5 - 6.6 m) have MI = 1.18 at 6.0 m depth. Skewness decreases down to 1.1 m, with coarsely-skewed samples dominating Unit 3. Below this depth and continuing to 3.7

m sediments are fine-skewed. Between 3.7 and 5.1 m (most of Unit 4) skewness is negative again, with finely-skewed sands below (Unit 5). In general, the range of skewness is much smaller throughout this core, compared to PO-3 (Fig. 4). The core bottomed in freshwater peat.

Core PO-7 is located on the landward margin of the Popham State Park dunefield - one of the largest coastal dunefields in the state of Maine (> 0.2 km^2; Nelson and Fink, 1978; Fig. 3). In this instance, a pulse auger core was used to extend the penetration of a 5 m-long vibracore PB-14 to 11 m (Figs. 3, 4 and 5). The upper portion of core PO-7 contains primary sedimentary structures in addition to textural and compositional attributes.

The finely laminated, well-sorted, fining-upward medium-to-fine sands extend from MHW down to about 2.7 m depth, with MI = 1.9 at 1.5 m (Unit 1, Fig. 4). This unit is underlain by cross-bedded, moderately sorted, medium-grained sand intercalated with organic sandy mud down to 4.2 m (Unit 2). Lithostratigraphic Unit 3 is absent from this core. Unit 4 (4.2 - 7.4 m) is characterized by moderately-to-poorly sorted, coarse-to-medium sand. At the top it is intercalated with thin (0.1 - 3 cm) layers of organic sandy mud. At a depth of 5.4 m the mean grain size is 0.64 φ and the sample contains 13.2% gravel. It is poorly sorted (σ_I = 1.44) and has MI of only 0.54. A thin layer below consists of poorly sorted fine-grained (M_Z = 2.13 ϕ) sand of Unit 4. Underlying these units the sediments become progressively finer (Fig. 4). Lithostratigraphic Unit 5 is characterized by fine-to very fine micaceous sands with a few granules and about 1% silt. Sample at 8.4 m depth is a well-sorted, very fine sand (M_Z = 3.38 ϕ) with 1.8% silt. Below this depth the sediments are progressively coarser and less well-sorted, terminating in moderately sorted fine-grained sands with MI of 1.32. Skewness does not show distinct trends in this core. In general, units 1, 4 and 5 show most negative values toward the middle parts of each unit, whereas Unit 2 has the opposite trend with the maximum of 0.54 at 3.5 m depth (Fig. 4). The general trends in the observed sediment characteristics from the three cores are: 1) an overall decrease in mean grain size from PO-3 to PO-7, away from the estuary mouth; 2) a distinct coarse, negatively-skewed, and relatively mineralogically immature Unit 4 between 2.5 (mean low water level) and 7.5 m below mean-high water (MHW) in all cores, and 3) a general downcore decrease in mean grain size and increase in mineralogical maturity index below this coarser unit in all cores (Unit 5).

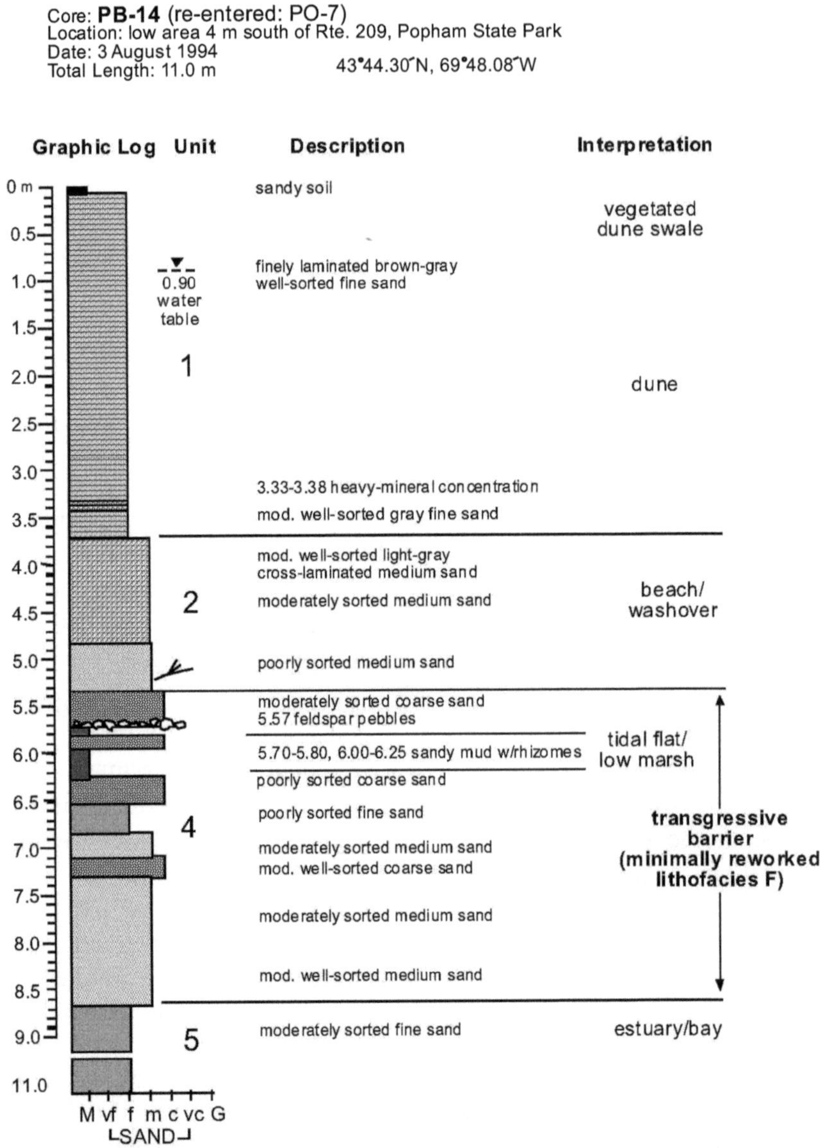

Figure 5. Lithologic log of vibracore PB-14 extended by pulse auger PO-7 (see Fig. 3 for core location). The coarse-grained Unit 4 below 5 m is interpreted as the core of a transgressive barrier/spit composed of minimally reworked fluvial sediments. For detailed textural and compositional characteristics of the core samples, rectified to mean high water elevation, refer to Fig. 4.

4.2 Major Lithostratigraphic Units – Interpretation

Downcore variations in textural characteristics and mineralogy can be a powerful tool in determining the energy conditions and, possibly, the depositional mode of the sedimentary units. The actualistic approach of studying sedimentary sequences in the rock record using physical characteristics from modern environments has been described by many authors (e.g., van Straaten, 1965; Davies *et al.*, 1971, and Mack *et al.*, 1983, among others). Downcore variations in mean grain-size and sorting of sediment samples can be the result of one or a combination of factors such as: 1) temporal changes in the texture of sediment delivered to the coastal area; 2) variation in transport distance and energy conditions (e.g., flow regime, degree of wave reworking, etc.), and 3) changes in depositional environment (Orton and Reading, 1993). In his classic work on sedimentology of barriers in Holland, van Straaten (1965) used a combination of sediment structures, grain-size characteristics, sediment color, and changes in mollusk shell content for stratigraphic correlation and interpretation. In many cases, however, only a limited number of the above characteristics are available for stratigraphic study (e.g., due to the type of coring operation, paleoecological conditions, and/or taphonomic peculiarities).

Textural (mean grain size, sorting, skewness) and compositional (mineralogical maturity index) characteristics of modern nearshore and beach environments (Fig. 2; Buynevich and FitzGerald, 2003a) form the basis for reconstructing the Holocene depositional environments. It is especially the deeper portion of the stratigraphic record, which cannot be sampled without destruction of primary sedimentary and biogenic structures, which relies for interpretation entirely on the knowledge of alternative environment-sensitive sedimentological characteristics.

In core PO-3, grain-size characteristics and high heavy-mineral content of Unit 3 suggest an upper beach environment. In turn, the relatively coarse-grained texture of Unit 4 is indicative of channel-margin facies. The lateral succession of beachface - channel margin - estuary channel environments exists today along the Riverside Beach. Furthermore, the mineralogical maturity of Unit 4 (MI = 0.28 - 0.47) is lower than even that of the present-day estuary channel (MI_{ave} = 0.70), suggesting that the Kennebec River once delivered coarser sediment than at present. Therefore, the lithostratigraphic Unit 4 is considered equivalent to lithofacies F. Sandy gravels and gravelly sands of this facies are also characteristic of the lowstand delta facies described by Barnhardt *et al.* (1995; 1996; 1997). The overall negative skewness of this unit also suggests partial wave reworking (Friedman, 1961). It should be noted that, in marked contrast to the present estuarine environments, the coarse unit is devoid of shell fragments throughout the barrier. This is likely due to a combination of lower abundance of marine macrofauna during more brackish estuarine conditions at lower sea-level, dissolution of carbonate material by rising acidic groundwater since the time of its emplacement in the barrier sequence, and/or other taphonomic processes. This conclusion is supported by the presence of considerable accumulations of shells of intertidal organisms in more recent, finer-grained sediments throughout the outer portion of the barrier. Finally, the fine-grained Unit 5, which is characterized by medium-to-fine, well-sorted, fine-skewed, mica-rich sands, suggests deposition on a tidal sandflat flanking the Kennebec River estuary at a lower sea level. Similar deposits are found in large tidal embayments adjacent to the study area (e.g., Sagadahoc Bay to the east, Small Point Harbor to the west; Fig. 1).

Unit 1 in core PO-4 is a modern interdunal aeolian deposit. Unit 3 extending from 0.6 to 2.5 m below MHW has similar characteristics to present day beachface sediments. This area is located along one of the former shorelines and predates the seaward portion of the barrier (Fig. 3), which formed by progradation through continued bar welding (FitzGerald *et al.*, 2000; Buynevich, 2001). The coarse grain-size of Unit 4 and low mineralogical maturity (MI_{ave} = 0.80) are similar to the estuarine channel characteristics (Fig. 4). However, since the location of core PO-4 is protected by bedrock ridges from all but the seaward direction, Unit 4 cannot be a channel-margin deposit. Rather, this unit could have formed in a channel-derived barrier spit 100-150 m offshore, when sediment was able to bypass the eastern bedrock ridge (Fig. 3). With continued sea-level rise, the sediment comprising the barrier was reworked onshore, forming a 3 m-thick transgressive unit. Due to limited wave reworking, Unit 4 in core PO-4 is finer than that in PO-3, but has a similar compositional signature and skewness, all typical of lithofacies F. The sedimentology and depth of Unit 5 are similar to the tidal flat sand in core PO-3. It could also represent a washover horizon deposited over a brackish-to-freshwater peat at 6.6 m below MHW, which formed in a boggy area of proto-Silver Lake. A similar sequence of dune, beach, transgressive/spit and washover facies overlying freshwater lake/bog deposits has been documented in a pulse auger core PO-5 obtained from the southwest shore of Silver Lake (Fig. 3) and has been dated at around 5,600 cal years BP (Buynevich and FitzGerald, 2003b).

In core PO-7, well-sorted, medium-fine (M_Z = 1.8 - 2.5 ϕ) laminated sands of Unit 1 are aeolian in origin and represent the parabolic dunes of State Park dunefield. The absence of Unit 3 (beach) is due to the proximity of the core site to the backbarrier. Underlying Unit 1, the 1.5 m-thick cross-bedded, moderately-sorted, finely-skewed, medium sands of Unit 2 are intercalated with organic sandy mud layers, and are interpreted to be washover deposits (Figs. 3 and 5). Similar to core PO-4, a 3.2 m-thick, coarse-grained, moderately-to-poorly sorted, negatively-skewed, submature sand of Unit 4 is interpreted as transgressive barrier/spit facies (lithofacies F; Fig. 5). It formed as part of the westward extension of a riverine-derived barrier complex. Therefore, facies F comprises much of the coarse-grained core of the barrier (Fig. 6) and its sedimentological signature is similar to that of modern estuarine channel.

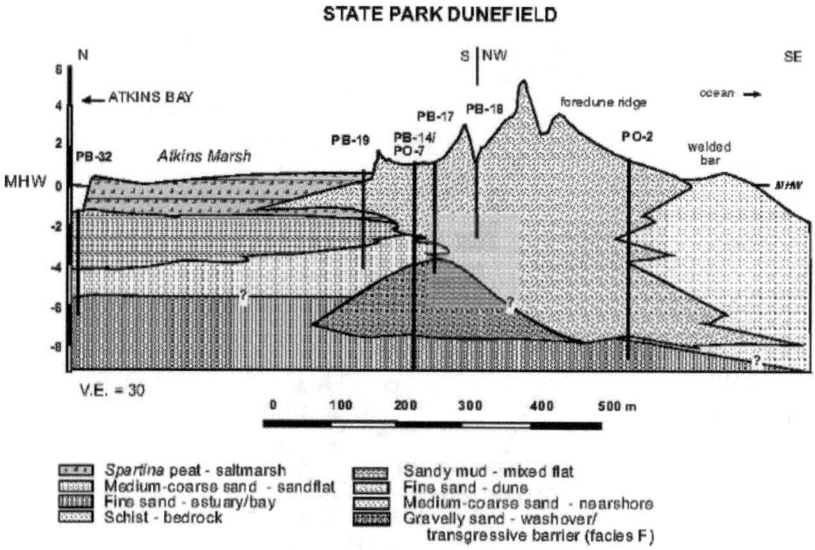

Figure 6. Shore-normal, composite stratigraphic section across the Popham State Park barrier system and adjacent Atkins Marsh (see Fig. 3 for transect location). Riverine-derived facies F occurs at depth and underlies the central portion of the Holocene barrier lithosome. This figure emphasizes the need for deep-penetrating cores (vertical bars) to be collected across the barrier, as the transgressive barrier core complex may often be missed when coring is confined to low elevations on the seaward and landward parts of the transect.

In addition, westward of Popham barrier complex (eastern Seawall Barrier), sediments in the lower portion of core SO-4 have relatively low quartz content in their coarse-sand fraction (Figs. 3 and 7). The distribution of the cores that penetrated this coarse-grained, mineralogically submature unit indicates that it is confined to the landward portion of the barriers and is bounded by a younger progradational sequence in a seaward direction (Fig. 3). This type of occurrence and preservation of transgressive barrier facies has been documented in many coastal lithosomes around the world (e.g., van Straaten, 1965; Kraft, 1971; Belknap and Kraft, 1981, 1985; Thom, 1984; Reinson, 1992; Roy *et al.*, 1994; van Heteren *et al.*, 1996; FitzGerald and van Heteren, 1999). In a seaward direction and extending onto the inner shelf, facies F grades into shell-rich Late Holocene estuary/nearshore deposits (facies E of Barnhardt *et al.*, 1997).

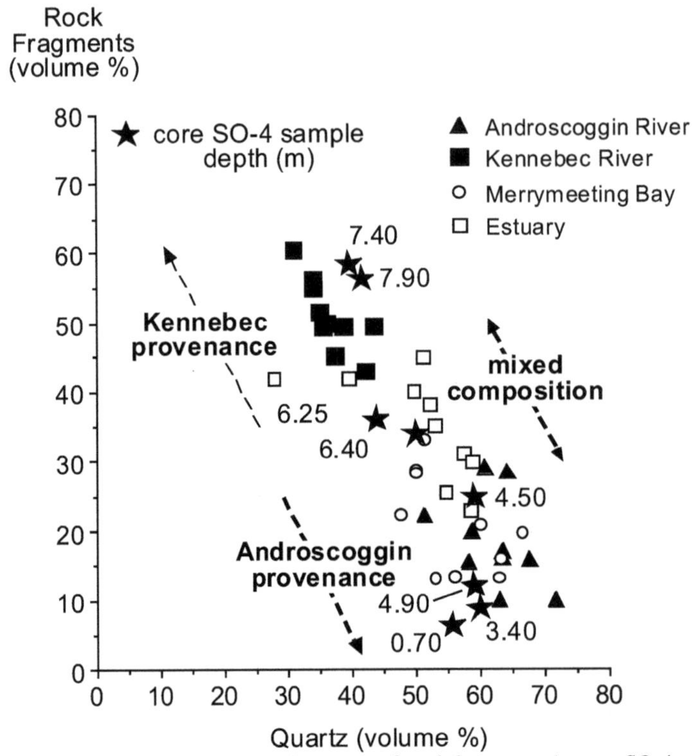

Figure 7. Relative proportions of quartz and rock fragments in core SO-4, eastern Seawall Barrier (see Fig. 3 for core location) compared to those of bottom sediment samples from the Androscoggin-Kennebec river systems and adjacent environments (from Kniskern et al., 1998). Note the compositional shift from the Kennebec affinity to mixed provenance within facies F (>5 m depth) and further increase in quartz content in the overlying beach sequence. The rock fragments in the lower portion of facies F are mostly low-grade metamorphic rocks. A coarse sand fraction was analyzed for all samples.

In their investigations in the Atkins Marsh, Nelson (1979) and Belknap *et al.* (1989) described a sandy unit at the base of the backbarrier sequence and interpreted it as an exposed intertidal sandflat and fluvial/subtidal flat deposit, respectively. During its earlier history, frequent flooding of the Kennebec River must have delivered large quantities of sand into the Atkins Bay prior to marsh aggradation. This scenario would explain the large areal extent of the coarse unit. Organic layers at the top of Unit 4 intercalated with coarse sands suggest intermittent marsh development between the flood events. Therefore, the top of this unit would represent the initial backbarrier development behind a prograding Popham State Park barrier system (Fig. 6). Below Unit 4 and down to 10.1 m below MHW, fine-grained and fine-skewed micaceous sands of Unit 5 suggest a tidal flat/estuarine environment, which existed behind a transgressive barrier spit, similar to that found in other cores (Fig. 4).

5. MID-HOLOCENE RIVERINE-DOMINATED PHASE

The comparison of textural and compositional sediment characteristics of three auger cores to those of modern nearshore and beach environments makes possible a reconstruction of the past depositional environments. Analysis of the distribution of the lithostratigraphic units provides a basis for stratigraphic correlation throughout the Holocene barrier sequence, from the present Kennebec River mouth westward toward Seawall Barrier (Fig. 3). The extent and depth range of Unit 4 (lithofacies F) found in all cores provide spatial information about an early transgressive barrier phase (Buynevich, 2001). An evolutionary model of this early phase envisions a transgressive, riverine-derived barrier/spit complex that extended for more than 4 km westward of the Kennebec River mouth approximately 5,600 years ago (Fig. 8). Where paleo-deltaic depocenters of the Androscoggin-Kennebec River system were abandoned on the inner shelf (10-30 m below present sea level; Belknap *et al.*, 1989; Barnhardt *et al.*, 1997), the entire shoreface and most of the barrier lithosome may consist of lithofacies F, such as the Mile Beach barrier located landward of the eastern lobe of the Kennebec River paleodelta (Fig. 1; Buynevich and FitzGerald, 1999; Belknap *et al.*, 2002).

Aside from the sedimentological evidence for the direct fluvial contribution to the early barrier complexes, there is preliminary evidence for the compositional shift within lithofacies F. Mineralogical analyses of sediments from the lower coarse-grained sequence of core SO-4 (eastern Seawall Barrier; Fig. 3) indicate a distinct downcore shift in lithological components. At the base, the quartz-poor, slate-dominated unit is compositionally similar to present-day sediments of the Kennebec River (Fig. 7; Kniskern *et al.*, 1998; Buynevich, 2001). In contrast, the overlying deposits possess a higher content of quartz and high-grade metamorphic rock fragments, which indicate an admixture of Androscoggin-derived lithologies (Kniskern *et al.*, 1998). This stratigraphic shift in the provenance of barrier sands suggests that the lowermost unit of core SO-4 was derived from the nearshore deposits laid down at the time when the Kennebec River may have been the primary, if not the sole, sediment supplier to the coast. This time period when the Androscoggin River was decoupled from the Kennebec and occupied the Thomas Bay - New Meadows valley could have been as early as 11.0 ka BP (Buynevich *et al.*, 1999; Buynevich, 2001). Alternatively, at any time during the Holocene, there may have been a time when Androscoggin-derived sands were sequestered in the Merrymeeting Bay region long enough to allow significant accumulation of sands of the Kennebec affinity west of the present river mouth.

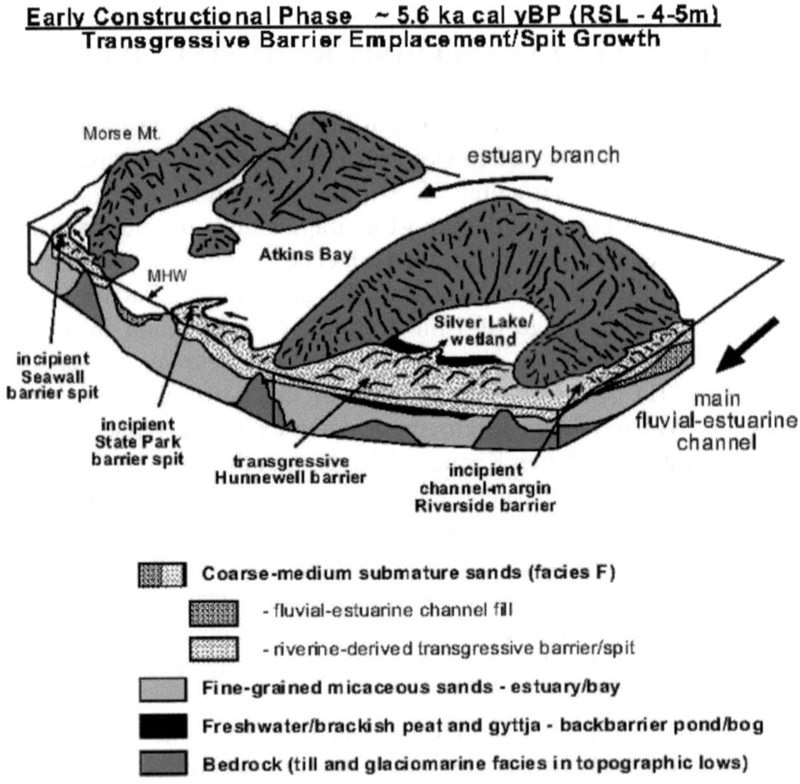

Figure 8. A schematic block diagram of an early constructional phase in coastal development at the mouth of the Kennebec River estuary. The coarse-medium, mineralogically submature lithofacies F are represented by fluvial-estuarine channel and channel-margin deposits, which supplied large portions of the proto-barriers west of the river mouth. The approximate age of barrier emplacement is based on radiocarbon-dated peat horizons underlying the transgressive barrier sands (Buynevich, 2001; Buynevich and FitzGerald, 2003b), with relative sea-level (RSL) position at that time based on data from Gehrels et al. (1996). The Atkins Bay region may have been a temporary branch of the main estuary prior to barrier growth, which led to the conversion of this area to a saltmarsh.

6. CONCLUSIONS

1. A distinct sedimentological signature of a mid-Holocene riverine-dominated phase in estuarine and barrier evolution is recognized along an embayed, west-central coast of Maine. Textural and compositional characteristics of recent channel sediments derived from the mouth of the Kennebec River estuary, such as coarse-medium mean grain size, moderate sorting, and relatively low mineralogical maturity, formed the basis for defining lithofacies F (Figs. 4 and 5). This facies is interpreted as minimally-reworked fluvial bedload, which contributes to a variety of depositional environments, from submerged early Holocene paleodelta deposits to a modern outer bar complex.

2. The diagnostic sedimentological signature and distribution of facies F along the coastal accumulation forms are used to identify and map the mid-Holocene (~4-5 ka cal BP) transgressive barrier lithosome as the core of modern Popham and eastern Seawall barriers (Fig. 8). The equivalent facies may comprise most of the barrier and shoreface deposits in areas where paleo-deltaic deposits were the sole sediment supply (e.g., eastern paleodelta lobe - Mile Beach barrier complex).

3. This study shows that bulk sedimentological properties of recent sediments, in combination with extensive coring efforts along adjacent barrier systems, can be used to examine the timing and nature of riverine bedload contribution to the Holocene coastal accumulation forms. In addition, there is also evidence for a downcore compositional shift within facies F, which may be attributed to changes in fluvial sediment supply and subsequent reworking by marine processes.

7. ACKNOWLEDGMENTS

This study was supported by the American Chemical Society Grant 32527-AC8 and Geological Society of America Grant 6398-99. We thank Sytze van Heteren, Paul McKinlay, and Amy Dougherty for their assistance in the field, Brent Taylor for processing the grain-size data, and Tara Kniskern and Eric Zamft for their help with mineralogical analyses. The comments by Jasper Knight, Helene Burningham, Elizabeth Pendleton, and

Emily Himmelstoss, and an anonymous reviewer helped to improve the manuscript. This is contribution number 11045 of the Woods Hole Oceanographic Institution.

REFERENCES

Ballantyne, C.K. 2002. Paraglacial geomorphology. *Quaternary Science Reviews*, 21, 1935-2017.

Barnhardt, W.A. 1994. *Late Quaternary relative sea-level change and evolution of the Maine inner continental shelf, 12-7 ka B.P.* Unpublished Ph.D. Dissertation, University of Maine, Orono, Maine.

Barnhardt, W.A., Gehrels, W.R., Belknap, D.F. and Kelley, J.T. 1995. Late Quaternary relative sea-level change in the western Gulf of Maine: evidence for a migrating forebulge. *Geology*, 23, 317-320.

Barnhardt, W.A., Kelley, J.T., Dickson, S.M., Belknap, D.F. and Kelley, A.R. 1996. *Surficial geology of the Maine inner continental shelf (map series).* Maine Geological Survey, Natural resources Information and Mapping Center, Augusta, Maine. Scale 1:100,000.

Barnhardt, W.A., Belknap, D.F. and Kelley, J.T. 1997. Stratigraphic evolution of the inner continental shelf in response to late Quaternary relative sea-level change, northwestern Gulf of Maine. *Geological Society of America Bulletin*, 109, 612-630.

Belknap, D.F. and Kraft, J.C. 1981. Preservation potential of transgressive coastal lithosomes on the U.S. Atlantic Coast. *Marine Geology*, 42, 429-442.

Belknap, D.F. and Kraft, J.C. 1985. Influence of antecedent geology on stratigraphic preservation potential and evolution of Delaware's barrier systems. *Marine Geology*, 63, 235-262.

Belknap, D.F., Shipp, R.C., Kelley, J.T. and Schnitker, D. 1989. Depositional sequence modeling of the late Quaternary geologic history, west-central Maine coast. In: Tucker, R.D. and Marvinney, R.G. (eds) *Maine Geological Survey, Studies in Maine Geology,* 5, 29-46.

Belknap, D.F., Kelley, J.T. and Gontz, A.M. 2002. Evolution of the glaciated shelf and coastline of the northern Gulf of Maine, USA. *Journal of Coastal Research*, SI 36, 37-55.

Borns, H.W., Jr. and Hagar D.J. 1965. Late glacial stratigraphy of a northern part of Kennebec River valley, western Maine. *Geological Society of America Bulletin*, 76, 1233-1250.

Buynevich, I.V., FitzGerald, D.M. and Parolski, K.F. 1999. *Geophysical investigation of the nearshore geologic framework, eastern Casco Bay - Reid State Park, Maine: Data analysis and implications for Late Quaternary coastal evolution.* United States Geological Survey Open-File Report 99-0380, 32pp.

Buynevich, I.V. 2001. *Fluvial-marine interaction and Holocene evolution of sandy barriers along an indented paraglacial coastline*. Ph.D. dissertation, Boston University.

Buynevich, I.V. and FitzGerald, D.M. 1999. Structural controls on the development of a coarse sandy barrier, Reid State Park, Maine. *ASCE, Coastal Sediments `99 Proceedings*, 2, 1256-1267.

Buynevich, I.V. and FitzGerald, D.M. 2003a. Textural and compositional characterization of recent sediments along a paraglacial estuarine coastline, Maine, USA. *Estuarine, Coastal, and Shelf Science*, 56, 139-153.

Buynevich, I.V. and FitzGerald, D.M. 2003b. High-resolution subsurface (GPR) profiling and sedimentology of coastal ponds, Maine, USA.: Implications for Holocene back-barrier evolution. *Journal of Sedimentary Research*, 73, 559-571.

Davies, D.K., Ethridge, F.G. and Berg, R.R. 1971. Recognition of barrier environments. *American Association of Petroleum Geologists Bulletin*, 55, 550-565.

Fenster, M.S. and FitzGerald, D.M. 1996. Morphodynamics, stratigraphy, and sediment transport patterns of the Kennebec River estuary, Maine, USA, *Sedimentary Geology*, 107, 99-120.

Fenster, M.S., FitzGerald, D.M., Kelley, J.T., Belknap, D.F., Buynevich, I.V. and Dickson, S.M. 2001. Net ebb sediment transport in a rock-bound, mesotidal estuary during spring freshet conditions: Kennebec River estuary, Maine. *Geological Society of America Bulletin*, 113, 1522-1531.

FitzGerald, D.M. and van Heteren, S. 1999. Classification of paraglacial barrier systems: coastal New England, USA. *Sedimentology*, 46, 1083-1108.

FitzGerald, D.M., Lincoln,J.M., Fink, L.K., Jr. and Caldwell, D.W. 1989. Morphodynamics of tidal inlet systems in Maine. In: Tucker, R.D. and Marvinney, R.G. (eds) *Maine Geological Survey, Studies in Maine Geology*, 5, 67-96.

FitzGerald, D.M., Rosen, P.S. and van Heteren, S. 1994. New England Barriers. In: Davis, R.A., Jr. (ed) *Geology of Holocene barrier island systems*. Springer-Verlag, 305-394.

FitzGerald, D.M., Buynevich, I.V., Fenster, M.S. and McKinlay, P.A. 2000. Sand circulation at the mouth of a rock-bound, tide-dominated estuary. *Sedimentary Geology*, 131, 25-49.

FitzGerald, D.M., Buynevich, I.V., Davis, R.A., Jr. and Fenster, M.S. 2002. New England tidal inlets with special reference to riverine-associated inlet systems. *Geomorphology*, 48, 179-208.

Forbes, D.L. and Syvitski, J.P.M. 1994. Paraglacial coasts. In: Carter, R.W.G. and Woodroffe, C.D. (eds) *Coastal Evolution: Late Quaternary Shoreline Morphodynamics*. Cambridge University Press, 373-424.

Friedman, G.M. 1961. Distinction between dune, beach, and river sands from their textural characteristics. *Journal of Sedimentary Petrology*, 31, 514-529.

Gehrels, W.R., Belknap, D.F. and Kelley, J.T. 1996. Integrated high-precision analyses of Holocene relative sea-level changes: lessons from the coast of Maine, *Geological Society of America Bulletin*, 108, 1073-1088.

Hannum, M.B. 1996. *Late Quaternary evolution of the Kennebec and Damariscotta River estuaries, Maine*. Unpublished M.S. Thesis, University of Maine, Orono, Maine.

Hayes, M.O. 1979. Barrier island morphology as a function of tidal and wave regime. In: Leatherman, S.P. (ed) *Barrier islands: from the Gulf of St. Lawrence to the Gulf of Mexico*. Academic Press, New York, 1-28.

Hussey, A.M., II. 1989. Geology of southwestern coastal Maine. In: Anderson, W.A. and Borns, H.W., Jr. (eds) *Neotectonics of Maine*. Maine Geological Survey, Bulletin, 40, 25-42.

Jensen, X.E. 1983. *Atlantic coast hindcasting shallow water significant wave information*. WIS Report, 8, Vicksburg, MS, 75pp.

Kelley, J.T., Belknap, D.F. and Shipp, R.C. 1987. *Geomorphology and sedimentary framework of the inner continental shelf of south central Maine*. Maine Geological Survey Open File Report 87-19, 76pp.

Kniskern, T.A., Buynevich, I.V., FitzGerald, D.M. and Peters, J.L. 1998. Sedimentological characteristics of fluvial-estuarine deposits in the Merrymeeting Bay, Maine: Implications for sediment transport. *GSA Northeastern Section Abstracts with Programs*, 30, 30.

Kraft, J.C. 1971. Sedimentary facies patterns and geologic history of a Holocene marine transgression. *Geological Society of America Bulletin*, 82, 2131-2158.

Kulm, L.D. and Byrne, J.V. 1967. Sediments of Yaquina Bay, Oregon. In: Lauff, G.H. (ed) *Estuaries.* AAAS Special Publication 83, 266-238.

Mack, G.H., Thomas, W.A. and Horsey, C.A. 1983. Composition of Carboniferous sandstones and tectonic framework of southern Appalachian-Oachita orogen. *Journal of Sedimentary Petrology*, 53, 931-946.

Malone, J.T. 1997. *Determining provenance of river sediment: Kennebec and Androscoggin Rivers, Maine*. Unpublished M.S. thesis. University of Maine, Orono, Maine.

Nace, R.L. 1970. World hydrology: Status and prospects. *IAS Symposium, Publication* No. 92, 1-10.

Nelson, B.W. 1979. *Shoreline changes and physiography of Maine's sandy coastal beaches.* Unpublished M.S. Thesis, University of Maine, Orono, Maine.

Nelson, B.W. and Fink, L.K., Jr. 1978. *Geological and botanical features of sand beach systems in Maine*. Critical Areas Program, Maine State Planning Office, Augusta, Maine, 269pp.

Orton, G.J. and Reading, H.G. 1993. Variability of deltaic processes in terms of sediment supply, with particular emphasis on grain size. *Sedimentology*, 40, 475-512.

Osberg, P.H., Hussey, A.M., II and Boone, G.M. 1985. *Bedrock geologic map of Maine, scale 1:500,000.* Maine Geol. Survey, Augusta, Maine.

Reinson, G.E. 1992. Transgressive barrier island and estuarine systems. In: Walker R.G. and James, N.P. (eds) *Facies models: response to sea level change.* Geological Association of Canada, 179-194.

Roy, P.S., Cowell, P.J., Ferland, M.A. and Thom, B.G. 1994. Wave-dominated coasts. In: Carter, R.W.G. and Woodroffe, C.D. (eds) *Coastal Evolution: Late Quaternary Shoreline Morphodynamics*. Cambridge University Press, 121-186.

Storlazzi, C.D. and Field, M.E. 2000. Sediment distribution and transport along a rocky, embayed coast: Monterey Peninsula and Carmel Bay, California. *Marine Geology*, 170, 289-316.

Thom, B.G. 1984. Transgressive and regressive stratigraphies of coastal sand barriers in southeast Australia. *Marine Geology*, 56, 137-158.

Thompson, W.B. and Borns, H.W., Jr. 1985. *Surficial geologic map of Maine, scale 1:500,000.* Maine Geol. Survey, Augusta, Maine.

van Heteren, S. 1996. *Preserved records of coastal-morphologic and sea-level changes in the stratigraphy of paraglacial barriers*. Unpublished Ph.D. Dissertation, Boston University, Boston, Massachusetts.

van Heteren, S., FitzGerald, D.M., Barber, D.C., Kelley, J.T. and Belknap, D.F. 1996. Volumetric analysis of a New England barrier system using ground-penetrating radar and coring techniques. *Journal of Geology*, 104, 471-483.

van Straaten, L.M.J.U. 1965. Coastal barrier deposits in South- and North-Holland. In: Schwartz, M.L. (ed) *Barrier Islands.* Dowden, Hutchinson and Ross, Stroudsburg, Pennsylvania, 171-217.

Chapter 16

PALEODELTAS AND PRESERVATION POTENTIAL ON A PARAGLACIAL COAST – EVOLUTION OF EASTERN PENOBSCOT BAY, MAINE

Daniel F. Belknap, Allen M. Gontz and Joseph T. Kelley
Department of Earth Sciences, University of Maine, Orono, ME 04469-5790, USA (belknap@maine.edu)

1. INTRODUCTION

1.1 Geologic Setting

The bedrock framework of the northern Gulf of Maine coast, USA (Fig. 1), controls the geometry of headlands and embayments (Shipp *et al.*, 1985, 1987; Kelley, 1987). Quaternary continental glaciers sculpted this paraglacial coast, culminating in the latest Wisconsinan Laurentide Ice Sheet, which reached its maximum extent in the region 20-22 ka (Hughes *et al.*, 1985). This ice sheet was marine-based in much of the Gulf of Maine 20-15 ka (Schnitker *et al.*, 2001) and during later stages of retreat through the Maine coastal lowlands (Stuiver and Borns, 1975; Dorion *et al.*, 2001). Sediments of a wide variety of (Thompson and Borns, 1985) were deposited duringeglacial retreat, interpreted in a sequence-stratigraphic model by Belknap and Shipp (1991) and Barnhardt *et al.* (1997). Sediment sources to the evolving Holocene coast included reworking from glacial and glaciomarine outcrops, as well as limited fluvial inputs.

D.M. FitzGerald and J. Knight (eds.),
High Resolution Morphodynamics and Sedimentary Evolution of Estuaries, 335-360.

Complex relative sea-level (RSL) changes (Fig. 2) near the former ice-sheet edge involved submergence to 70-130 m above present sea level 15-13 ka during deglaciation. The coast emerged rapidly during continuing isostatic rebound 13-11 ka, with relative sea-level fall to 60 m below present. Submergence and transgression occurred 10.8 ka to present as isostatic rebound slowed and eustatic sea-level rise predominated (Belknap *et al.*, 1987a; Barnhardt *et al.*, 1997). The flat early to mid-Holocene sea-level record is interpreted as the passage of a marginal forebulge (Barnhardt *et al.*, 1995).

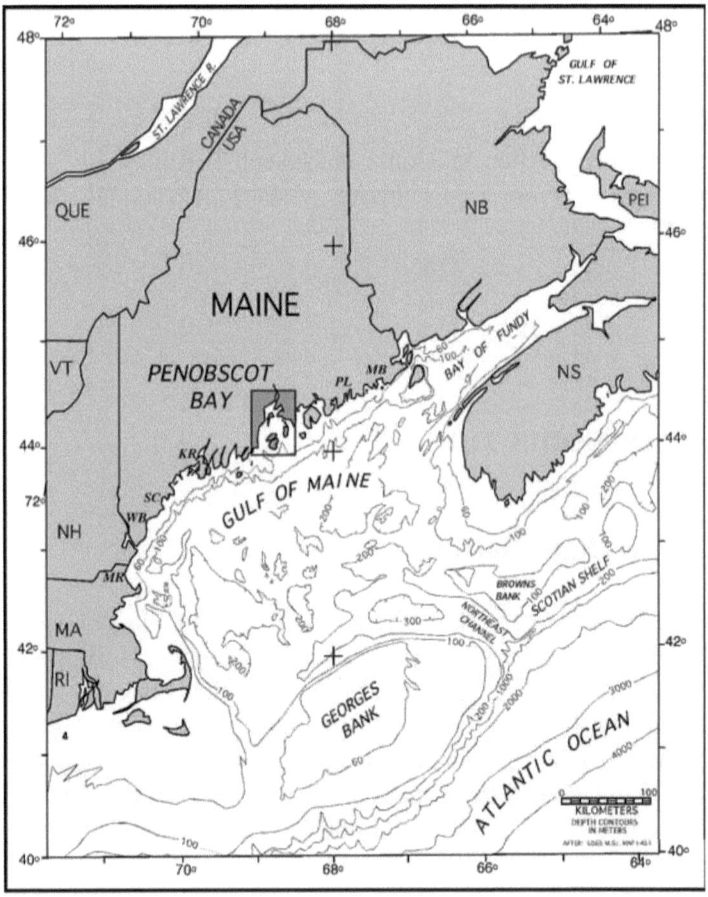

Figure. 1 Location map, Gulf of Maine, USA, with Penobscot Bay (PB) study area outlined in box. MR – Merrimack River, WB – Wells Bay, SC – Saco Bay, KR – Kennebec River, PL – Pleasant Bay, MB – Machias Bay.

Inner shelf stratigraphy has been studied with high-resolution seismic profiling, vibracoring, side-scan sonar, sediment grab samples, and submersible investigations. Distinct differences among embayments and along the coast reflect the various weightings of geomorphic and process controls. Three typical environments are: (1) barriers in open embayments (Wells Bay, Saco Bay), (2) large rivers with lowstand paleodeltas (Kennebec River, Merrimack River), and (3) estuaries (Penobscot Bay, Pleasant Bay, Machias Bay). Stratigraphy of the inner shelf reflects the interplay of the bedrock and glacial framework, sediment supplies, rate of sea-level change, and the direct effects of tidal, wave, and mass-wasting processes.

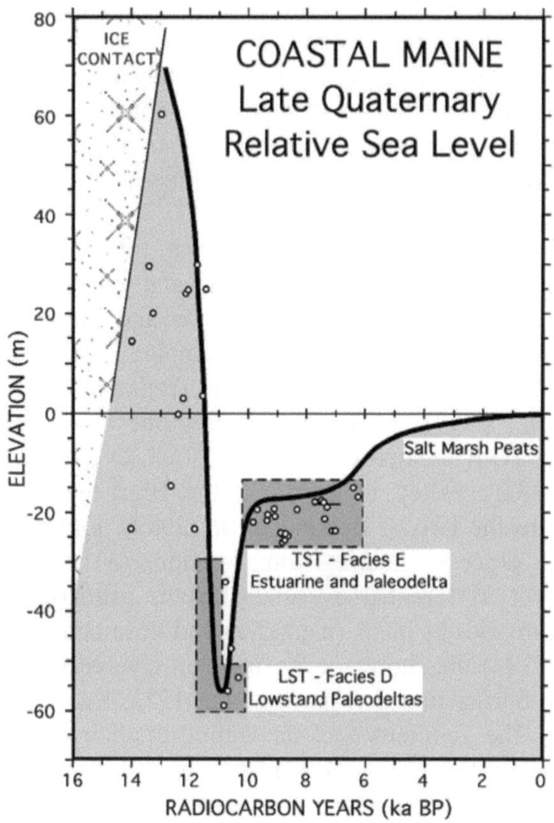

Figure. 2 Local relative sea-level curve, northern Gulf of Maine, based on radiocarbon-dated marine fossil shells from glaciomarine and post-glacial sediments, as well as more than 100 salt-marsh peat samples ca. 6 ka BP to present (see Belknap et al., 2002, their Fig. 2). Uncorrected radiocarbon years BP, after Belknap et al. (1987a), Kelley et al. (1992, 1995) and Barnhardt et al, (1995, 1997).

1.2 Previous Studies

Marine stratigraphic analysis of Penobscot Bay began with Ostericher's (1965) early seismic profiling and short piston cores. More advanced studies followed, by USGS (Knebel and Scanlon, 1985a,b; Knebel, 1986; Scanlon and Knebel, 1989) and University of Maine researchers (Kelley and Belknap, 1989; Barnhardt, 1994), culminating in a 1:100,000 scale geologic map (Barnhardt *et al.*, 1996) and detailed stratigraphic model (Barnhardt *et al.*, 1997) of Penobscot Bay. Several studies of pockmarks in western and eastern Penobscot Bay demonstrated distributions and abundance of gas-related features, and a continuing debate over their origins and evolution (Belknap, 1991; Kelley *et al.*, 1994; Rogers, 1999; Gontz, 2002; Gontz *et al.*, 2002; Ussler *et al.*, 2003). A newly discovered paleodelta of the Penobscot River (Belknap *et al.*, 2002) has stimulated analysis of Gulf of Maine coastal and shelf systems with regard to a possible episode of rapid sea-level rise that overtopped features at –30 m. It also allows a new system for comparison to lowstand deltas and shorelines (-55-65 m) as well as late Holocene nearshore sand wedges.

Evolution of Penobscot Bay from 10-8 ka, following lowstand, included tidal channel migration, infill, avulsion, and overstepping. A complete understanding of the stratigraphic evolution and preservation potential of nearshore and shelf facies of this complex glaciated coast requires identification of primary erosional surfaces (Belknap and Kraft, 1981, 1985; Belknap *et al.*, 1994). In Penobscot Bay these include: (1) the Basal Unconformity (Ub) created during falling sea level by littoral and fluvial erosion, including valley incision, (2) the Bluff-toe Unconformity (Ubt) (relabeled from the Urb of Belknap *et al.*, 2002), formed by wave, ice and mass-wasting processes during transgression, (3) the Tidal Ravinement Unconformity (Urt) created by tidal currents primarily on the flanks of channels in estuaries or inlets (e.g. Allen and Posamentier, 1994; Belknap *et al.,* 1994), and (4) the shoreface Ravinement Unconformity (Ur) created by wave erosion during transgression (Stamp, 1922; Swift, 1968; Belknap and Kraft, 1985). The geometry and the timing of their creation may in some cases be distinguish Basal (Ub) regressive and bluff-toe (Urb) transgressive unconformities, but these erosional surfaces often occupy the same position in the stratigraphic sequence and are easily misinterpreted. Similarly, distinction between wave- and tide-formed erosional surfaces may be difficult at some locations in an embayment, and erosional surfaces may reoccupy and further erode older surfaces (Belknap and Kraft, 1985; Belknap *et al.*, 1994). Facies may have higher or lower preservation potential between these surfaces depending on paleotopography, sediment type, exposure to waves and currents, and rate of sea-level change.

1.3 Hypotheses

No lowstand paleodelta is recognized at the mouth of Penobscot Bay, however, there is the mid-estuary Penobscot Paleodelta (Belknap *et al.*, 2002). This represents a pause in transgression in the mud-dominated estuary, punctuated by sandy deltaic progradation. Unlike the lowstand Kennebec Paleodelta, which was reworked during transgression (Belknap *et al.*, 1989; Belknap and Shipp, 1991; Barnhardt *et al.*, 1997), the Penobscot Paleodelta appears to be intact below an estuarine mud cap. For these reasons, it is an important marker for comparing rate of RSL change and rate of sediment input. Hypotheses for this distinctive change in estuarine accumulation and preservation include: (1) a rapid, brief acceleration of RSL rise, possibly caused by a meltwater pulse (e.g. Fairbanks, 1989; Bard *et al.*, 1996), (2) a plateau followed by acceleration in the RSL curve caused by passage of a marginal forebulge (Barnhardt *et al.*, 1995; Balco *et al.*, 1998), (3) changes in local sediment source, (4) a major channel avulsion by bluff erosion at a former isthmus between Islesboro Island and Sears Island (Fig. 3), and/or (5) changes in the upper Penobscot River drainage system.

1.4 Purpose

The purpose of this paper is to document seismic facies and cores from the Penobscot Paleodelta, and to develop a sequence stratigraphic model of Penobscot Bay related to RSL change. This detailed analysis of a small sand body within a major estuary provides comparisons to lowstand paleodeltas elsewhere in the Gulf of Maine. In addition, this local example can be compared to larger-scale and longer-term models of transgressive shelf evolution.

2. METHODS

2.1 Seismic Reflection Profiling

Seismic reflection profiles (Fig. 3) were collected with the Triton-Elics digital acquisition and processing system, using an Applied Acoustics Engineering boomer at 100 J, peak frequency ca. 1.5 kHz, and a 20-element hydrophone. Digital data were geo-referenced with a differential GPS navigation system, nominal accuracy of 1-3 m. The system was employed from 10-15 m research vessels.

2.2 Vibracoring

Vibracores were collected using a Rossfelder P3 underwater vibracorer with a 10 cm diameter by 5.5 m length steel barrel, with 9.5 cm diameter plastic core liners. Cores locations (Fig. 3) were selected from the seismic survey, and precisely placed along a repeat seismic line on the day of the coring using a temporary marker buoy. Cores were described, photographed, sampled, and analyzed for standard sedimentological parameters (grain size, water content, bulk density, magnetic susceptibility, and shear strength) at the University of Maine sedimentology laboratory.

2.3 Analysis

Belknap and Shipp (1991) and Barnhardt *et al.* (1997) provide a detailed discussion of the interpretation of seismic facies for the Maine coast. Facies are delineated on the basis of intensity of acoustic contrast at facies boundaries, external geometry of the reflection unit, geometry and intensity of internal reflectors, relationship to adjacent facies and tied to numerous outcrops and cores. These seismic facies form the basis of a sequence stratigraphic interpretation for the late Quaternary evolution of the northern Gulf of Maine (Belknap *et al.*, 1987b, 2002; Belknap and Shipp, 1991; Barnhardt *et al.*, 1997). Ostericher (1965) initiated study of seismic facies in Penobscot Bay, and the high-resolution seismic stratigraphy developed by Knebel and Scanlon (1985a,b) and Knebel (1986) followed similar interpretation methods. All sedimentary thicknesses described below are related to a simplifying assumption of water velocity for the P-wave, 1.5 km/sec. This has proven reasonably accurate in most local studies when compared to cores, bridge borings and refraction profiles on land, but may underestimate thickness of till.

The seismic reflection at the base of the sequence in Penobscot Bay is a sharp return with long-lasting echo. It commonly has a spiky to hyperbolic upper surface, and is clearly linked to rock outcrops at islands and on the margins of embayments. It is interpreted as bedrock, represented on figures as BR. Knebel (1986) and Knebel and Scanlon (1985a,b) concur in this interpretation (their unit Pz).

A common but discontinuous facies non-conformably overlies BR. It has is a strong upper surface return and chaotic internal reflections. It forms blankets a few meters thick as well as mounds of a few meters to greater than 15 m thickness. Its external geometry is often lenticular or mounded. When traced to outcrops on the sea floor, seen in sidescan sonar (Barnhardt

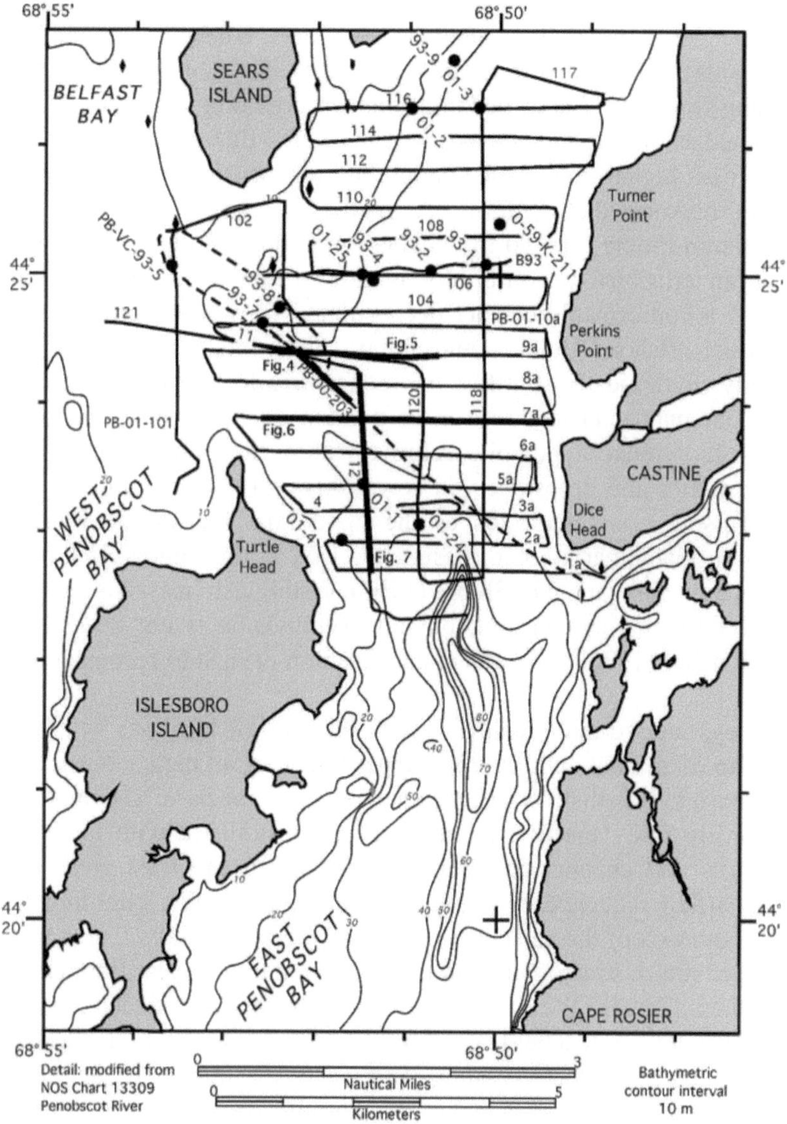

Figure. 3 Location of seismic reflection profiles and vibracores, northern Penobscot Bay. Barnhardt et al. (1997, their Fig. 5) location indicated by "B97." Seismic reflection profiles PB-00-200 through 203 collected 10/13/00, shown as dashed line. PB-01-1 through 01-12 collected 01/11/01, and PB-01-101 to 121 collected 09/07/019, shown as solid lines. Locations of figures in this paper are highlighted in bolder lines. Cores are solid dots: piston core O-59-K-211 from Ostericher (1965); vibracores PB-VC-93-1 through 5 from Barnhardt et al. (1997), collected 8/30/93; vibracores PB-VC-01-1, 2, 3, 4, collected 09/14/01; 01-24 and 25 collected 09/17/01.

et al., 1996, 1998) and submersible observations (Belknap *et al.*, 1988; Belknap, 1991), this unit has abundant boulders. Knebel (1986) and Knebel and Scanlon (1985a,b) recognize unit Qdu as moraines and outwash, including some better-stratified units than are found in this study. This unit is interpreted as till, represented as T on interpreted figures.

The next higher unit in the stratigraphy is a nearly continuous blanket of variable thickness that conformably overlies BR and/or T. It ranges from a few meters to more than 50 m thickness in bedrock valleys. The top of this unit is an erosional unconformity in this present study area, although it becomes a conformable contact deeper than 70 m below present sea level and farther offshore (Belknap and Shipp, 1991). Internal reflectors are often distinctly rhythmically stratified and concentrically draped over underlying layers, but may tend toward a more ponded geometry near the top of the section. Individual hyperbolic returns within the unit are interpreted as iceberg dumps and dropstones. In cores, this unit is a stiff, blue-gray sandy mud. This unit is interpreted as the uppermost Pleistocene Presumpscot Formation glaciomarine mud, designated GM in the figures. Belknap *et al.* (1989) and Belknap and Shipp (1991) distinguish massive, draped and ponded subfacies in this unit, but that subdivision is not needed for this study. Knebel (1986) and Knebel and Scanlon (1985a,b) recognize this unit as Qp.

Valleys incised into the GM were filled in a few locations by a lenticular unit up to 35 m thick, with a sharp upper boundary of high acoustic contrast, and a sharp to less-distinct impedance contrast at the base. The upper surface is relatively flat. This unit is distinctly cross-stratified with large sigmoid clinoforms and channel cut-and-fill structures in east-west cross sections, and an offlap sequence of regular southerly prograding clinoforms in the north-south axis of the paleochannel of East Penobscot Bay. Cores from near its surface return coarse sand and gravel with marine shells (Barnhardt *et al.*, 1997, their Fig. 5). We interpret this unit as fluvial and deltaic sand and gravel (labeled S). Knebel (1986) and Knebel and Scanlon (1985a,b) found this unit on a single crossing, and used the term Qf for the same interpretation. In the deep center of this fill are a series of diffractions suggesting interference at a rough surface, and/or individual coarse cobbles and boulders (labeled Gr). Shipp (1989) and Barnhardt (1994) found similar reflections in the Kennebec Paleodelta and Gouldsboro Bay, interpreted as thin gravel lenses (TGL). All are too deep for our present sampling techniques, and would require a drill rig to sample at 30 m or more depth in the sediment column.

The uppermost seismic facies is transparent to weakly stratified, with a draping geometry conformably overlying the Quaternary units and bedrock.

Cores confirm that this is Holocene estuarine mud, shown as M in the figures (Qh of Knebel (1986) and Knebel and Scanlon, 1985a,b).

3. RESULTS

Earlier seismic data (PB-93-01) and vibracores (Barnhardt *et al.*, 1997) provided evidence for a sand deposit in mid bay. The 1993 line was a successful attempt to parallel Knebel and Scanlon's (1985a, their Fig. 6) seismic line, which they had interpreted as fluvial channel infill. Core PB-VC-93-04 (Barnhardt *et al.*, 1997, their Fig. 5) recovered two articulated, growth-position *Mya arenaria* shells giving identical 8.7 ka radiocarbon dates in sand and gravel facies on the eastern flank of the thickest clinoform unit. These dates provide critical timing for the construction of sandy fluvial infill facies of Eastern Penobscot Bay, and these intertidal to shallow subtidal soft-shell clams are an approximate indicator of paleo-sea-level.

In a trial of new seismic reflection equipment on October 13, 2000, we crossed a distinctive set of prograding clinoforms under the eastern channel of Penobscot Bay (Fig. 4). For many years we had been exploring outer Penobscot Bay for an analogue to the sandy Kennebec paleodelta of the west-central Maine coast (Belknap *et al.*, 1986). The work by Knebel (1986), Knebel and Scanlon (1985a, b), and Barnhardt *et al.* (1997) had uncovered only limited evidence for sandy units in the mid and upper bay. Transect PB-00-203 (Fig. 4) demonstrates six distinct seismic facies. The bedrock basement (BR) is overlain by till (T) in several locations. The deepest portions of the bedrock channel/basin (near time mark 15:17) are obscured, probably by diffractions and losses in the overlying incised valley fill. Total sediment thickness is greater than 85 milliseconds, or 64 m. BR and T are overlain by up to 35 m of glaciomarine mud (GM), which is rhythmically stratified throughout, with less variability at the bottom and a very strong rhythmic acoustic contrast at its top. Continuation of GM to the east of time mark 15:17 is unclear, and the GM interpreted above till in the eastern portion of the profile may actually be the base of Unit S. Lowered sea level allowed valley incision by the paleo-Penobscot River. In the thalweg 55-55 m below present SL is a strong set of diffractions, interpreted as gravel channel lag (Gr). This unit terminates abruptly on its margins, and appears to be lenticular in form. Unit S contains a distinctive set of clinoforms, almost conformable with GM at the base, but steeply inclined in the middle and top of the unit. Acoustic impedance contrast, and correlation to Barnhardt *et al.* (1997) core PB-VC-93-04 (1.5 km to the northeast) allow confidence in the interpretation as sand and gravel. There appears to be a single set of prograding and aggrading channel-fill structures making up Paleochannel A.

A second channel cut-and-fill structure, Paleochannel C, occurs west of time mark 15:14. The top of these channel units is relatively flat. The uppermost unit is 5-6 m of more acoustically transparent Holocene mud (M).

Seismic line PB-01-09a (Fig. 5) contains the same sequence of seismic facies as in figure 4. The deep portion of the center of the figure is difficult to interpret – we show either GM or a weakly reflective till mound below -60 m. The transition from GM to S east of time mark 19:48 is not as clear as to the west – the strata appear nearly conformable. This may represent a transition from rapid accumulation in deglacial meltwater-fed flows that provide a characteristic draped geometry for GM, to the channel cut-and-fill of post-glacial river-valley sedimentation. Similar ambiguities are evident in the Kennebec Paleodelta (Belknap *et al.*, 1989; Barnhardt *et al.*, 1997). The interpretation of the sandy channel fill unit suggests several phases of progradation and aggradation in Paleochannel A. Paleochannel C is a distinctive east-to-west prograding infill above a distinct disconformity. As in figure 4, these sandy units are conformably overlain by Holocene estuarine mud.

Another representative W-E line, PB-01-07a (Fig. 6), covers the four identified paleochannels, and is a complete cross-section through the Penobscot Paleodelta. The stratigraphy is similar to the examples discussed previously, but is complicated by possible natural gas wipeout (NG) near time mark 18:26, and complex diffractions at the base of Paleochannels A and D. Our preferred interpretation is gravel lag (Gr), but gas-enhanced reflectors or seismic side-echoes are alternative possibilities.

Figure 7 is a N-S line parallel to the axis of the paleodelta. Unlike the paleochannels described above, Unit S here displays a monotonic offlap and downlap geometry, prograding out over GM into a topographic low. Strong hyperbolic diffractions at the base of S are interpreted as gravel lag (Gr). This facies is interpreted as the southern terminus of a paleodelta lobe, which was prograding into an estuary (or lake?) when sea level was near -30 m. To the south of the terminus is mud charged with natural gas, but there is also a small sand unit near time mark 21:11. The latter may be a portion of another delta lobe. The gas-charged mud is conformable with the overlying estuarine mud. Figure 8 provides a close-up view of the paleodelta terminus. Note that the progradation was accompanied by aggradation of several meters. This may suggest building in a time of rapidly rising sea level. Alternatively, this may represent a channel avulsion event. In either case, the clear, near 100% preservation of topset and foreset units suggests that later wave or tidal current erosion was not able to rework this terminus. The exquisite preservation suggests an unusual mechanism for preservation.

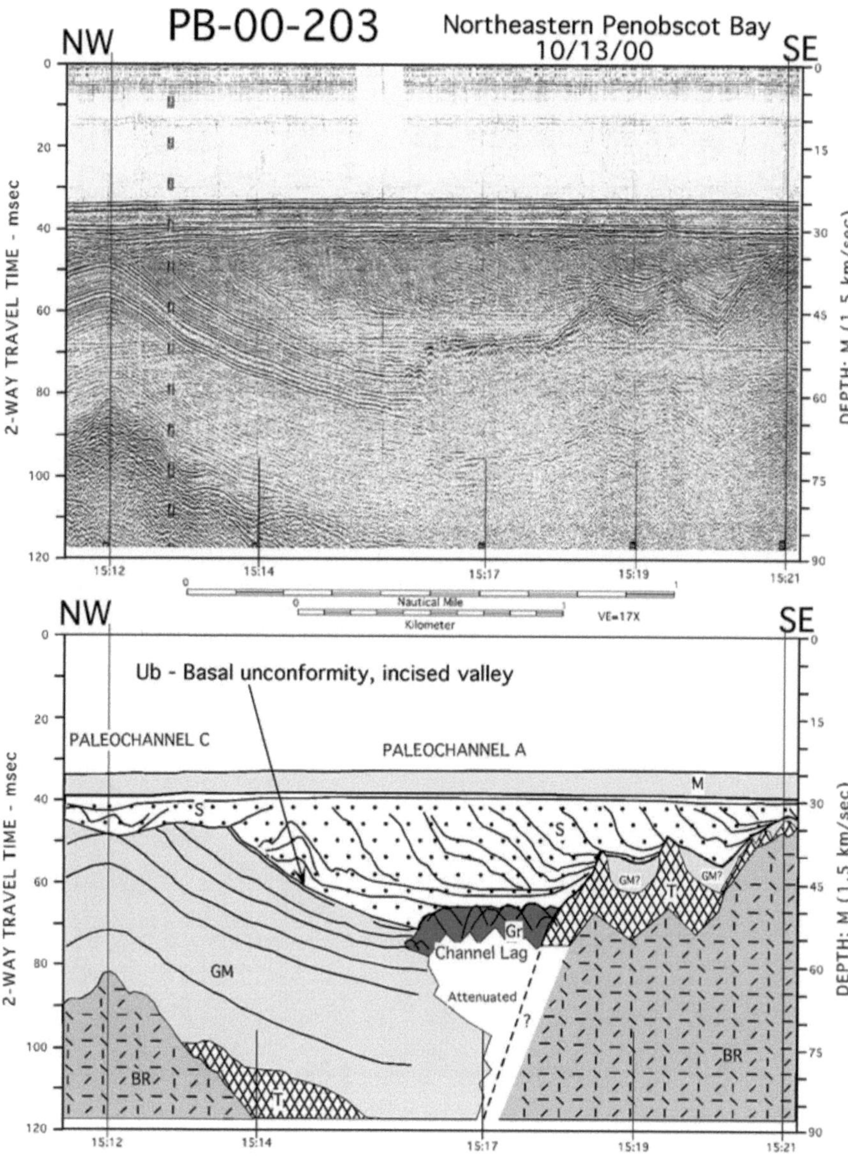

Figure. 4 Detail of Applied Acoustic Engineering (AAE) digital boomer seismic reflection profile PB-00-203 in Eastern Penobscot Bay, see Fig. 3 for location. Clinoforms identify infill of two (of four identified) channels incised into glaciomarine mud at lowstand. Infill began with gravel lag and terminated by channel avulsion. M = Holocene estuarine mud, S = sand, Gr = gravel, GM = glaciomarine mud, T = till, BR = bedrock, Ub = basal unconformity. After Belknap et al. (2002, their Fig. 10).

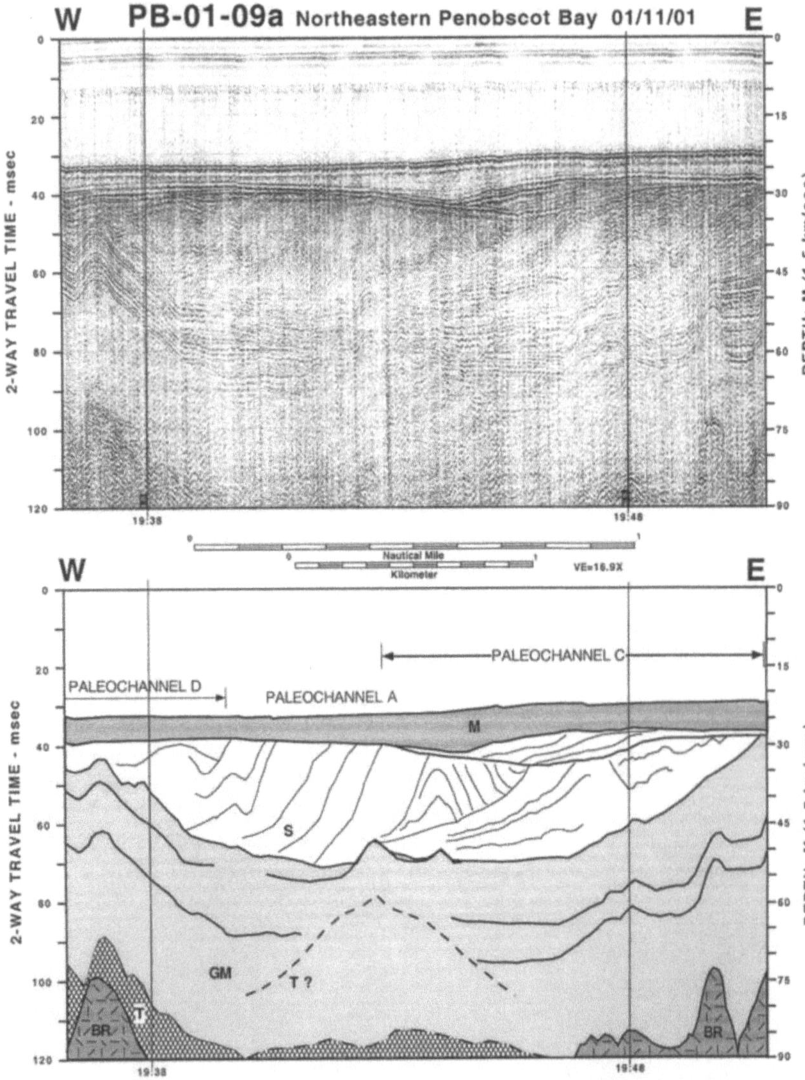

Figure. 5 Detail of AAE digital boomer seismic reflection profile PB-01-09a in Eastern Penobscot Bay, see figure 3 for location. Clinoforms identify infill of three (of four identified) channels incised into glaciomarine mud at lowstand. Infill of channel A on the east side is nearly conformable with underlying GM. Two alternative interpretations (till or glaciomarine) are shown in the lower central portion of the profile where reflections may be obscured by attenuation by overlying thick sand and gravel units. Legend is in caption for Figure 4.

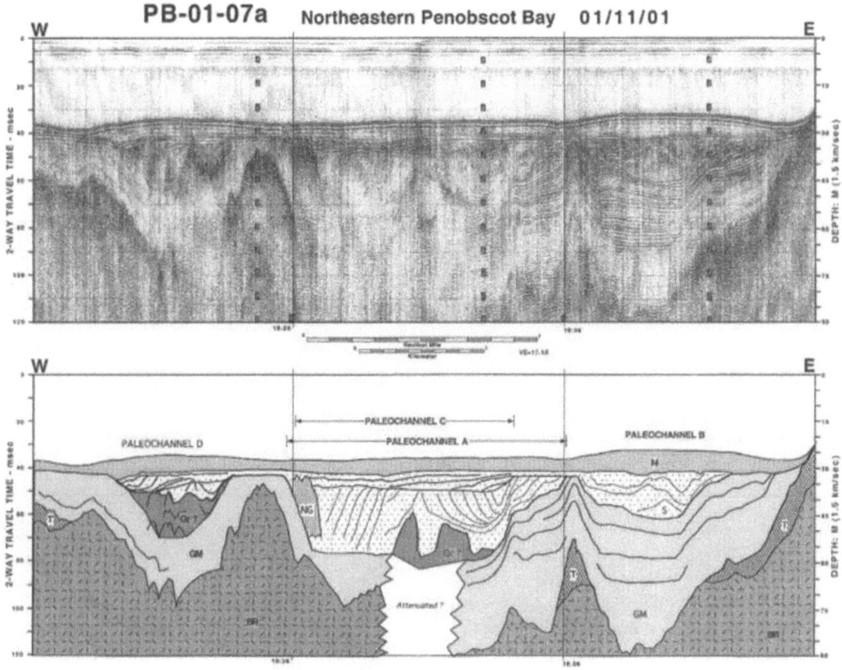

Figure. 6 Detail of AAE digital boomer seismic reflection profile PB-01-07a in Eastern Penobscot Bay, see figure 3 for location. Clinoforms identify infill of all four identified paleochannels. A complex set of reflectors is interpreted as gravel lag within the deeper portions of channels A and D, however gas-enhanced reflections or other interpretations are also possible. Glaciomarine mud is clearly identified by rhythmically draped stratification in the eastern half of the profile, while it is more indistinct to the west, possibly due to signal attenuation by the thick channel sands and gravels. Legend is in caption for Figure 4.

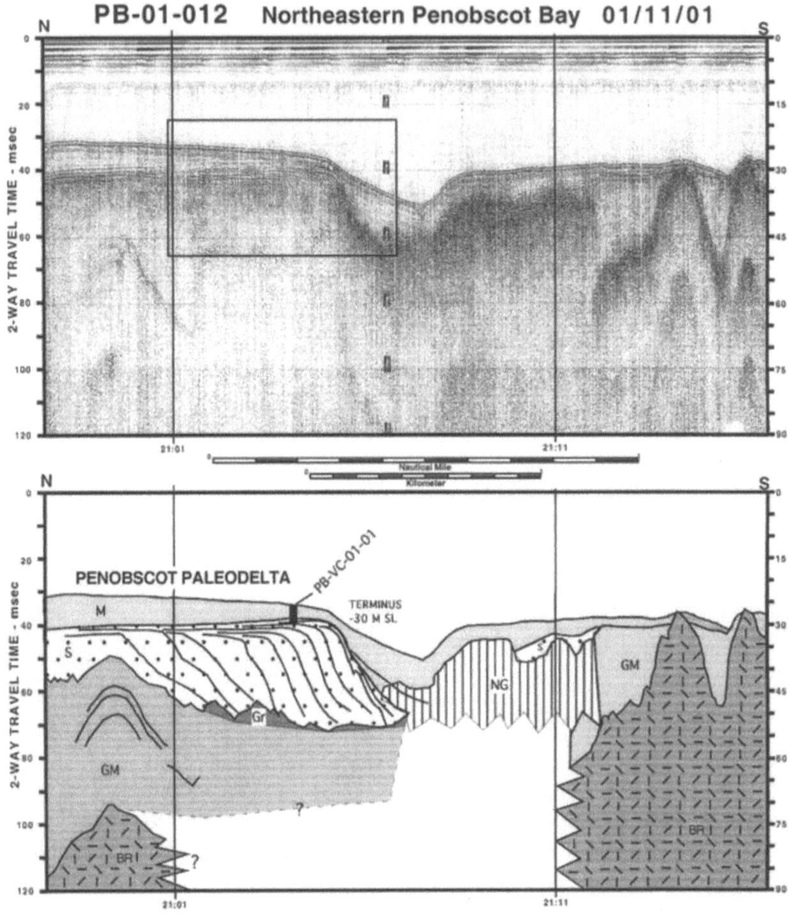

Figure. 7 Detail of AAE digital boomer seismic reflection profile PB-01-12 in Eastern Penobscot Bay, north to south down the axis of the paleodelta, see figure 3 for location. Clinoforms indicate progradation of a delta front south into a 20-m deep basin. Progradation was accompanied by a 3 m rise in the topset beds to a terminus at approximately 30 m below present sea level. Box is closer view (Fig. 8). Legend is in caption for figure 4 (after Belknap et al., 2002, their Fig. 11).

Vibracore PB-VC-01-01 (Fig. 9) was a top priority for our sampling. It penetrated through the Holocene estuarine mud and recovered sand, gravel and *Mya arenaria* shells from the paleodelta terminus. Unfortunately, these shells are not as well preserved as in the 1993 core, and have not been dated. However, we feel confident that the dates from PB-VC-93-04 provide an approximate age of formation of at least the final stages of this delta front 9-8 ka.

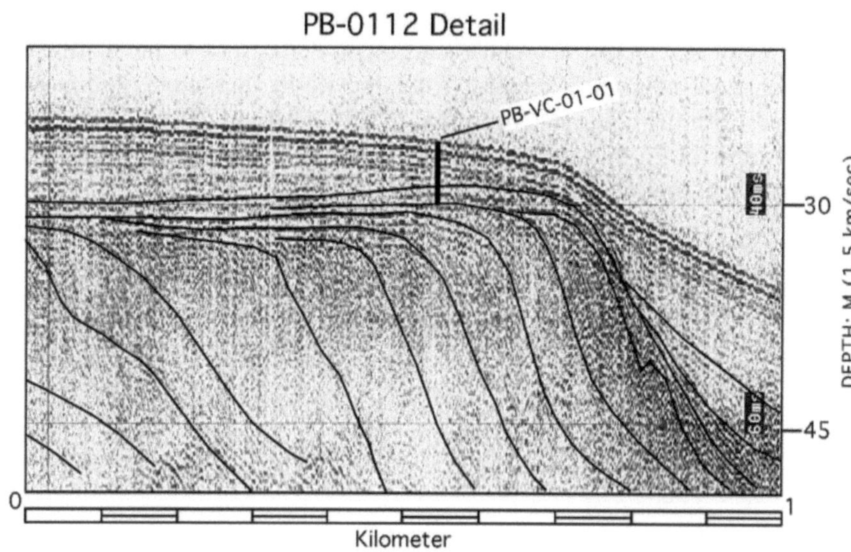

Figure. 8 Close-up view of Penobscot Paleodelta terminus, PB-01-12. See figure 7 for full view.

4. DISCUSSION

Three W-E boomer profile cross-sections (Figs. 4-6) near the center of the paleodelta, and one (Fig. 7) N-S down the axis (PB-01-012) to the terminus, sketch out the geometry of the Penobscot Paleodelta. Multiple channels demonstrate avulsion and infill by prograding sigmoid clinoforms. The terminus progrades nearly a kilometer while aggrading 2-3 m at its top to –30 m, most likely during sea-level rise. By correlation to *Mya arenaria* shells in core PB-VC-03-04, giving 8.7 ka radiocarbon dates on the flank of the oldest, largest paleochannel (A), we suggest construction of this sand body ca. 9-8 ka. The delta built into a narrow muddy embayment or an estuary, as indicated by the marine fossils. The earliest phases of progradation might possibly have been into a lake. Penobscot Bay was tightly constricted south of Islesboro Island at this time, based on paleogeographic reconstruction from bathymetric contours.

An isopach map (Fig. 10) of the sand and gravel channel-delta facies (S) was constructed from all the seismic lines available (Fig. 3). The thickest sands and gravels (30-35 m) are found near the southwest terminus of the feature. The southern terminus is clearly defined, but the sand body most likely extends north beyond the data grid. Distinct lobes and paleochannel fills are evident. Overall geometry suggests a delta front to the south, with fluvial channel sources extending from the north. Sediment volume was integrated from the contours for a total of 290 x 10^6 m^3. There is no formal estimate of error on this value, presented here as two significant digits, due to the nature of interpolation between seismic lines.

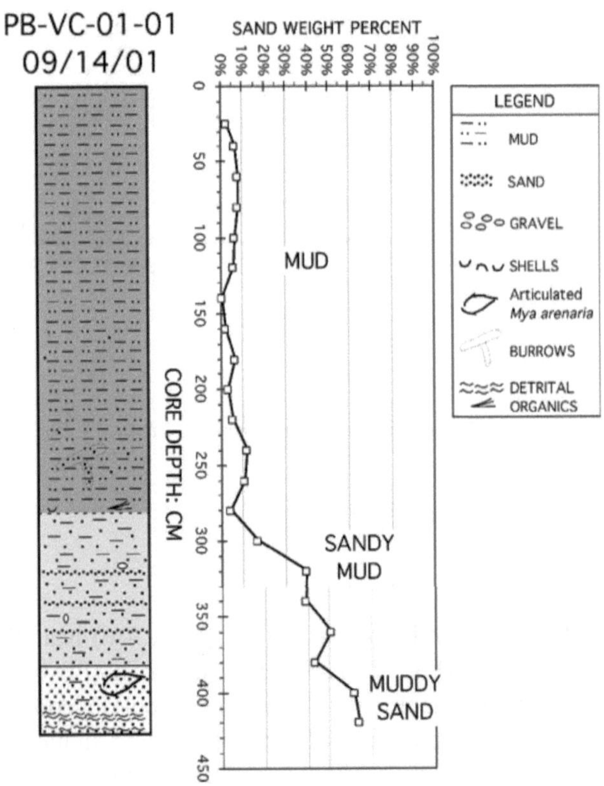

Figure. 9 Core log, vibracore PB-VC-01-01, Penobscot Paleodelta terminus. See figure 3 for location and figures 7 and 8 for relationship to seismic stratigraphy.

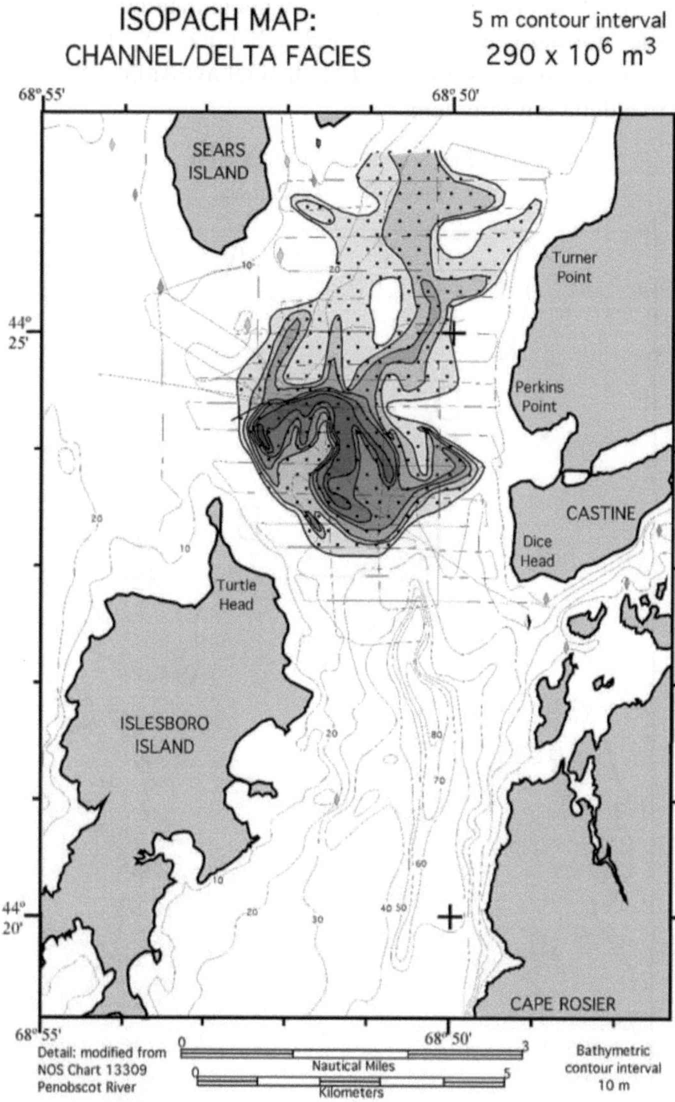

Figure. 10 Isopach map of Penobscot Paleodelta, ca. 8-9 ka BP delta complex, buried beneath modern estuarine mud, between Castine, Islesboro Island and Sears Island, shown as stippled contours. Isopachs in 5 m contour intervals, thickest unit > 30 m.

The relatively close spacing of the seismic lines permitted confident correlation of four distinct paleochannels (Fig. 11). The sequence of channels is labeled A (oldest) to D (youngest). The modern channel of the Penobscot River (20 m contour of figures 3 and 11) may be reworking some sand from the northern portions of Paleochannels A, C and D. The paleochannel pattern suggests a sequential channel avulsion pattern as would be expected in a delta environment.

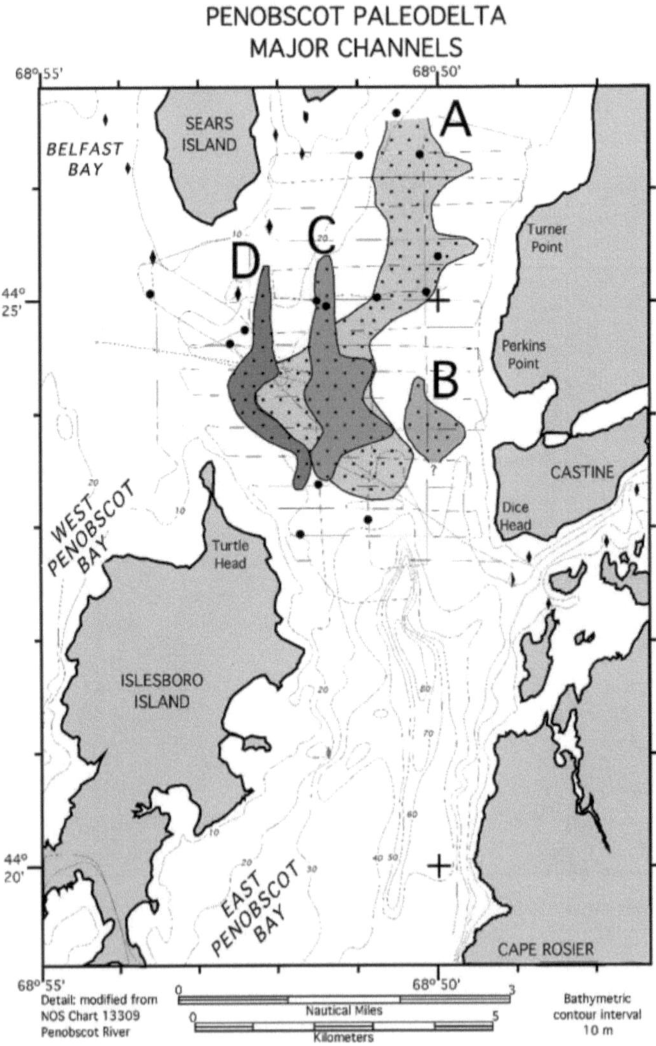

Figure. 11 Channels and sub-lobes of the Penobscot Paleodelta, lettered sequentially from oldest (A) to youngest (D) on the basis of cross-cutting relationships.

Bedrock topography provides the primary control on morphology in coastal Maine, and in turn, influences sedimentary processes. Penobscot Bay and other deeply indented estuaries experience steep gradients in wave energy (high at the mouth) and tidal currents (strongest in constricted channels in the middle bay) (Belknap *et al.*, 1986; Dalrymple *et al.*, 1992). Erosion during post-glacial sea-level fall incised more than 70 m below present sea level into the glaciomarine Presumpscot Formation, outwash and till. We have found no evidence for a distinct lowstand paleodelta of the Penobscot River, but there are extensive plains of gravel at –50 to –70 m (Barnhardt *et al.*, 1996). Depocenters for estuarine and embayment fill gradually moved up the incised valleys during transgression (Fig. 12).

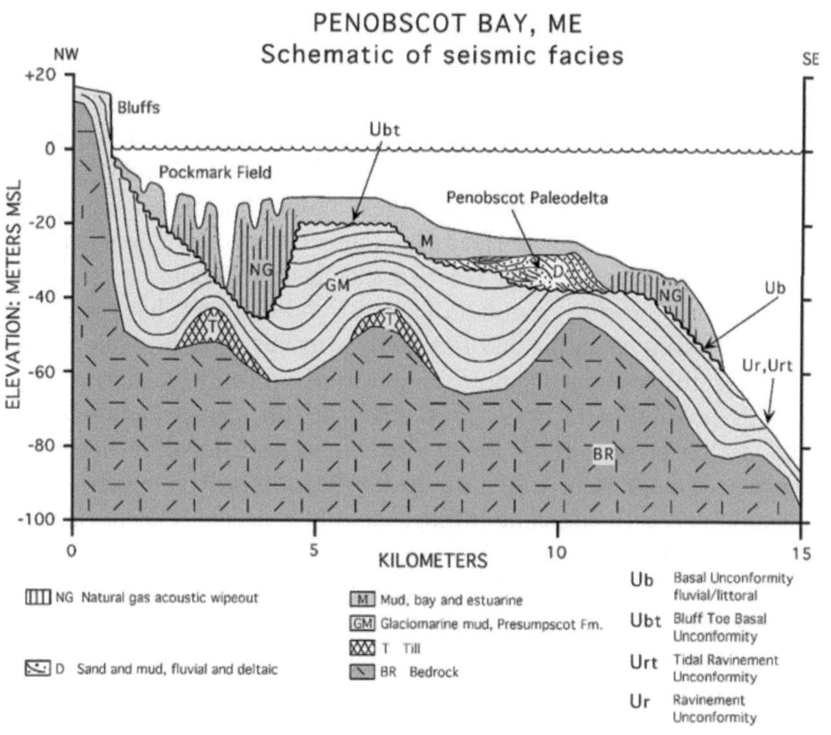

Figure. 12 Schematic of seismic facies, Penobscot Bay, ME. The Penobscot Paleodelta (D) progrades and downlaps onto the basal unconformity Ub at the top of glaciomarine Presumspscot Fm. (GM). Modified from Belknap et al. (2002, their Fig. 12).

Our first hypothesis is that a distinct paleodelta formed during the slowdown in sea-level rise 9-8 ka, at –30 m. Shipp *et al.* (1991) found a possibly correlative –30 m terrace grouping on the Kennebec Paleodelta.

This may point toward a regionally significant event. Renewed rapid sea-level rise and burial 7-6 ka (Fig. 2) may explain the excellent preservation of the Penobscot Paleodelta. We have suggested migration of a glacial forebulge as the reason for the distinctive shape of the local relative sea-level curve between 10 and 7 ka (Barnhardt *et al.*, 1995). This is further supported by glacioisostatic tilting of lake paleoshorelines nearby (Balco *et al.*, 1998).

A second hypothesis concerns possible evidence for a global meltwater pulse. Evidence for a eustatic meltwater pulse at this time is equivocal. The 7-6 ka time period is 1000-2000 years too late for Fairbank's (1989) meltwater pulse IB (although IB could help explain the rapid rise prior to 9 ka). Other well-constrained sea-level curves, such as in Delaware (Belknap and Kraft, 1977; Nikitina *et al.*, 2000) show little evidence for accelerated rate of RSL rise at that time.

A third possibility is that a pulse of new sediment became available to the Penobscot River, allowing rapid progradation. There are local sources of sand in gravel in eskers and outwash in bluffs near the present river (Thompson and Borns, 1985) that would have been suitable sources. The briefness of the pulse of sediment input is more difficult to explain, however. Alternatively, (hypothesis four) thick Presumpscot Formation and till found in the saddle between Turtle Head (Islesboro Island) and Sears Island may indicate an interfluve between Western and Eastern Penobscot Bay prior to ca 9-8 ka. Bluff erosion from both the west and east may have opened this isthmus, allowing estuarine and fluvial flow to the west, possibly stranding the Penobscot Paleodelta and protecting it from tidal current erosion.

A fifth possibility is drainage changes in the upper Penobscot River system. Balco *et al.* (1998) found that Moosehead Lake in northwestern Maine was a major source of flow to the Penobscot River prior to 8.4 ka (radiocarbon years). Isostatic tilting, possibly during the passage of a decaying glacial forebulge, shifted its outlet to the Kennebec River after that time. Isostatic tilt and larger-than-present runoff resulted in downcutting of terraces on the middle and lower Penobscot River prior to 8 ka (A.R. Kelley *et al.*, 1994), a close coincidence with the timing of deposition of the Penobscot Paleodelta.

5. CONCLUSIONS

Penobscot Bay provides a high-resolution example of sequence stratigraphy (Fig. 13), focusing on the erosional unconformity surfaces that define the boundaries of depositional sequences, and the bundles of facies that form systems tracts related to positions and rates of change of sea-level (e.g. Posamentier and Vail, 1988). The basal unconformity (Ub), as defined here, was created by littoral and fluvial erosion during rapidly falling postglacial sea level ca. 12-10 ka, creating a sequence boundary between Pleistocene glaciomarine and Holocene shelf and coastal systems. Locally, rivers incised valleys into glacial and glaciomarine sediments, creating accommodation space for Holocene estuarine systems. Lowstand was reached 10.8 ka at –55 to –65 m (Belknap *et al.*, 1989; Barnhardt *et al.*, 1997). Lowstand systems tracts comprising paleodelta, shoreline, and basin deposits occur in some locations (Belknap and Shipp, 1991; Barnhardt *et al.*, 1997), but Penobscot Bay appears to not preserve a thick lowstand deposit, just a widespread gravel lag at this level.

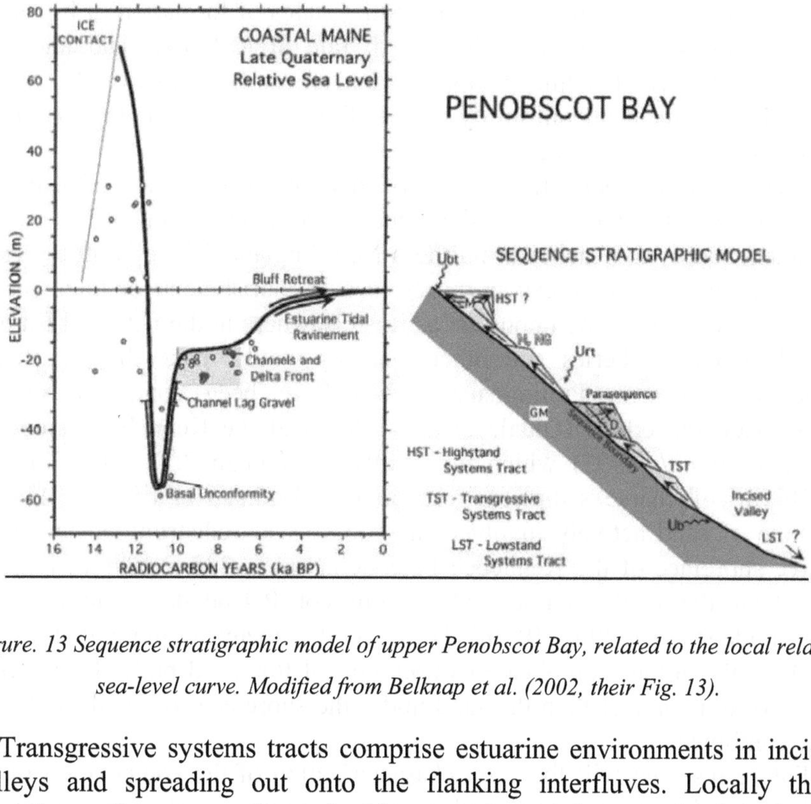

Figure. 13 Sequence stratigraphic model of upper Penobscot Bay, related to the local relative sea-level curve. Modified from Belknap et al. (2002, their Fig. 13).

Transgressive systems tracts comprise estuarine environments in incised valleys and spreading out onto the flanking interfluves. Locally these estuarine sediments are charged with gas and many have evolved pockmark fields (Kelley *et al.*, 1994; Rogers, 1999; Gontz *et al.*, 2002) (Fig. 12). A plateau in the rate of sea-level rise 9-8 ka resulted in a parasequence expressed as a paleodelta of the Penobscot River. This Penobscot Paleodelta was later preserved, possibly by an increased rate of sea-level rise (e.g. Belknap and Kraft, 1981) caused by passage of a marginal forebulge or by a global meltwater pulse. This event may correlate with a similar parasequence and shorelines of the Kennebec Paleodelta (Belknap *et al.*, 2002), and may be regionally significant. Alternative or contemporaneous causes for this shift in depositional systems and preservation may include changes in the Penobscot River drainage and/or sediment sources, or erosion of the isthmus between Sears Island and Islesboro resulting in a channel avulsion.

A transgressive basal unconformity (Ubt) is created at the toe of bluffs, reworking glacial and paraglacial sediments as a source for Holocene environments. This unconformity has the same stratigraphic position as the regressive basal unconformity (Ub), but in some cases can be distinguished as a planar, near horizontal surface, in contrast to the incised valley geometry formed during sea-level fall and lowstand. Reworking of estuarine and embayment sediments occurs in channels and channel margins at the tidal ravinement unconformity (Urt). Initial states of formation of highstand systems tracts result from a slowing of rate of RSL rise, and increased relative influence of sediment supply at barriers and in estuaries. The wave-formed ravinement unconformity (Ur) is found in the outer, higher energy portions of the embayment.

One reason for conducting studies of this type is the survey of nearshore sand and gravel resources. Nearshore coarse sediments are natural sources of sediment supply to beaches in northern New England. There is little present exploitation for commercial purposes, beach nourishment, or engineering, but this may come to be important as it is elsewhere in the U.S. and Europe. Conflicts with fisheries and other interests would make this a difficult process at present. Preliminary studies have been conducted on the Merrimack Paleodelta (Oldale *et al.*, 1983) and the Kennebec Paleodelta (Barnhardt *et al.*, 1997), which are estimated to contain 2.1×10^9 m^3 and 1.3×10^9 m^3 of coarse sediments respectively (Belknap *et al.*, 2002). (The volume of sand actively involved in processes of redistribution under the active shoreface of the Kennebec Paleodelta, some 335×10^6 m^3, is much less than the total volume). The Penobscot Paleodelta is an order of magnitude smaller, 290×10^6 m^3. The cover of 5 meters or more of Holocene mud would further complicate exploitation of this sand body. However, it has a volume greater than the sand under the shorefaces of Wells and Saco Bay combined.

Future research will focus on characterization of the composition of the sediments in these stratigraphic units, timing and mechanism of emplacement, and further refinement of understanding of preservation potential within embayments and on the inner shelf. This research on the glaciated shelf of northern New England has applications to similar shelves in Atlantic Canada, northern Europe and many other locations worldwide.

6. ACKNOWLEDGMENTS

We gratefully acknowledge funding by the National Science Foundation, Maine-New Hampshire Sea Grant, NOAA for ship-time additions, and NOAA's National Undersea Research Program at the University of Connecticut, Avery Point for submersible support. We specifically acknowledge NSF Major Research Instrumentation grant OCE-9977367 for equipment used in this project, and NOAA-ME-NH Sea Grant project R/CE-235 for other research funding. We wish to acknowledge technical support by Geoff Shipton of Triton Elics Company. We specifically thank our former students and continuing colleagues who helped in the collection of these data and in discussions of the concepts over a number of years: Gregory A. Balco, Walter A. Barnhardt, Stephen M. Dickson, Robert A. Johnston and Alice R. Kelley, as well as fieldwork assistance by numerous other University of Maine undergraduate and graduate students. We thank Captain Tony Codega of the *R/V Friendship*, Maine Maritime Academy, and Captain Corey Roberts of the *R/V Alice Siegmund*, The Island Institute, for help with the geophysical profiling, and Captain Randy Flood, Don Bradford, and the crew of the *R/V ARGO Maine* for able assistance in collecting the vibracores. We thank Woodrow B. Thompson and Jasper Knight for helpful reviews.

REFERENCES

Allen, G.P. and Posamentier, H.W. 1994. Transgressive facies and sequence architecture in mixed tide- and wave-dominated incised valleys: examples from the Gironde Estuary, France. In: Dalrymple, R., Boyd, R. and Zaitlin, B. (eds) *Incised-Valley Systems: Origin and Sedimentary Sequences*. SEPM Special Publication 51, 225-240.

Balco, G., Belknap, D.F. and Kelley, J.T. 1998. Glacioisostasy and lake-level change at Moosehead Lake, Maine. *Quaternary Research*, 49, 157-170.

Bard, E., Hamelin, B., Arnold, M., Montaggioi, L., Cabioch, G., Faure, G. and Rougerie, F. 1996. Deglacial sea-level record from Tahiti corals and the timing of global meltwater discharge. *Nature*, 382, 241-244.

Barnhardt, W.A. 1994. *Late Quaternary relative sea-level change and evolution of the Maine inner continental shelf 12-7 ka BP*. Unpublished PhD Dissertation, University of Maine, Orono, 196 pp.

Barnhardt, W.A., Belknap, D.F. and Kelley, J.T. 1997. Stratigraphic evolution of the inner continental shelf in response to late Quaternary relative sea-level change, northwestern Gulf of Maine. *Geological Society of America Bulletin*, 109, 612-630.

Barnhardt, W.A., Belknap, D.F., Kelley, A.R., Kelley, J.T. and Dickson, S.M. 1996. *Surficial geology of the Maine inner continental shelf: Rockland to Bar Harbor, Maine*, Geologic Map No. 96-11, Maine Geological Survey, Augusta, 1:100,000.

Barnhardt, W.A., Gehrels, W.R., Belknap, D.F. and Kelley, J.T. 1995. Late Quaternary relative sea-level change in the western Gulf of Maine: evidence for a migrating glacial forebulge. *Geology*, 23, 317-320.

Barnhardt, W.A., Kelley, J.T., Dickson, S.M. and Belknap, D.F. 1998. Mapping the Gulf of Maine with side-scan sonar: a new bottom-type classification for complex seafloors. *Journal of Coastal Research*, 14 , 646-659.

Belknap, D.F. 1991. *Submersible investigation of seafloor pockmarks, central Maine coast: geology and biology of natural gas (?) craters.* Final Report to NOAA National Undersea Research Center, Univ. Connecticut, Avery Point, 11 pp.

Belknap, D.F., Andersen, B.G., Anderson, R.S., Anderson, W.A., Borns, H.W., Jr., Jacobson, G., Jr., Kelley, J.T., Shipp, R.C., Smith, D.C., Stuckenrath, R. Jr., Thompson, W.B. and Tyler, D.A. 1987a. Late Quaternary sea-level changes in Maine. In: Nummedal, D., Pilkey, O.H., Jr. and Howard, J.D. (eds) *Sea-Level Fluctuation and Coastal Evolution.* SEPM Special Publication 41, 71-85.

Belknap, D.F., Kelley, J.T. and Gontz, A.M. 2002. Evolution of the glaciated shelf and coastline of the northern Gulf of Maine, USA. *Journal of Coastal Research Special Issue*, 36, 37-55.

Belknap, D.F., Kelley, J.T. and Robbins, D.H.W. 1988. Sediment dynamics of the nearshore Gulf of Maine: Submersible experimentation and remote sensing. In: Babb, I. and DeLuca, M. (eds) *Benthic Productivity and Marine Resources of the Gulf of Maine.* NOAA National Undersea Research Program, Research Report 88-3, 143-176.

Belknap, D.F., Kelley, J.T. and Shipp, R.C. 1987b. Quaternary stratigraphy of representative Maine estuaries determined from high resolution seismic reflection profiling: preliminary report. In: FitzGerald, D.M. and Rosen, P.S. (eds) *Glaciated Coasts.* Academic Press, San Diego, 177-207.

Belknap, D.F. and Kraft, J.C. 1977. Holocene relative sea-level changes and coastal stratigraphic units on the northwest flank of the Baltimore Canyon Trough geosyncline. *Journal of Sedimentary Petrology*, 47, 610-629.

Belknap, D.F. and Kraft, J.C. 1981. Preservation potential of transgressive coastal lithosomes on the U.S. Atlantic Shelf. *Marine Geology*, 41, 419-442.

Belknap, D.F. and Kraft, J.C. 1985. Influence of antecedent geology on evolution of barrier systems. *Marine Geology*, 63, 235-262.

Belknap, D.F., Kraft, J.C. and Dunn, R.K. 1994. Transgressive valley-fill lithosomes: Delaware and Maine. In: Boyd, R., Zaitlin, B.A. and Dalrymple, R. (eds) *Incised Valley Fill Systems.* SEPM Special Publication 51, 303-320.

Belknap, D.F. and Shipp, R.C. 1991. Seismic stratigraphy of glacial-marine units, Maine inner shelf. In: Anderson, J.B. and Ashley, G.M. (eds) *Glacial-Marine Sedimentation; Paleoclimatic Significance.* Geological Society of America Special Paper 261, 137-157.

Belknap, D.F., Shipp, R.C. and Kelley, J.T. 1986. Depositional setting and Quaternary stratigraphy of the Sheepscot Estuary, Maine. *Géographie physique et Quaternaire*, 40, 55-69.

Belknap, D.F., Shipp, R.C., Kelley, J.T. and Schnitker, D. 1989. Depositional sequence modeling of late Quaternary geologic history, west-central Maine coast. In: Tucker, R.D. and Marvinney, R.G. (eds) *Studies in Maine Geology - Vol. 5: Quaternary Geology.* Maine Geological Survey, Augusta, 29-46.

Dalrymple, R.W., Zaitlin, B.A. and Boyd, R. 1992. Estuarine facies models: conceptual basis and stratigraphic implications. *Journal of Sedimentary Petrology*, 62, 1130-1146.

Dorion, C.C., Balco, G.A., Kaplan, M.R., Kreutz, K.J., Wright, J.D. and Borns, H.W., Jr. 2001. Stratigraphy, paleoceanography, chronology, and environment during deglaciation of eastern Maine. In: Weddle, T.K. and Retelle, M.J. (eds) *Deglacial History and Relative Sea-Level Changes Northern New England and Adjacent Canada.* Geological Society of America Special Paper 351, 215-242.

Fairbanks, R.G. 1989. A 17,000-year glacio-eustatic sea level record: influence of glacial melting rates on the Younger Dryas event and deep-ocean circulation. *Nature*, 342, 637-642.

Gontz, A.M. 2002. *Evolution of seabed pockmarks in Penobscot Bay, Maine.* Unpublished MS Thesis, University of Maine, Orono, 118 pp.

Gontz, A.M., Belknap, D.F. and Kelley, J.T. 2002. Seafloor features and characteristics of the Black Ledges area, Penobscot Bay, Maine, USA. *Journal of Coastal Research Special Issue*, 36, 333-339.

Hughes, T., Borns, H.W., Jr., Fastook, J.L., Hyland, M.R., Kite, J.S. and Lowell, T.V. 1985. Models of glacial reconstruction and deglaciation applied to Maritime Canada and New England. In: Borns, H.W., Jr., LaSalle, P. and Thompson, W.B. (eds) *Late Pleistocene History of Northeastern New England and Adjacent Quebec.* Geological Society of America Special Paper 197, 139-150.

Kelley, A.R., Kelley, J.T. and Belknap, D.F. 1994. Geomorphology and surficial geology of the lower Penobscot River Valley. In: Hanson, L.S. (ed) *New England Intercollegiate Geological Conference Guidebook to Field Trips in North-Central Maine.* 85th Annual Meeting, Trip A-1, 1-13.

Kelley, J.T. 1987. An inventory of coastal environments and classification of Maine's glaciated shoreline. In: FitzGerald, D.M. and Rosen, P.S. (eds) *Glaciated Coasts.* Academic Press, San Diego, 151-176.

Kelley, J.T. and Belknap, D.F. 1989. *Geomorphology and sedimentary framework of Penobscot Bay and adjacent inner continental shelf.* Open File Report 89-3, Maine Geological Survey, Augusta, ME, 35pp.

Kelley, J.T., Dickson, S.M., Belknap, D.F., Barnhardt, W.A. and Henderson, M. 1994. Giant sea-bed pockmarks: evidence for gas escape from Belfast Bay, Maine. *Geology*, 22, 59-62.

Kelley, J.T., Dickson, S.M., Belknap, D.F. and Stuckenrath, R. Jr. 1992. Sea-level change and the introduction of late Quaternary sediment to southern Maine inner continental shelf. In: Fletcher, C.H. and Wehmiller, J.F. (eds) *Quaternary Coastal Systems of the United States, Marine and Lacustrine Systems.* SEPM Special Publication 48, 23-34.

Kelley, J.T., Gehrels, W.R. and Belknap, D.F. 1995. Late Holocene relative sea-level rise and the geologic development of tidal marshes at Wells, Maine, U.S.A. *Journal of Coastal Research*, 11, 136-153.

Knebel, H.J. 1986. Holocene depositional history of a large glaciated estuary, Penobscot Bay, Maine. *Marine Geology*, 73, 215-236.

Knebel, H.J. and Scanlon, K.M. 1985a. Sedimentary framework of Penobscot Bay, Maine. *Marine Geology*, 65, 305-324.

Knebel, H.J. and Scanlon, K.M. 1985b. *Maps showing sea-floor topography, depth to bedrock, and sediment thickness, Penobscot Bay, Maine.* Miscellaneous Field Studies Map MF-1751, U.S. Geological Survey, 1:100,000.

Miller, G.T. 1998. *Deglaciation of Wells Embayment, Maine: interpretation from seismic and side-scan sonar data.* Unpublished MS Thesis, University of Maine, Orono, 210pp.

Nikitina, D.L., Pizzuto, J.E., Schwimmer, R.A. and Ramsey, K.W. 2000. An updated Holocene sea-level curve for the Delaware coast. *Marine Geology*, 171, 7-20.

Oldale, R.N., Wommack, L.E. and Whitney, A.B. 1983. Evidence for postglacial low relative sea-level stand in the drowned delta of the Merrimack River, western Gulf of Maine. *Quaternary Research*, 33, 325-336.

Ostericher, C. 1965. *Bottom and subbottom investigations of Penobscot Bay Maine, 1959.* U.S. Naval Oceanographic Office Tech. Report 173, 177pp.

Posamentier, H.W. and Vail, P.R. 1988. Eustatic controls on clastic deposition II - sequence and systems tract models. In: Wilgus, C.K., Hastings, B.S., Kendall, C.G.St.C., Posamentier, H.W., Ross, C.A. and Van Wagoner, J.C. (eds) *Sea-level changes - an integrated approach.* SEPM Special Publication 42, 125-154.

Rogers, J.N. 1999. *Mapping of subaqueous, gas-related pockmarks in Belfast Bay, Maine using GIS and remote sensing techniques.* Unpublished MS Thesis, University of Maine, Orono, ME, 139pp.

Scanlon, K.M. and Knebel, H.J. 1989. Pockmarks on the floor of Penobscot Bay, Maine. *Geo-Marine Letters*, 9, 53-58.

Schnitker, D., Belknap, D.F., Bacchus, T.S., Friez, J.K., Lusardi, B.A. and Popek, D.M. 2001. Deglaciation of the Gulf of Maine. In: Weddle, T.K. and Retelle, M.J. (eds) *Deglacial History and Relative Sea-Level Changes, Northern New England and Adjacent Canada.* Geological Society of America Paper 351, 9-34.

Shipp, R.C. 1989. *Late Quaternary sea-level fluctuations and geologic evolution of four embayments and adjacent inner shelf along the northwestern Gulf of Maine.* Unpublished PhD Dissertation, University of Maine, 832pp.

Shipp, R.C., Belknap, D.F. and Kelley, J.T. 1991. Seismic-stratigraphic and geomorphic evidence for a post-glacial sea-level lowstand in the northern Gulf of Maine. *Journal of Coastal Research*, 7, 341-364.

Shipp, R.C., Staples, S.A. and Adey, W.H. 1985. Geomorphic trends in a glaciated coastal bay: a model for the Maine coast. *Smithsonian Contributions to the Marine Sciences*, No. 25, Smithsonian Institition Press, Washington, DC, 76pp.

Shipp, R.C., Staples, S.A. and Ward, L.G. 1987. Controls and zonation of geomorphology along a glaciated coast, Gouldsboro Bay, Maine. In: FitzGerald, D.M. and Rosen, P.S. (eds) *Glaciated Coasts.* Academic Press, San Diego, 209-231.

Stamp, L.D. 1922. An outline of the Tertiary geology of Burma. *Geological Magazine*, 59, 481-501.

Stuiver, M. and Borns, H.W., Jr. 1975. Late Quaternary marine invasion in Maine: Its chronology and associated crustal movement. *Geological Society of America Bulletin*, 86, 99-104.

Swift, D.J.P. 1968. Coastal erosion and transgressive stratigraphy. *Journal of Geology*, 77, 444-456.

Thompson, W.B. and Borns, H.W., Jr. 1985. *Surficial Geologic Map of Maine.* Maine Geological Survey, Augusta, ME, 1:500,000.

Ussler, W., III, Paull, C.K., Boucher, J., Friederich, G.E. and Thomas, D.J. 2003. Submarine pockmarks: a case study from Belfast Bay, Maine. *Marine Geology*, 202, 175-192.

Index

Coastal Systems and Continental Margins

1. B.U. Haq (ed.): *Sequence Stratigraphy and Depositional Response to Eustatic, Tectonic and Climatic Forcing*. 1995 ISBN 0-7923-3780-8
2. J.D. Milliman and B.U. Haq (eds.): *Sea-Level Rise and Coastal Subsidence*. Causes, Consequences, and Strategies. 1996. ISBN 0-7923-3933-9
3. B.U. Haq, S.M. Haq, G. Kullenberg and J.H. Stel (eds.): *Coastal Zone Management Imperative for Maritime Developing Nations*. 1997 ISBN 0-7923-4765-X
4. D.R. Green and S.D. King (eds.): *Coastal and Marine GeoInformation Systems: Applying the Technology to the Environment*. 2001 ISBN 0-7923-5686-1
5. M.D. Max (ed.): *Natural Gas Hydrate in Oceanic and Permafrost Environments*. 2000 ISBN 0-7923-6606-9; Pb 1-4020-1362-0
6. J. Chen, D. Eisma, K. Hotta and H.J. Walker (eds.): *Engineered Coasts*. 2002 ISBN 1-4020-0521-0
7. C. Goudas, G. Katsiaris, V. May and T. Karambas (eds.): *Soft Shore Protection*. An Environmental Innovation in Coastal Engineering. 2003 ISBN 1-4020-1153-9
8. D.M. FitzGerald and J. Knight (eds.): *High Resolution Morphodynamics and Sedimentary Evolution of Estuaries*. 2005 ISBN 1-4020-3295-1

MIX
Papier aus verantwortungsvollen Quellen
Paper from responsible sources
FSC® C105338

If you have any concerns about our products,
you can contact us on
ProductSafety@springernature.com

In case Publisher is established outside the EU,
the EU authorized representative is:
Springer Nature Customer Service Center GmbH
Europaplatz 3, 69115 Heidelberg, Germany

Printed by Libri Plureos GmbH
in Hamburg, Germany